Cave and Karst Systems of the World

Series editor

James W. LaMoreaux, Tuscaloosa, USA

More information about this series at http://www.springer.com/series/11987

Martin Knez • Tadej Slabe
Editors

Cave Exploration in Slovenia

Discovering Over 350 New Caves During Motorway Construction on Classical Karst

Co-authors: Martin Knez, Tadej Slabe, Franci Gabrovšek, Janja Kogovšek, Andrej Kranjc, Andrej Mihevc, Janez Mulec, Bojan Otoničar, Matija Perne, Metka Petrič, Tanja Pipan, Mitja Prelovšek, Nataša Ravbar, Stanka Šebela, Nadja Zupan Hajna, Pavel Bosák, Petr Pruner, Hong Liu

Springer

Editors
Martin Knez
Karst Research Institute
Research Centre of the Slovenian Academy of Sciences and Arts
Postojna
Slovenia

Tadej Slabe
Karst Research Institute
Research Centre of the Slovenian Academy of Sciences and Arts
Postojna
Slovenia

Technical editing done by Alenka Možina

ISSN 2364-4591 ISSN 2364-4605 (electronic)
Cave and Karst Systems of the World
ISBN 978-3-319-21202-9 ISBN 978-3-319-21203-6 (eBook)
DOI 10.1007/978-3-319-21203-6

Library of Congress Control Number: 2015946082

Springer Cham Heidelberg New York Dordrecht London

Printed on acid-free paper

Springer International Publishing AG Switzerland is part of Springer Science+Business Media (www.springer.com)

Preface

Thanks to the successful cooperation with companies and institutes in charge of planning and building motorways, i.e. the Motorway Company of the Republic of Slovenia (DARS), the Engineering Company for Public Roads that was renamed DRI Investment Management and colleagues from the former Institute for the Protection of Natural and Cultural Heritage of Slovenia–the Nova Gorica Regional Office and the Institute of the Republic of Slovenia for Nature Conservation–the Nova Gorica Regional Unit, karstologists are able to study and record fascinating karst phenomena discovered in the course of motorway construction, which represent an important part of the Slovenian natural and cultural heritage. Furthermore, such collaboration enables up-to-date application of the fundamental knowledge of karstology to the planning of life in this sensitive region.

During construction, the earthmoving works have revealed a cross-section of the original karst surface, of the low and mostly covered karst of the Dolenjska region featuring subsoil stone forests and of the unique karst that can be observed in the breccia of Vipavska dolina (Vipava Valley). Over 350 new caves were opened in the karst region, including unroofed caves.

Research results have led to a number of new findings on the manner of the karst surface and underground formation, on the flow of water through the karst aquifers and on the evolution of our karst on various types of rocks and in various conditions.

The first part sums up the research results obtained during construction on the Classical Karst, newly discovered karst phenomena, research of sediments and flowstone, and dating thereof. It continues with presentation of construction on low and covered karst of the Dolenjska region and on the breccia of the Vipava Valley. The next section is dedicated to the planning of traffic roads. The book concludes with studies on karst waters, their protection and biological characteristics.

It is our aim to preserve as many karst phenomena as possible, which is quite a challenge given that many caves are hidden beneath roadways and behind the rims of tunnels.

The collaboration described above can serve as an example for planning and implementing various activities in the karst landscape, for familiarization with and for protection of the Slovenian natural and cultural heritage.

The book was written by researchers of the ZRC SAZU Karst Research Institute. Most of the authors also teach the Karstology doctoral study programme at the University of Gorica Graduate School and are members of the Unesco Chair on Karst Education. The study findings are effectively used in planning and implementing the programme in terms of the course content covering the development of caves and karst aquifers as well as development challenges on the karst.

Leon Drame, Franjo Drole and Jure Hajna took part in the field research and the compilation of documentation, and Mateja Zadel helped with laboratory work. Matters of administration were handled by Sonja Stamenković.

The palaeomagnetic research concerning the age of the cave sediments was conducted in collaboration with colleagues from the Institute of Geology, Academy of Sciences of the Czech Republic. The subsoil shaping of the karst surface was developed also within the Yunnan University International Joint Research Centre for Karstology and the Yunnan International Karst Environmental Laboratory.

Technical and text editing was provided by Alenka Možina, and translations by Milena Djokić, Andreja Golob, Darja Mevlja and David Bošković. Pictures were made by Iztok Sajko and Tamara Korošec.

The book was produced with the support of the Slovenian Research Agency.

We would also like to extend our gratitude to Acad. Prof. Andrej Kranjc, Ph.D., and Assist. Prof. Marko Komac, Ph.D., for reviewing our work.

Martin Knez
Tadej Slabe

Contents

Part V Planning

Part VI Construction and Use of Motorways with Regard to Karst Waters

1 Introduction

Slovenia lies at a crossroads of major traffic routes connecting Central Europe with the Mediterranean Area, the Mediterranean Area with the Pannonian Basin, and the Pannonian Basin with the Alps. Slovenian motorways entail an approx. 615 km long network of roads. The motorway network lies in the direction of two major European traffic corridors, i.e. the X. and the V. Pan-European corridors.

Almost a third (167 km) of motorway sections runs across the karst surface, particularly in the southern half of the country where they cross the Primorska, Notranjska, and Dolenjska karst area (Fig. 1.1). Connecting important parts of the country and opening them up to Europe, the construction of modern motorways is considered to be one of the biggest construction undertakings in Slovenia in the last two decades.

In the period from 1970–1972, the first Slovenian (then Yugoslavian) four-lane motorway from Vrhnika to Postojna was built; it was 30 km long and ran across the so-called Primorska motorway section. Two years later, an 11 km long motorway section from Postojna to Razdrto was built. The Primorska section which runs mostly on the karst surface was extended until the year 2004. At Gabrk near Divača, a 16 km long section branches off the Primorska section, it runs across the surface of the Classical Karst in the direction towards Sežana and Italy. It was built in the years 1994–1998. Below Mount Nanos, a motorway towards Vrtojba and Italy and completed in 2002 branches off and crosses the karstified areas. South of Ljubljana, there is the Dolenjska motorway section that branches off towards Dolenjska and Croatia. It was built progressively in the period between 1987 and 2007. Almost 70 km of its entire length runs across karstified surfaces. The entire motorway network encompasses 22 supply stations with all the required infrastructure (filling stations, restaurants, parking areas), fourteen rest areas (parking lots, benches, tables, sanitary fittings), and nine service stations with small parking lots (DARS 2015).

Almost half of Slovenia consists of karst and more than half of the country's water supply comes from karst aquifers. Slovenia is the country of the original Karst, which gave its name to a peculiar landscape on carbonate rocks in numerous languages, and is, in fact, the cradle of karstology. The delicate karst landform requires thorough knowledge of its characteristics, and concerted preservation efforts for it is an integral part of our natural and cultural heritage.

This volume is the result of years of experience which we have obtained from studying karst features in the course of motorway construction where we are still involved in the planning and monitoring of the construction works. In addition, it references some of the most recent findings made in karstology. It was established that some, otherwise well-familiar, karst features have been neglected in karst studies. Therefore, we have highlighted relevant examples from the original Karst, the low karst of the Dolenjska region and the karst in the breccias of the Vipava valley.

M. Knez and T. Slabe (eds.), *Cave Exploration in Slovenia*,
Cave and Karst Systems of the World, DOI 10.1007/978-3-319-21203-6_1

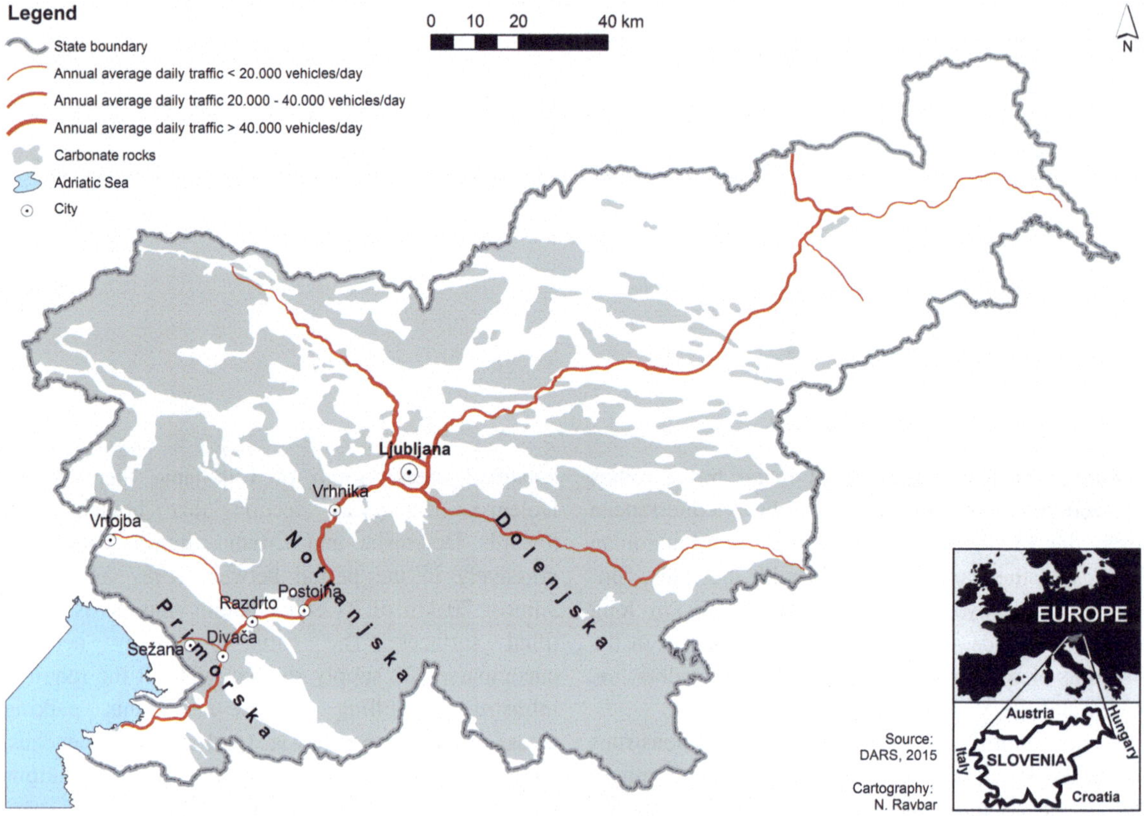

Fig. 1.1 National motorway routes and the distribution of carbonate rocks in Slovenia

In particular, our research focused on the Karst region in Slovenia, rising above the north-easternmost part of the Adriatic Sea and lined by an extensive flysch area in the southeast with altitudes of over 600 m. On a broader scale, this plateau with a total surface of 440 km^2 and altitudes ranging from 200 to 500 m is part of the Outer Dinarides. In terms of the plate tectonics theory, it is located on the northern deformed margin of the Adriatic plate and emerged as the result of thrust tectonics. Here we find only Cretaceous and Palaeogene rocks. They are characterized by a great variety of limestones which were mostly formed in relatively shallow sedimentation basins featuring diverse fauna and lush flora. In the Karst region, there is no trace of surface streams which were the go-to theory to explain the plateau development in the past. Initially, the region was surrounded and covered by flysch and therefore subject to floods. Vertical percolation was held at a minimal level. Later, the soil water dropped several hundred metres down into the karst. At the contact between carbonates and flysch, surface watercourses created a typical and vast contact karst. Today, it is typical for karst rivers to sink underground as soon as they reach underlying limestone beds instead of flysch beds. Underground, the water runs towards the Timavo Springs in Italy. The largest watercourse is the Reka River that sinks into the Škocjanske jame Caves (Škocjan Caves), with 65 % of the water percolating dispersedly from the surface. In terms of ecology, Karst is one of the most vulnerable ecosystems in the country.

The low karst of the Dolenjska region is mostly covered with various sediments over a distinct karst surface with very prominent stone forests (Knez et al. 2003). Underground water is often found shallowly beneath the surface, while the valley systems are occasionally flooded.

Karst also developed in breccias consisting of the rubble from the slopes of the Nanos plateau. Breccias lie on more or less impermeable flysch. The water running over the contact area has carved out the biggest caves around.

For a number of years now, karstologists have been involved in the planning and construction of motorways in the Karst (Bosák et al. 2000a, b; Knez and Slabe 1999b, 2000, 2001c, 2002a, 2004a, b, 2005, 2006a, b, 2007, 2008, 2009a, b, c, d, 2010a, b, c, d, e, f, 2011a, b, 2012a, b, c, d, 2014; Knez and Šebela 1994; Knez et al. 1994, 2003, 2004a, b; Kogovšek 1993, 1995a, b, c; Kogovšek et al. 1997; Mihevc 1996, 1999a; Mihevc and Zupan Hajna 1996; Mihevc et al. 1998; Slabe 1996, 1997a, b, 1998a; Šebela and Mihevc 1995; Šebela et al. 1999). When outlining the route of the motorways and railway lines, we make an effort to not interfere with the integrity of the karst landscape, and to bypass the more important surface karst features (i.e. dolines, poljes, collapse dolines, karst walls) and the caves which have already been discovered. The impact of the construction process and the use of motorways on karst waters are examined in greater depth. Motorways are supposed to be impermeable, which is why waters from the roadway are first collected in oil separators and released into the karst in a purified state.

We have studied the impact of traffic roads on karst water. Kogovšek (1993, 1995a) has examined the composition of the contamination of the daily runoff from the motorways. The stagnant waters, which were found in smaller quantities in the caves along traffic roads, contained traces of mineral oils (Knez et al. 1994).

In the scope of motorway construction, we carry out karstological supervision (Fig. 1.2). Additionally, we study newly-discovered karst features that make up an important part of our natural heritage, we provide advice on their preservation (as far as possible due to construction works), and help the construction team with new insights. We have made a series of new findings on the formation and development of the karst surface, epikarst, and the cavernosity of the aquifer.

Fig. 1.2 Exploring the cave with a collapsed roof due to the mining during construction of the motorway Klanec–Sermin

1.1 Researching the Karst Surface and New Caves in the Course of Construction

Surface landforms, epikarst and subsoil karst features were revealed following the removal of soil and vegetation from the karst surface and, of course, the large-scale earthmoving works during the digging of roadcuts and tunnels. Our task is to study these features in the context of natural heritage, to provide advice as to their preservation and, naturally, bring construction workers up to date with the new findings. These findings will be used to overcome obstacles arising during construction.

The karst surface is dissected by dolines (Fig. 1.3) and unroofed caves (Fig. 1.4). Today, dolines reflect the way the surface is shaped by precipitation water which percolates vertically through it, passing through the non-flooded part of the aquifer down to the groundwater. Dolines are filled up with soil, some more prominently than others. Doline floors feature shafts and crevices which drain away the water. First, the dolines must be cleared from the soil and next the floors must be reinforced with rocks stacked into arches (Fig. 1.5), since the mouths of shafts are often

Fig. 1.3 Dolines dissect karst surface

Fig. 1.4 Karst surface dissected by an unroofed cave

Fig. 1.5 Closing up a fissure cave

Fig. 1.6 Filling a doline (which now lies below the motorway) with rubble

smaller than the nearby cavities beneath them. Next, the dolines must be filled up with layers of rubble (Fig. 1.6). The unroofed caves are of similar shapes or more elongated. They are old caves that emerged to the surface due to the lowering of the karst surface, i.e. they no longer have the upper parts of the rim. In this case too, it is necessary to clear away the fine-grained fillings, which happen to be old cave sediments, and then fill up the caves with rocks and rubble. Otherwise, water could gradually carry away these sediments and subsidence could occur on the surface in response.

The epikarst is crisscrossed with crevices; the most prominent are found in the Cretaceous limestone and less so in the Palaeogene one. More crevices opened up at the bottoms and slopes of dolines. They are mostly filled with soil and their walls are covered with subsoil rock forms. Due to the lowering of the karst surface, many shafts are already immediately below the surface.

On a 75 km stretch of the motorway, which has been built in the Karst region in recent years, more than 350 caves had opened up (Fig. 1.7). In terms of aquifer development, caves can be divided into old caves through which water flowed when the karst aquifer was surrounded and covered with flysch higher up, and into shafts, through which water is percolating vertically from the permeable karst surface to the groundwater (Fig. 1.8). The deepest shaft measured 110 m. Old caves are either empty or filled with sediments, the latter accounting for almost two thirds of all the caves, while one third of the caves are already unroofed.

The caves opened up as a result of the removal of vegetation and soil from the surface, with an exceptionally large number having been opened up during roadcut digging (Fig. 1.9). When the rock was blasted, their roofs caved in, the cross-sections of the passages having been preserved in the banks. The majority of the shafts opened up on doline bottoms, after the soil and sediments had been removed.

We proceeded to examine all of the caves, map them out, define their shapes and rock relief, we took samples of the sediments for palaeomagnetic and pollen research and samples of flowstones for mineralogical research and dating. Based on the cave shapes and the geological features we have predicted their extension, which will be of particular use to the builders as the construction goes on.

1.2 Research Led to New Insights About Karst Development

A peculiar and common karst feature is the unroofed cave (Fig. 1.10). This surface karst feature, which bears relevance still today, is a rather familiar phenomenon, which, however, has not been subject to any comprehensive studies until now. In fact, it was pretty much neglected up until now when it turned out that the number of such surface features is much greater than originally presumed. The number of papers published on unroofed caves correlates with the construction of new motorway sections (Knez and Slabe 1999a, b, 2000, 2001c, 2002a, 2004a, b, 2005, 2006a, 2008, 2009b, d, 2010b, c, d, f, 2011, 2012b; Knez and Šebela 1994; Knez et al. 2012; Kogovšek et al. 1997; Mihevc 1996; Mihevc and Zupan Hajna 1996; Mihevc et al. 1998; Slabe 1996, 1997a, b, 1998a; Šebela and Mihevc 1995; Šebela et al. 1999). The shape of an unroofed cave is the result of the type and shape of the original cave and of the development of the karst aquifer and its surface under various geological, geomorphologic, climatic and hydrological conditions. The distinctiveness of the surface shape of an unroofed cave is dictated by the velocity with which the sediments

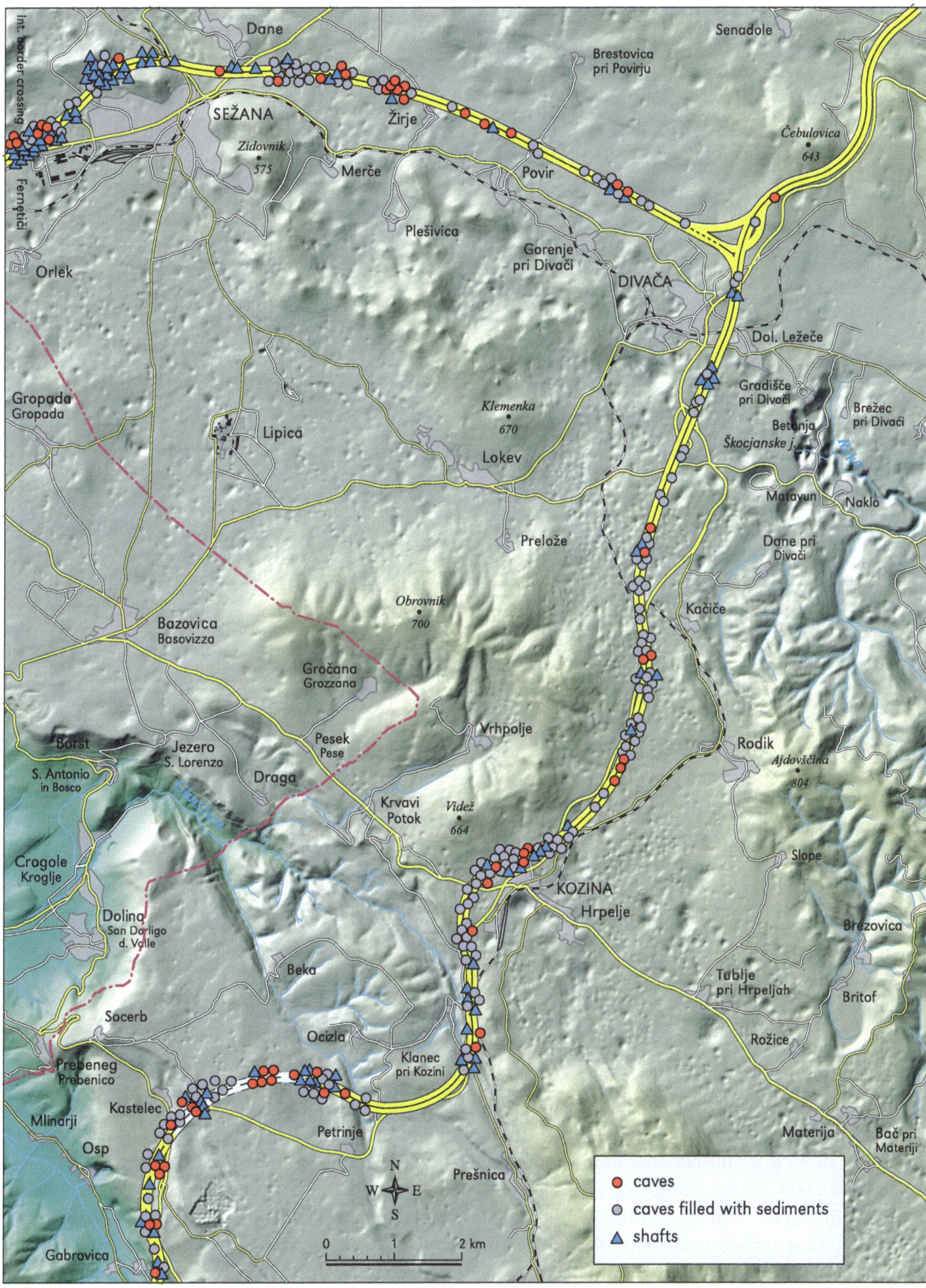

Fig. 1.7 Caves that have opened during construction works between Razdrto, Fernetiči, and Črni Kal (the motorway in the sketch is 15-times wider)

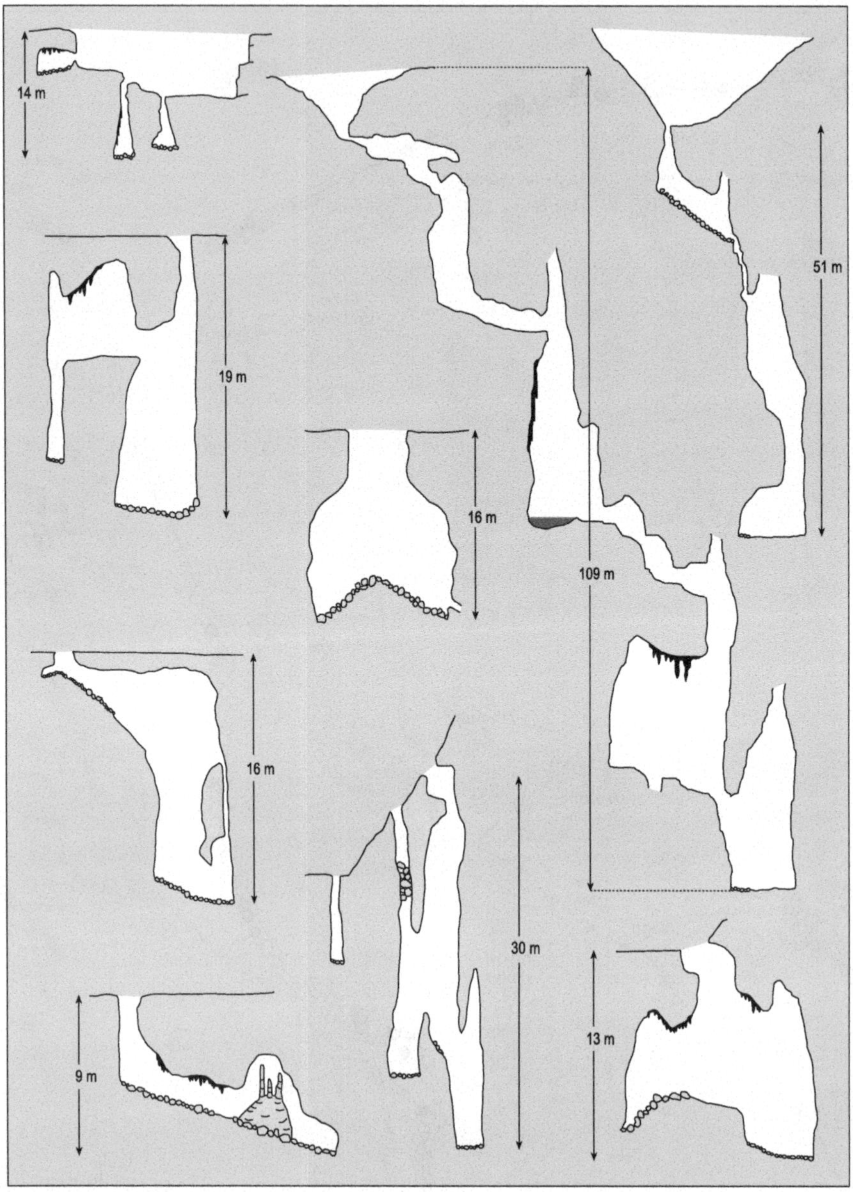

Fig. 1.8 Various caves that have opened during motorway construction

were carried off from the cave in comparison with the subsidence of the surrounding surface. If the process was slow, it is still possible to make out the soil and vegetation on the surface or zones of sediments and flowstone; if it was faster, the unroofed caves on the karst surface look more like dolines or strings of dolines or occur as elongated notches. They are often made up from a network of various old forms, i.e. caves, and of the present-day karst forms featuring dolines and shafts.

A great portion of the caves was filled up with sediments (Fig. 1.11). In most cases it is fine-grained flysch alluvium with interbedded gravel. We also took samples of sediments for palaeomagnetic research. It was

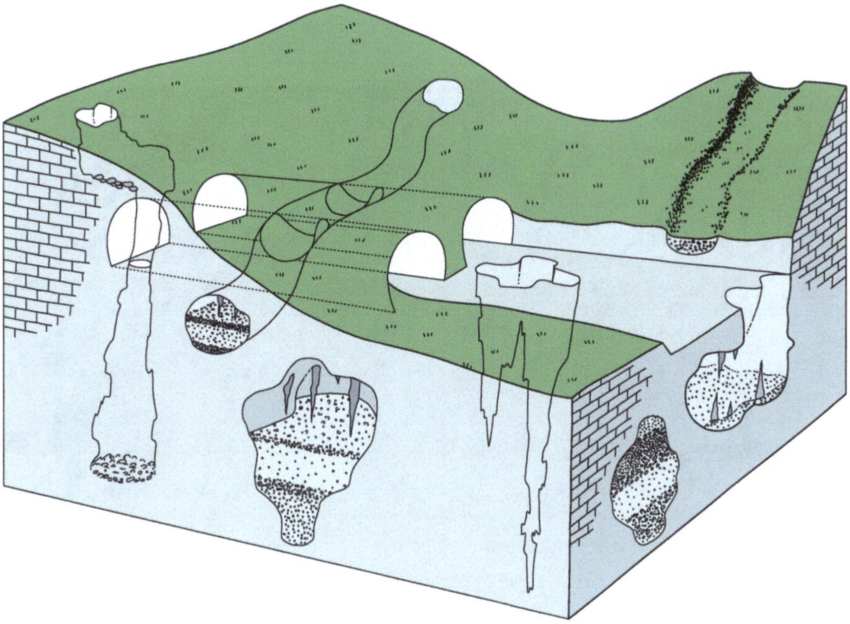

Fig. 1.9 Caves of various shapes and sizes opened during construction works which dictated all subsequent construction works

Fig. 1.10 Brezstropa jama unroofed cave near Povir

Fig. 1.11 A cave that revealed in a roadcut and has been completely filled with sediments

Fig. 1.12 Roof of a cave which collapsed during construction

established that the sediments in the caves near Kozina and Divača are of the older Olduvai period. This led to the conclusion that the caves must have been filled following the Messinian Salinity Crisis, i.e. approximately 5.2 million years ago (Bosák et al. 2000a, b).

Hence, unroofed caves are increasingly turning into a rather distinct feature of the karst surface, constituting an important part of epikarst and providing an intriguing clue to the development of the karst aquifer.

Dating the sediments gives us the chance to identify the oldest periods in the karstification, giving rise to conclusions that the oldest caves in the Karst region are much older than initially assumed by karstologists.

1.3 Road Planning

Before the construction works are to commence, an on-site check of the accuracy of the current cave data is carried out, supplementing it with any new measurements, and providing an interpretation as to their development. The existing knowledge on the cavernosity of the aquifer is presented to aid understanding, supported with prognostic subsurface maps with special emphasis on the anticipated lithological and tectonic changes in the rock. Karst cavernosity is to be demonstrated in detail prior to the actual construction works. The location of underground caves is determined by drilling. In this process, we also determine measuring indicators and the possible fill type (flowstone, alluvium). The shape, type and frequency of caves in the vicinity are predicted based on the existing knowledge from examined surface and subsoil features.

Motorway construction in the Karst region is heavily influenced by the cavernosity of this landscape (Fig. 1.12). In the Slovenian Karst region, which is strongly marked by very lively tectonics and lithostratigraphic diversity, coupled with diverse development, it is hard to predict where or when a cave will open. As a rule of thumb, they generally occur in places where flysch and limestone come in contact. In order to establish the extent of the karst aquifer cavernosity it is essential to possess solid and comprehensive knowledge of karst and be thoroughly and regularly involved in the road planning and construction process.

In view of the connection of surface and underground karst features, the road planning process must include a karstological evaluation of both the karst surface and the karst underground, the special features in their hydrology, and finally an assessment of the presented variants. It is safe to say that road construction in the Karst region will invariably lead to the uncovering of numerous karst features, i.e. dolines, filled or empty caverns and parts of old and present-day drainage paths dissecting the karst. Many karst caves have been denuded by the subsiding karst surface and are easily identifiable on the surface of the Karst region. Recently, unroofed caves, which came to light in the course of motorway construction, are attracting special attention. We are aware that an underlying thorough karstological study of the area envisaged to feature the road is a prerequisite for good route selection and that it is one of the basic starting points for construction planning amidst such a unique and delicate landscape.

We collect information on the surface karst features first by consulting literature published on this topic, archives and various collections; we make sure to particularly single out dolines, examples of collapse and other morphological forms. After the reconnaissance of the route, we lay down starting points to commence mapping along the selected route. We assess different rocks in terms of their karstological features. The known entrances to the underground areas are shown on thematic maps, and maps are supplemented with potential new entrances. Next, the branching of underground cavities is predicted based on the results of surface mapping and on the interpretation of the development of unroofed caves which are expressed in the morphology and can be made out in the relief. If necessary, we can predict the possibility of excess material deposits based on surface mapping.

We know from experience that underground cavities and parts of cave systems will invariably come up on every route that runs across karst. The shape and type of cavities can be partially predicted, drawing on the existing knowledge of surface and underground features. We determine the type of the caves found in the broader area surrounding the route, their position and role in the aquifer, as well as their form, rock relief, featured flood sediments and flowstone, and display them on relevant maps. For easier understanding, we present the insights on aquifer cavernosity obtained thus far and design a prognosis which highlights the anticipated litho-tectonic changes in the rock.

Due to the specific properties of carbonate rock, the karst waters, which sink down the examined area, easily pass into the underground (karst aquifer); they can pass through 100 m thick rocks in roughly an hour. Even though flysch rocks—in the Karst region they have constant direct contact with carbonates—are often presented as strictly impermeable strata, it should be pointed out that flysch (in many places in thinner layers) is merely an isolated lenticular bedding overlying permeable carbonate rock. It should also be noted that underground trunk conduits, though in smaller numbers, are also formed in flysch rocks and that the precipitation water accumulated on flysch runs off into the karst. To this end, we have to perform hydrogeological mapping of the site. This entails delineating and determining the underlying features of the hydrogeological units in the area next to the route, listing hydrological objects (captured and non-captured springs, surface flows, water caves, boreholes, measuring stations, and so on) and determining the physical and chemical properties of the springs. If necessary, we conduct two tracer tests at low and high water levels, mostly for determining the course and speed of the underground flow in the broader area surrounding the route. The results of the site mapping and the tracer tests help us design and update the existing hydrogeological maps, compile a database on the status of the environment and assess the impact of the construction on karst waters.

Fundamental guidelines applying to planning traffic roads can be summed up as such:

- route selection is based on an overall assessment of karst with emphasis on the local characteristics;
- the selected course of the route makes sure to bypass individual distinct karst features;
- one of the priority objectives in the course of planning is to preserve the karst aquifer.

1.4 Cave Preservation

The shafts were the easiest to preserve. Their smaller entrances were sealed with concrete slabs (Fig. 1.13). It was also possible to preserve old caves that had a firm rim. The caves located in crushed rock that opened as a result of the blasting had to be filled up. The caves that had been cut off by the roadcuts with entrances in their banks were covered with rock walls (Fig. 1.14). Their rim was too broken, rendering the caves unsuitable for further inspection, while from sediment-filled caves water might wash the clay out onto the roadway. We left one well-preserved cave open as a tourist attraction for passengers crossing the border with Italy (Fig. 1.15). The most interesting and well-preserved caves were, however, fully protected and left open to access even though they are located beneath the roadway or—as in the case of the Kastelec tunnel—wind around the tunnel tube. They can be accessed via concrete tubes, which end in a closed shaft next to the road (Fig. 1.16) and, in the tunnel, with a door (Fig. 1.13c).

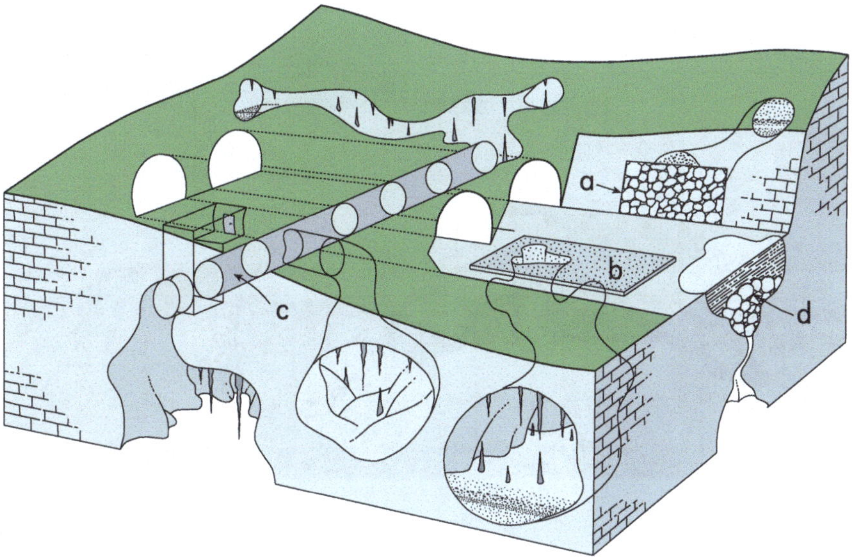

Fig. 1.13 Closing and preserving of caves. *Legend* **a** in road cuts the caves are hidden behind rocky scarps; **b** the caves lying below the road are covered by concrete lids; **c** entrance into the cave meandering around the tunnel; **d** bottom of karst fissures and tops of shafts are often closed by arches of rocks

Fig. 1.14 Closing a cave in the embankment of a roadcut

Fig. 1.15 A cave that has remained open before the Fernetiči border crossing

Fig. 1.16 Construction of a road-side artificial entrance

Fig. 1.17 Preparation works for laying pipes for road drainage

We have also studied the consequences of different blasting activities inside the caves, which will aid in further construction and in the preservation of karst features.

1.5 Karst Protection in Light of Motorway Construction and Usage

Our experience with tracing water and with accidental spills of different substances in karst, suggested that we must keep in mind the high degree of cavernosity of the karst aquifer. Numerous caverns that we found during the construction corroborated our assumption. Poor permeability characterizes only individual relatively small areas found either at doline floors which are covered with larger quantities of washed off soil (dolines on Palaeogenic limestone of the Divača lowland are dotted with puddles in the wet seasons) or yet even smaller areas of loam which fill old caves. This lends itself to the conclusion that road construction and road usage must be approached with circumspection. Daily traffic can leave many environmentally harmful substances on the road

Fig. 1.18 Drainage channels at the side of the road

Fig. 1.19 Construction of an oil separator

surface (Kogovšek 1993), and mineral oils were found in stagnant waters in caves next to the roads (Knez et al. 1994). As a result of these insights and our relentless efforts, motorway designs aim at providing impervious road surfaces. Pipes (Fig. 1.17) and drains (Fig. 1.18) are provided along the road, leading down to wastewater separators (Fig. 1.19). Ideally, untreated water does not reach the permeable karst surface. However, the technical aspect of the drainage system must be brought in line with our wishes. As it is, separators are often too small, and heavy rainfall can flush the sediments out.

1.6 Conclusion

It was established that the involvement of karstologists in the road construction process in the Karst region is useful. However, it is vital that we are involved both in the planning and construction process and, later on, in the monitoring of the impact that motorways have on the environment. Karstologists should therefore be involved in the full process of human intervention in the delicate karst landscape, from start to finish. Through meaningful collaboration we can preserve our cultural heritage and deepen our fundamental knowledge on the formation and development of karst and on the construction of motorways in a very distinct setting. Seeing as there are several different types of karst, each requiring a unique approach, the co-operation between the construction crew and karstologists should be permanent and regular. The co-operation between motorway engineers, constructors and karstologists produced new findings which are applicable in the planning process and realization of other human interventions in karst.

Part I

Classical Karst: Construction Monitoring

2 Development and Karstification of the Karst Aquifer as Discovered Between Klanec and Črni Kal

Between Klanec and Črni Kal, on a 6.5 km route of the motorway, 67 caves opened during earthworks, road cutting and tunnel digs (Fig. 2.1). The majority consisted of old caves, i.e. caves which once had water passing through. Two thirds of these caves were filled with alluvium. Research carried out in these caves augmented our knowledge about the development of this part of the Karst region. A cave system extending more than 500 m, which we tried to preserve in full, opened in the Kastelec tunnel near the Brezno na Škrklovici Shaft. Underneath the road, the passages of this system are connected with concrete tubes which are accessible via a gully at the side of the road.

Due to the construction of the tunnel and extensive roadcuts, the impact of the activities affecting the Karst Edge and its hinterland was significant, in several places leading to the unearthing of today's epi-karst and a part of the vadose zone, both of which are also intersected with old caves. Besides the palaeo-karst, these caves provide the oldest traces of karst development in this area. The caves which were discovered during construction gave us new insights about the cavernosity of the Karst and its development.

2.1 Karst Surface and Karstification

The route runs through alveolinid-nummulitid limestone and, to a smaller extent, flysch rock. It crosses several thrusting deformations between carbonate rock and Eocene flysch.

After Eocene flysch had been deposited in the Pyrenean phase, the rock underwent NW–SE folding. Later, the folds were deformed due to thrust, normal and longitudinal faults. The area between Petrinje and Črni Kal belongs to the imbricate structure of Čičarija thrust unit (Pleničar et al. 1969). The motorway route crosses folds, thrusts and fault deformations (mostly oriented NW–SE and NE–SW). Up to 100 m wide fissured zones striking N–S are mostly made up of open fissures along which shafts and karren systems have developed. These fissures are highly conducive to karstification (Fig. 2.2).

2.1.1 Alveolinid-Nummulitid Limestones

The alveolinid-nummulitid limestones representing the end of carbonate sedimentation in southwest Slovenia lie concordantly on miliolid limestone. The limestone is dominated by the fauna from the Alveolinidae and Nummulitidae families to such a degree that it has given this type of limestone its very name.

The bedding thickness changes in a lateral direction, with medium and thick beds being predominant. In some places, the bedding cannot be determined. The upper part of the alveolinid-nummulitid limestone in particular is more compact and homogenous; the bedding is also less prominent.

The biomicritic and biosparitic limestone, mostly packstone, is usually light brown, light grey or yellowish white in colour. It features numerous fossils of

M. Knez and T. Slabe (eds.), *Cave Exploration in Slovenia*,
Cave and Karst Systems of the World, DOI 10.1007/978-3-319-21203-6_2

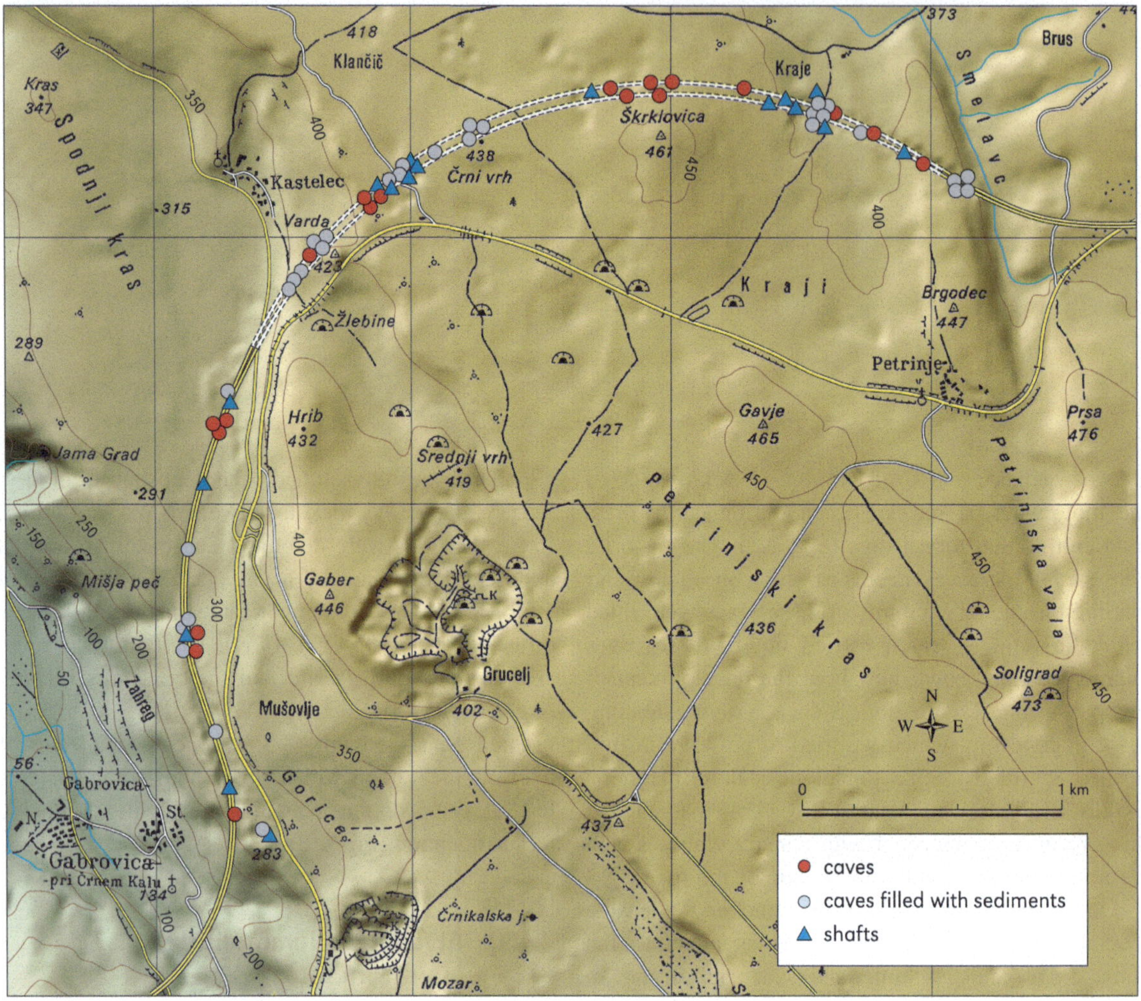

Fig. 2.1 Caves in the route uncovered during motorway construction

alveolinids, nummulitids and discocyclinids. Typically, one can locally also find in this limestone brachiopods, echinoids, corals, lithothamnians, various lamellibranchs, etc.

In most cases, the alveolinid-nummulitid limestone has deposited on the miliolid limestone continuously. Considering the limestone type, it was established that the sedimentation took place in an open and shallow shelf, while elsewhere the sedimentation conditions were more subtle. There, sedimentation took place in sheltered areas of the open shelf or, perhaps, in small lagoons (Jurkovšek et al. 1996).

Nummulites are the most common fossil remains in the alveolinid-nummulitid limestone with beds varying laterally from a few metres and up to 300 m. Usually, however, the genera Nummulites, Operculina and Assilina occur in a mix with all three genera present. There are areas where nummulites are predominant and others where operculines prevail. It follows that the former limestones could be termed nummulitid and the latter operculinid limestone. Assilines are usually featured in the rocks to a smaller degree.

2.1.2 Flysch

Along the border with flysch, we observe transitional beds of carbonate and non-carbonate rock, of mainly marl and marly limestone, often containing

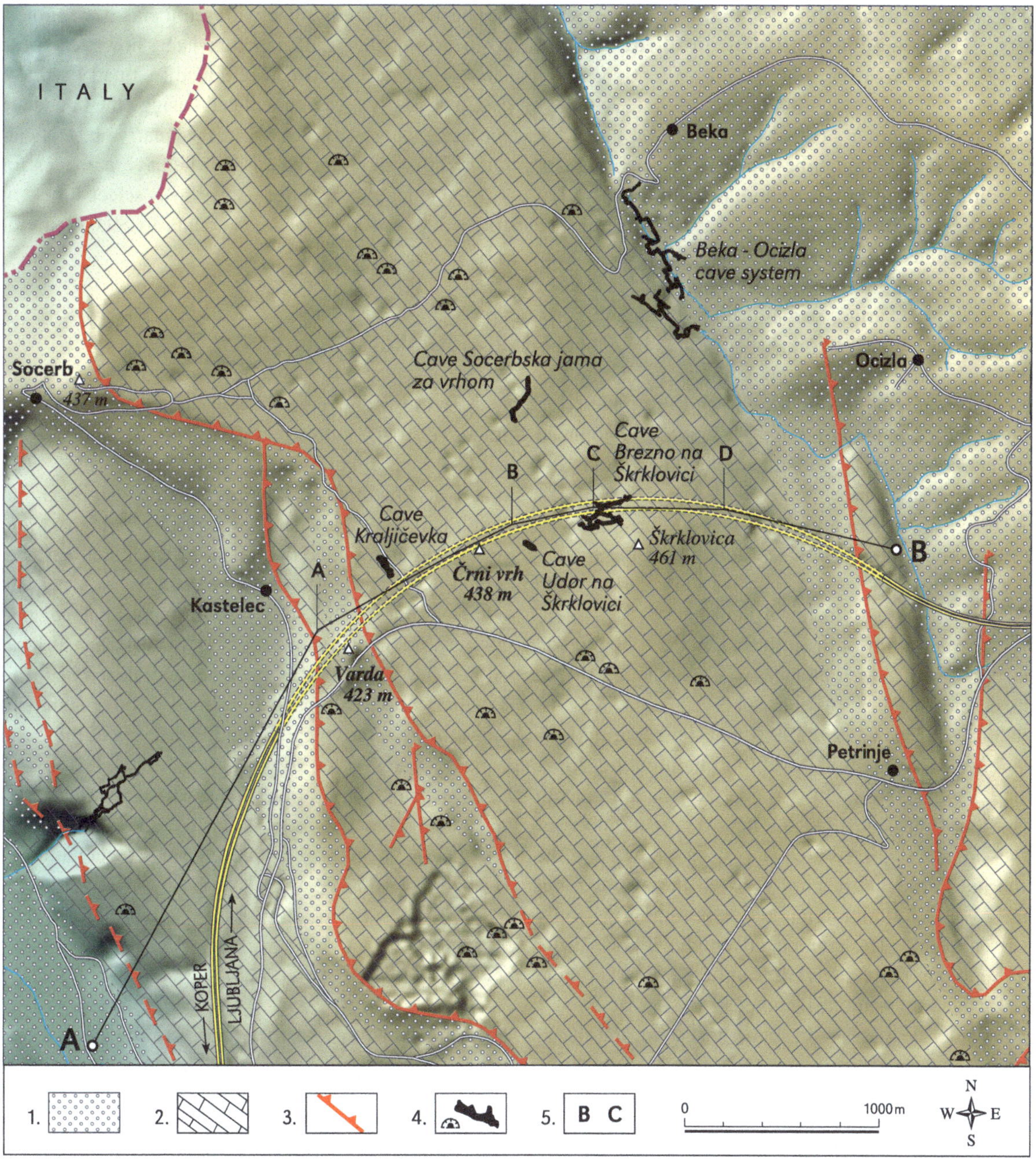

Fig. 2.2 Fault zones and already known caves in the studied area (the legend provided in Fig. 2.9)

abundant pelagic Foraminifera. Nevertheless, a longer or shorter hiatus may occur laterally between the two formations.

The flysch, marked by alternating sequences of marl, sandy siltstone, and coarse-grained carbonate sandstone with intermittent thicker or thinner insertions of breccia and conglomerates, was thrust-imbricated onto the alveolinid-nummulitid limestone. The area of contact was important for the typical speleogenetic development. It was established that in contact with impermeable rock, limestone is not just a water barrier as it is where runoff collects, carving from larger drainage channels through which the material is washed off below the ground.

2.1.3 Karstification of the Alveolinid-Nummulitid Limestone

In light of current research, it was found that alveolinid-nummulitid limestone is more resistant to erosion compared to the stratigraphically closer adjacent limestone. In most cases, it was only the surface that was subject to karstification, and the area directly below. Only on rare occasion does karstification occur deeper down. Surface karstification is heavily pronounced on inclined terrain where diluvial sediments occur. One can however topographically distinguish alveolinid-nummulitid limestone from the underlying miliolid limestone based on the more pronounced surface karstification and the numerous karren dissecting the rock surface.

The weathering of alveolinid-nummulitid limestone is platy; the limestone breaks up into rubble or displays irregular disintegration. The soil overlying the bedrock is often only a few 10 cm thick.

A detailed study of the influence of the rock type on the number of cave entrances was conducted in the immediate vicinity of the surveyed area. It was established (Knez 1995, 1996) that the Liburnia Formation beds, featuring alveolinid-nummulitid limestone, have a considerably smaller karstification depth than the Cretaceous beds. The average number of caves in the examined area is 1.01 cave/km^2, and the Liburnia Formation reaches 75 % of the average, but the alveolinid-nummulitid limestone only 0.43 cave/km^2. The significantly higher figures from the Senonian (2.42 cave/km^2) and from the Turonian (2.18 cave/km^2) should be noted.

In order to assess the cavernosity of the Karst region between Petrinje and Črni Kal, we measured the karst caves, uncovered in the course of quarry works, in the Črnotiče quarry with a laser theodolite. This involved shafts, horizontal caves and vertical fissures, mostly completely filled with cave sediment and flowstone. The passages were up to 220 m long with a diameter of up to 15 m. Some caves were situated immediately below ground, while others had opened a few metres deeper. Also, some were filled with sand and alluvial clay, while other instances featured rubble, and sometimes flowstone, deposited over sand and clay. Some of the karst caves which had not been completely filled with sediments were subject to measurements and mapping, later, however, these caves were destroyed and removed by quarry work.

Cavernosity was calculated using the geodetic measurements of the caves in the quarry. It was calculated that the total volume of the cave passages for 13 geodetically measured caves in a select block of rock (300 × 400 × 19 m) amounts to 89,074 m^3 or 3.9 %.

2.2 Epikarst

Underlying a thin layer of soil and overlying poorly fractured rock, subsoil karren with a characteristic subsoil rock relief (Slabe 1998a) had formed. The massively fractured rock has already disintegrated into individual, mainly smaller, pieces which were reshaped, i.e. rounded off, below ground; their surface, which was exposed to even weathering, indicates the composition of the rock. As it were, Palaeogenic fossils are often found sticking out of the surface. Along the contact of limestone and flysch, subsoil channels and scallops had formed. Above-sediment anastomoses are typical of the lower faces of the basal conglomerates in the flysch.

In Kozina, the former Palaeokarstic surface was unearthed directly below the present surface and revealed remains of dinosaurs and other animals (Debeljak et al. 1999).

2.3 Discovered Caves

On a 6.5 km route of the motorway, 67 caves were opened during earthworks, roadcuts and tunnel excavation. In terms of numbers, old caves—once marked by through-flowing water—were predominant (49); two thirds of these were filled with fine-grained alluvium and occasionally with coarse rubble.

Old caves opened as unroofed caves (Knez and Slabe 1999a, b, 2000, 2001c, 2002a, 2004a, b, 2005, 2006a, 2008, 2009b, d, 2010b, c, d, f, 2011, 2012b; Knez and Šebela 1994; Knez et al. 2012; Kogovšek et al. 1997; Mihevc 1996; Mihevc and Zupan Hajna 1996; Mihevc et al. 1998; Slabe 1996, 1997a, b, 1998a; Šebela and Mihevc 1995; Šebela et al. 1999).

Fig. 2.3 Walls of a cave network that was filled with alluvia and flowstone

Caves also opened at the perimeter of the roadcuts and during the excavation of the tunnel. Their passages reached up to 8 m in diameter.

A larger system of caves opened on the eastern side of the south entrance to the Kastelec tunnel. A part of the system was already missing the roof, one part consisted of hollow passages, and the other part was filled with sediments and flowstone (Figs. 2.3, 2.4 and 2.5). The rim of the passages had been characteristically reshaped above the sediments (Slabe 1995). From the largest passage, 18 alluvium samples were collected for palaeomagnetic study (Fig. 2.6). With regard to the location within the region and the development of this part of the Karst region, it was established that the deposits, much like other cave deposits in the immediate vicinity, consist of flysch deposits (Bosák et al. 2000a, b). The uppermost part of the profile is made up of a 3 m thick rubble layer. Below lies a 2 m layer which is sandier, whereas the lowermost layer is argillaceous. The sediment is yellowish brown (10YR 5/6) to light olive grey (5Y 6/2) in colour. From the bottom of the profile up, at a depth of 4.5–5 m, lies a 10–20 cm thick connective layer of clay that is of a darker shade (brownish yellow 10YR 6/8) than other sediments in the profile. The altitude of the deeper samples is 395 m a.s.l. Individual, unroofed parts of the cave, which were doline-like, were clearly distinguishable on the surface even before works began.

However, the largest cave system, stretching over 500 m (Figs. 2.7, 2.8 and 2.9), opened in the tunnel not far from the already known Brezno na Škrklovici Shaft (Reg. No. 1391; The Cave Registry of the Speleological Association of Slovenia and the Karst Research Institute ZRC SAZU) has distinct signs (large scallops) indicating that the water flow in the water-filled cave must have been slow. Three large passages opened in the tunnel, however they were not

Fig. 2.4 Section of the cave network that was filled with alluvia and flowstone

interconnected. Still, it appears that they are all part of a single cave system. The shape of the inclined passages (up to 8 m in diameter) suggests that the cave was shaped by the flow of water and that its reshaping by percolating water was less pronounced. The large scallops are the result of a slow water current, which was in fact the main culprit for the present shape of the cave, while the ceiling cups indicate that the cave was once completely filled with water. The ceiling cups are of various shapes, they run narrow and extremely high (up to several metres) along the cracks; they have relatively level tops and are often clustered into smaller or larger ceiling cupolas. Originally, the passages were filled with fine-grained alluvium, which had mostly been flushed out by now.

At the surface, specifically at the entry shaft to the Brezno na Škrklovici Shaft, the beds of alveolinid-nummulitid limestone dip 20° to the southeast. In the cave, in the left tunnel tube, the beds dip 30° or 40° to the east or northeast. At a vertical distance of 80 m, the bedding dip direction has thus slightly changed.

While making way for the roadcut, an old cave opened next to the steep karst edge, which was in fact, a system of minor passages with a round cross-section, their roofs criss-crossed by cups and cupolas—features dating back to when the cave had been shaped below the ground water level. Occasionally, a small conduit can be found on the bottom of the passage, created by the through-flow of small quantities of water over the rock bed—suggesting a more recent cave reshaping.

Many of the caves are characterized by cracks along vertical fissures, either hollow or filled with alluvium. Often they were 1–2 m, rarely 3 m, wide but they could run up to several tens of metres deep. Those that were filled with fine-grained deposits had supra-alluvial scallops along their rim.

Fig. 2.5 The wall of the passage that has formed along the alluvium, with the passage being filled with fine-grained alluvium

Fig. 2.6 Alluvia in an unroofed cave

In addition, there was a shaft with a diameter of 4 m located within the perimeter of the described cave system, before the tunnel (Fig. 2.10). It was completely filled with fine-grained alluvium. Its walls showed traces of percolating water and were only occasionally reshaped under the alluvium.

Caves also emerged along the contact of limestone and flysch. In places where limestone overlies flysch, most of the respective caves had formed in flysch. The caves had a diameter of 2 m, and they were completely filled with flowstone and fine-grained alluvium.

Eighteen shafts had opened, the deepest two exceeding 60 m each. Shafts were either simple or gradient. Some displayed old flowstone which had been corroded by water running down the walls.

2.4 Aquifer Development

Caves, which were opened during the construction works, reveal the most important periods in the development of this part of the Karst. The epikarst and the upper part of today's vadose zone are intersected by traces from different periods of karst development. The oldest caves demonstrate distinct traces of cavernosity of the aquifer, suggesting that the preserved part of the Karst was entirely shaped below the ground water table—which is today 230 m deep underground. As a result of the lowering of the karst surface, some of the caves have already lost their roofs (Knez and Slabe 2002a). The shapes of the other caves suggest formation in a phreatic zone. However, most

Fig. 2.7 A cave that was uncovered during the tunnel digging (*Photo* Feruccio Hrvatin)

of them were filled with fine-grained deposits which were later in part washed out. The alluvial fill in the caves is associated with ground water level rises following the Messinian Event (Bosák et al. 2000a, b). It appears that the shaft, which was filled with fine-grained alluvium, reveals a distinct period of formation of the upper part of the aquifer in vadose conditions before it had been reached by floodwaters.

A relatively swift transition of the formation of this part of the Karst region into permanent vadose conditions followed. Apart from alluvial traces, particularly deposits and rare traces of faster water currents, the caves show no distinct signs of formation in epiphreatic or vadose conditions. Today, this part of the Karst region is subject to formation with dispersed water percolating down from the karst surface. Water caves develop only at the contact area of flysch and limestone, which is crossed and entered by streams. It is safe to say that the development of this part of the Karst is closely connected with the rapid decrease of the ground water level.

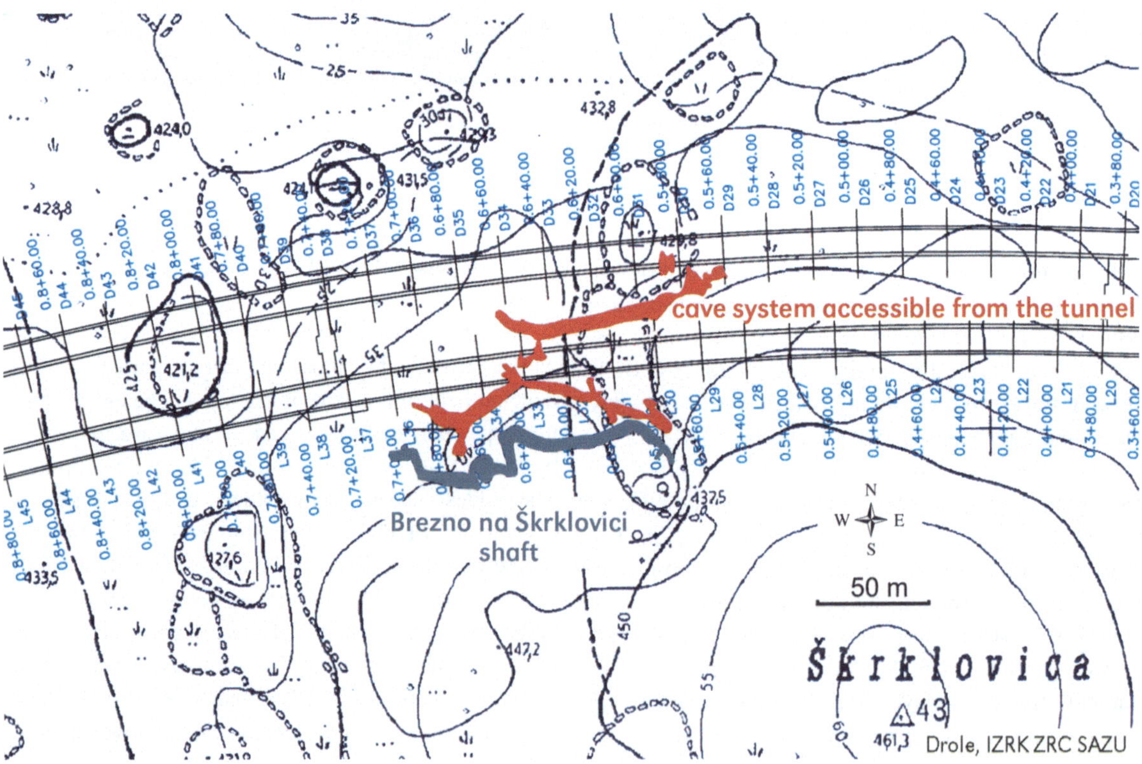

Fig. 2.8 A preserved cave network accessible from the tunnel

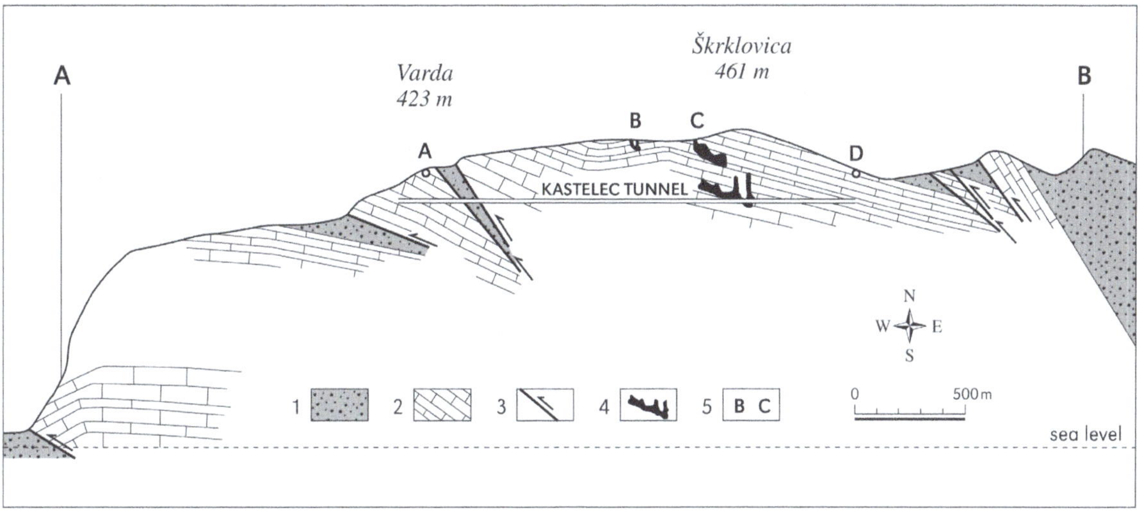

Fig. 2.9 Cave cross-section AB. The height of profile is enlarged by a factor of 3 relative to the length. Legend: *1* Eocene flysch, *2* Eocene alveolinid-nummulitid limestone, *3* thrust, *4* karst cave, *5* labelled points on the profile (B—Cave Udor na Škrklovici, C—Cave Brezno na Škrklovici)

Fig. 2.10 An old shaft filled with fine-grained alluvium

A part of the old caves is filled with coarse rubble which came about as a result of rock disintegration in the cold Pleistocene periods.

In several instances, it was observed that the water, percolating from the surface and trickling down the walls of the shafts—either independent shafts or such dissecting older caves—dissolved the flowstone which had once covered these walls. Will further research reveal that this is merely the result of the widening of cracks, of changes affecting the karst surface, or perhaps even the result of human impact due to the removal of vegetation?

2.5 Cave Preservation

All the caves discovered during the motorway construction were measured, mapped out (Fig. 2.11) and explored. Supported by road builders, we tried to preserve as many as possible. Now they are hidden behind rock embankments at the edges of respective roadcuts, and behind concrete rims inside the tunnel. Caves located under the road, with narrower openings and, despite blasting, a relatively unfractured rock rim, were covered with concrete slabs. We made efforts to preserve the largest cave system in the tunnel in full. The passages of this system are connected by concrete tubes running under the road, accessible through a shaft located at the side of the road in the tunnel.

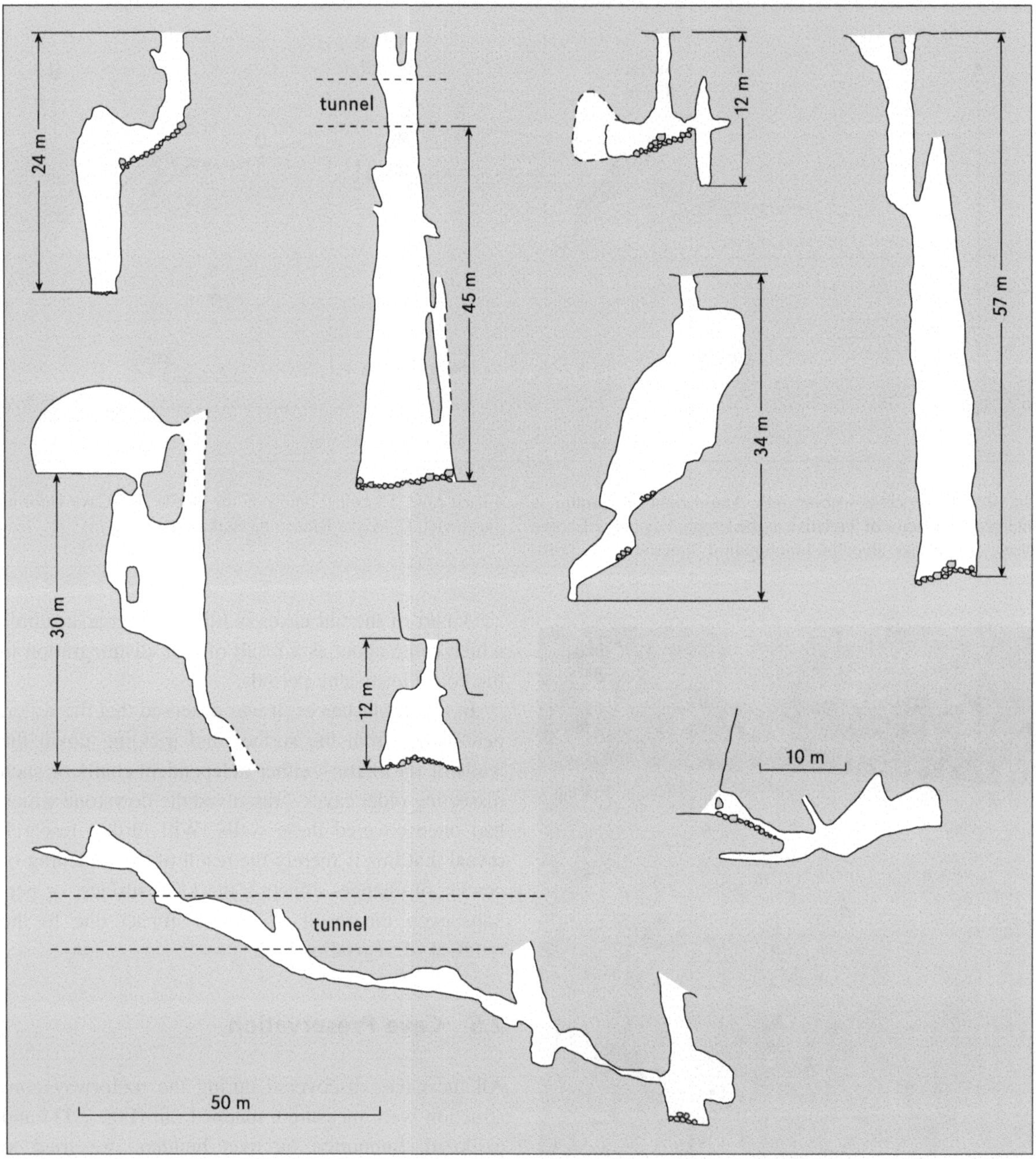

Fig. 2.11 Cross-sections of caves uncovered during tunnel-digging

2.6 Conclusion

On a 6.5 km stretch of the motorway, a total of 67 caves opened during the construction works, of which 49 caves, i.e. the majority, showed signs of having had water flowing through at one point. Inside the Kastelec tunnel a large cave system was discovered in the NW part of the Škrklovica hill (461 m), and proven to be genetically connected to the previously discovered Brezno na Škrklovici Shaft. A great portion of the caves was filled with fine-grained alluvium. Most of the motorway route runs over alveolinid-nummulitid limestone which is relatively resistant against karstification, accounting for the fact that in most cases only its surface and the parts close to the surface are distinctly karstified. The signs of karstification from various periods are evident in the epikarst in the upper part of today's vadose zone. The oldest caves still contain evidence of the cavernosity of the aquifer, when the part of the karst, which is still preserved today, was shaped entirely in the phreatic zone. Some of the caves consisting of a system of smaller passages with a circular cross-section, have ceilings that feature ceiling cups and cupolas, i.e. shapes dating back to the time when the caves had been formed below the former ground water table. This was followed by a relatively swift drop of the ground water level which led to the reshaping of the older signs of development in vadose conditions.

Unroofed Caves Near Kozina and Their Identification

3

Unroofed caves were old caves that became exposed by the lowering of the karst surface. They are in fact preserved by their fill—mostly fine-grained alluvium. It is also often that they feature preserved flowstone and an intact rock rim.

During earthworks preceding the motorway construction, this important karst feature, also characterizing the surface, attracted special attention (Knez and Slabe 1999a, b, 2000, 2001c, 2002a, 2004a, b, 2005, 2006a, 2008, 2009b, d, 2010b, c, d, f, 2011, 2012b; Knez and Šebela 1994; Knez et al. 2012; Kogovšek et al. 1997; Mihevc 1996; Mihevc and Zupan Hajna 1996; Mihevc et al. 1998; Slabe 1996, 1997a, b, 1998a; Šebela and Mihevc 1995; Šebela et al. 1999). Earthworks revealed that the karst surface is scattered with several distinct types of unroofed caves, which in itself are not an uncommon phenomena. We were able to single out the typical shapes of unroofed caves found on karst terrain, i.e. individual doline-like forms that occur in strings, and oblong notches.

Because the surface of the Karst region has lowered so dramatically (Kranjc 1997), there are old caves and shafts opening up all the time in the course of the construction of the motorways. Old caves are either void or filled with alluvium. The caves were formed as a part of a system of cavities in a period when impermeable rocks had enclosed the aquifer higher up, causing the ground water in the aquifer to be at a higher level. But karstification gave rise to a drop in the water table in the aquifer—today it is 200 m and more below ground, and the karst surface is still lowering.

Unroofed caves are therefore regarded as distinct surface karst forms which were in part reshaped by surface processes that make up an important part of epikarst.

In the course of the earthworks for the construction of the Kozina motorway, we were able to make out the typical shapes of unroofed caves. We have also discovered passages and large cave systems carved in horizontal or inclined karst surfaces. The area is mainly dotted with old caves which had through-flowing water at a time when this portion of the aquifer was shaped in the phreatic or epiphreatic zone. Three caves were empty; the rest was filled with fine-grained flysch alluvium, gravel and rubble. The passages measured up to 5 m in diameter. To a large extent, the old caves had missing ceilings as a result of the lowering of the karst surface. The largest cave system that had been discovered at the start of the route near Kozina (Figs. 3.1 and 3.2) was made up of empty passages, one of which had been already known before the earthworks, and alluvium-filled passages with thin ceilings, as well as passages with no ceilings whatsoever. At the bottom of

M. Knez and T. Slabe (eds.), *Cave Exploration in Slovenia*,
Cave and Karst Systems of the World, DOI 10.1007/978-3-319-21203-6_3

Fig. 3.1 Unroofed caves at Kozina (*Photo* Alma Bavdek)

soil- and alluvium-covered dolines, we discovered vertical shafts draining the water from the permeable surface, but their openings were small and the shafts could not be entered.

3.1 Unroofed Caves and the Karst Surface

The karst surface along the motorway route between Kozina and Klanec has a notably diverse geological structure. The road runs relatively transversely over different lithological and stratigraphic belts of Cretaceous and Palaeogene carbonates and Eocene non-carbonates.

In the northsouth direction toward Klanec, we can trace Turonian and Senonian Cretaceous limestones followed by beds of Liburnia Formation. Among them, in the lower part, are Cretaceous Vreme beds followed by Palaeocene Kozina beds, Thanetian miliolid limestone and Eocene alveolinid, nummulitid and alveolinid-nummulitid limestone. West of Klanec, and for the first time in this section, the motorway crosses a narrow impermeable belt of flysch rock. After less than 100 m, the route resumes its course over alveolinid-nummulitid limestone. The final section of the route no longer runs over karst landscape, but instead over exposed marlstone and sandstones (flysch).

It is highly likely that the surface karst morphology and morphogenesis of this part of the Karst region also reflect its lithostratigraphic foundation. In the southern section of the route, which is characterized by the prevailence of Palaeogene limestone, the karst relief is relatively levelled with slight waves, and marked by the absence of sharp karst forms indicating surface erosion. By contrast, karst relief on the northern part of the route, north and east of Kozina, has intensively eroded the lower Cretaceous rock surface with numerous karren and various surface notches.

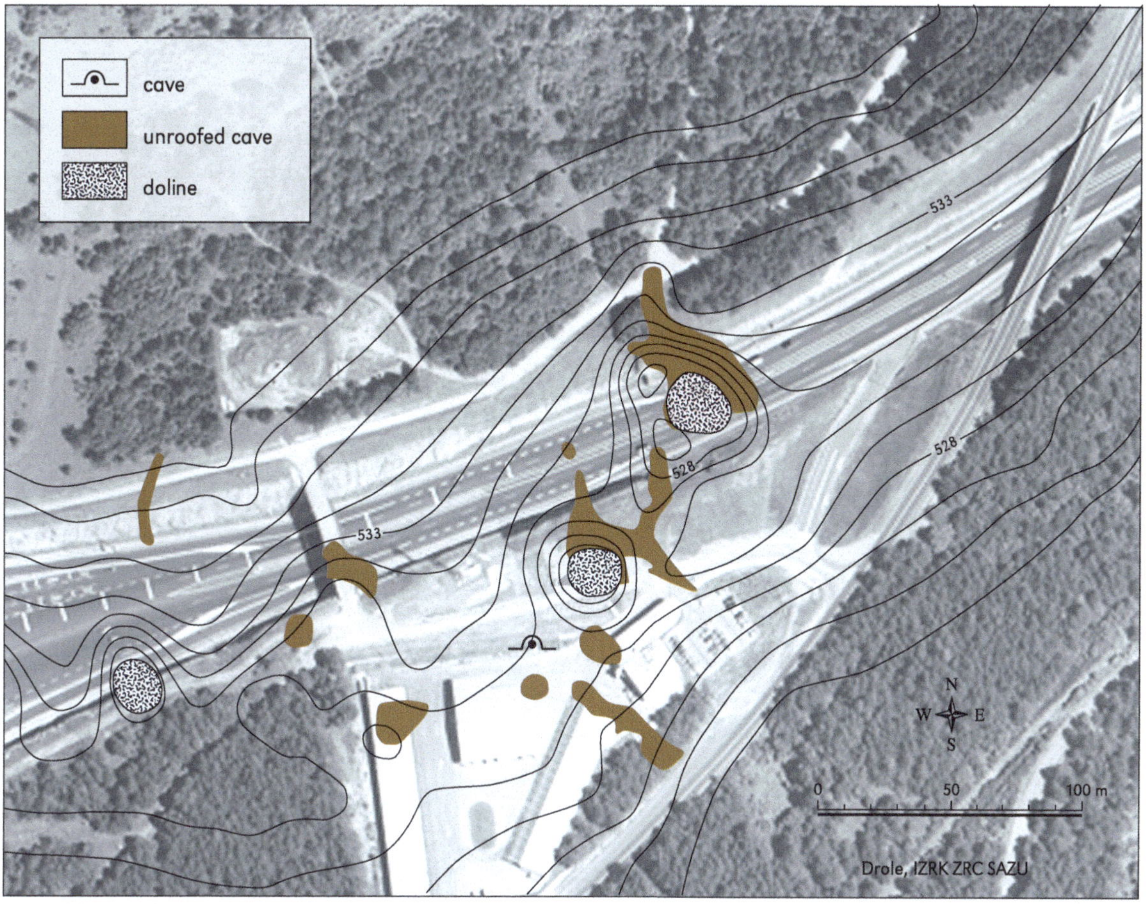

Fig. 3.2 Karst surface with unroofed caves and dolines

As early as the planning stage for the motorway route, we had identified the notches that dissect the slopes of the dolines, stretching several tens of metres around, as unroofed passages (Fig. 3.3). Most of our assumptions were confirmed by earthworks.

On the northern slope of a large doline, a narrow and shallow notch was found behind the service station, which we had already identified during fieldwork and from the surface map. Earthworks revealed that the cave was filled with fine-grained alluvium. One part of the passage was also uncovered at the edge of the doline floor, while a cross-section of a passage was found on the opposite slope. Both were filled with fine-grained alluvium. It thus appears that the doline developed in the middle of a cave system.

On the relatively steep hillside overlooking Klanec, the beginning section of a horizontal passage, which was filled with fine-grained alluvium, appeared as a small half-doline on the surface. Earthworks for the roadcut revealed two more passages—situated higher and lower—and belonging to the cave system, mentioned before.

The largest cave system (Figs. 3.2, 3.4 and 3.5), which was unearthed at the new, beginning section of the route, stood out on the surface as a network of more or less prominent notches, the most prominent ones dotting the slopes of the dolines and connecting dolines, and of smaller, relatively shallow dolines. The cave system was made up of unroofed sections as well as a smaller hollow passage which we had known about before initiating earthworks, and roofed passages. To a large extent the cave was filled with fine-grained alluvium and pebbles (Fig. 3.6). The bottom of the southwestern part was almost completely covered with heaps of flowstone, stalactites and stalagmites. Even flowstone was covered with fine-grained alluvium. Coarse rubble (Figs. 3.7 and 3.8) occasionally covered the surface or filled the

Fig. 3.3 An unroofed cave—an elongated indentation in the karst surface

Fig. 3.4 An unroofed cave at Kozina

Fig. 3.5 An unroofed part of a cave network after clearing out the alluvia

Fig. 3.6 Gravel alluvium in an unroofed cave

Fig. 3.7 A rubble-filled cave

caves; it was commonly overlaying old alluvial deposits or filling the space below the shelves which were created in the beginning section of passages. The rubble came about as the result of rock disintegration during the cold Pleistocene epoch. The bottoms of the two larger dolines that had formed at the perimeter of the old cave system were covered with layers of brown and red soil a few metres deep. Their rim featured rock formations suggesting notable water percolation. Also at their bottoms were entrances into narrow vertical shafts. More distinct notches develop from unroofed caves, if the high water velocity facilitates sediment transport from the surface, often the result of another doline being located close by.

On a section of the route between Kozina and Klanec the dolines are lined up. We found pieces of flowstone at the bottom of these dolines. At the edge of a doline, next to a belt of exposed flowstone, we also unearthed a passage with a diameter of over 5 m. It was filled with large pebbles, stalactites and stalagmites (Fig. 3.9). Further research will show whether such line-up of dolines may be brought in connection with the old denuded cave. Drilling works uncovered flowstone in one of the notches on the slope of one of the dolines.

Fig. 3.8 An old rubble-filled cave

Fig. 3.9 A cave filled with gravel, fine-grained alluvium, and flowstone

Flowstone was not an uncommon find along the route. The flowstone, which we discovered on the rock surface following the removal of turf and soil layers, was one of the first signs alarming us that any further excavations must be carried out under strict professional supervision. Based on this, we had expected to find other vertical or lateral sediment in the vicinity. The most common discovery in this area were cave sediments.

The largest heaps of flowstone in terms of surface area and thickness were determined in the northern part of the route, on Cretaceous rock. After it was dug out, it was identified almost exclusively as variously bedded flowstone. There were almost no stalactites and stalagmites, as they were probably removed together with the sediment or crushed and washed away at an earlier stage. Flowstone in this section of the route is mainly orange-brown, whereas white, meat-red and grey-yellow flowstone occurs significantly less. It should also be noted that the flowstone undergoes changes, rather rapid and in short succession, in terms of morphology, colour and preservation level. In general though, the flowstone is well preserved, mainly because it was well-shielded against surface influences, blocking the onset of the weathering process. The most common findings were column-shaped calcite crystals up to a few centimetres in size, occasionally exceeding 10 cm.

Near the Kozina motorway exit, we discovered completely black flowstone of a smaller scale; it was partially weathered and non-homogeneous. It contained a large amount of non-carbonate inclusions. A significant portion of yellow-grey cave sediment was also found in a larger flowstone deposit in the cave close by. The basic structure of the flowstone was not weathered. Subsequently, cave sediment filled the space in the otherwise highly porous flowstone.

North of the motel near Kozina, still in the northern section of the route, a highly homogenous and compact bed of orange-brown flowstone was found when drilling. With regard to the group of active dolines in that area, it was assumed that the excavation will unearth an old cave system as well.

In the central and south section of the route, flowstone occurs in considerably smaller amounts. Individual locations are less frequent, although not to be disregarded. Some of the subterranean karst forms, which we came across later in the more than 20 m deep roadcut, could be explained based on the expert examination of the flowstone which had been found beforehand on the surface.

The central and south section of the route displayed a variegated composition and form of flowstone. We would especially like to point out the discovery of a large heap of yellow-white bedded flowstone in alveolinid-nummulitid limestone. This flowstone stands out because of its colour, the crystallization of the calcite crystals, and its preservation level. It was discovered only a few metres below the surface, almost on top of the hill.

3.2 Conclusion

Unroofed caves are an important karst form that makes up a part of the karst surface and epikarst, and provides us with valuable evidence of the karst development. The form of unroofed caves can be traced back to the cross-section of the lowering karst surface with its cave systems. Water, draining from the surface, carries away cave sediment, which is why unroofed caves often occur as hollows in the karst surface. In many cases they are complex karst formations. Dolines occur in the centre of old passages or at their perimeter. In the latter case, unroofed caves resemble dolines. Several unroofed caves may be lined up in a row. Open semi-circular doline-like forms emerge when the slope is cut off with an old sediment-filled passage. They may also occur as individual notches on flat terrain or riddling the sides of dolines. In most cases they feature no rocks and may be distinguished from the remaining terrain based on the vegetation, provided, of course, there is no cave sediment or flowstone readily identifiable on the surface. Besides the lithostratigraphic composition of the epikarst, which is a major influencing factor on surface morphogenesis, the occurrence of flowstone on the surface and on the rock relief can in many ways suggest the possible presence of an old (and present) underground morphology.

The typical, and increasingly familiar, karst forms such as notches, doline-like forms on flat or inclined terrain and strings of dolines are all indicators which make unroofed caves an increasingly distinguishable and easily identifiable karst form.

4 The Section Between Divača and Kozina Revealed Many Characteristics of Karst Development

The motorway section from Divača to Kozina stretching 7.5 km, which is described in this document, revealed 50 old caves (Fig. 4.1), most of them filled with alluvium, many of which unroofed, and 6 of them qualifying as shafts. Old caves contain the oldest geological traces dating back to the very early periods of the karst aquifer development. We were able to preserve the more significant caves. Today, these are either hidden below the road or accessible via man-made entrances.

4.1 Karst Surface

From Divača the road runs over the Divača lowland, rising, at its edge, to the karst plain of Kozina, and running across it almost to Kozina. The first section of the motorway route overlies rudist Cretaceous limestone of the Lipica Formation and continues over Cretaceous and Palaeocene limestone of the Liburnia Formation that contains coal; it is followed by a belt of miliolid limestone of the Slivje Formation and Eocene Operculina limestone, next comes an extensive belt of alveolinid-nummulitid limestone which precedes the Kozina hill and runs across it all the way to the beginning of the karst plain. Due to the anticline-synclinal structure of this part of the Karst region, this particular geologic structure gets repeated along the route to Kozina in a similar yet reverse sequence.

The surface of this area was dotted by dolines (Fig. 4.2). The largest dolines have a diameter of 200 m and a depth of 35 m. The bottoms of some cultivated dolines featured fields. Old cultivated dolines were revealed in the course of archaeological excavations. The doline bottoms often displayed rocks and stones covered with a layer of soil; alternatively, the soil was overlying thick layers of loam that showed no indication of human intervention. We proceeded with measurements as to the amount of soil that was transported into dolines after humans have begun interfering in this setting. It was established that a 0.14 m thick layer of soil was transported from the surface down into the dolines. Macroscopic descriptions of sediments in the dolines were carried out and the sediment layers were sampled for mineralogical analyses.

Under, occasionally relatively thick, layers of brown and red soil, especially along distinctive fissures, the surface is often dissected, whereas in other places the epikarst part of the aquifer is shallow. Typical subsoil rocky features have developed underground. The subsoil rocky surface of alveolinid-nummulitid limestone is unique. Along their rims, dolines which have developed along faults, have walls with a typically rocky relief. Karren and solution pans are the dominating forms. Unroofed caves make up another important element of the karst surface.

M. Knez and T. Slabe (eds.), *Cave Exploration in Slovenia*,
Cave and Karst Systems of the World, DOI 10.1007/978-3-319-21203-6_4

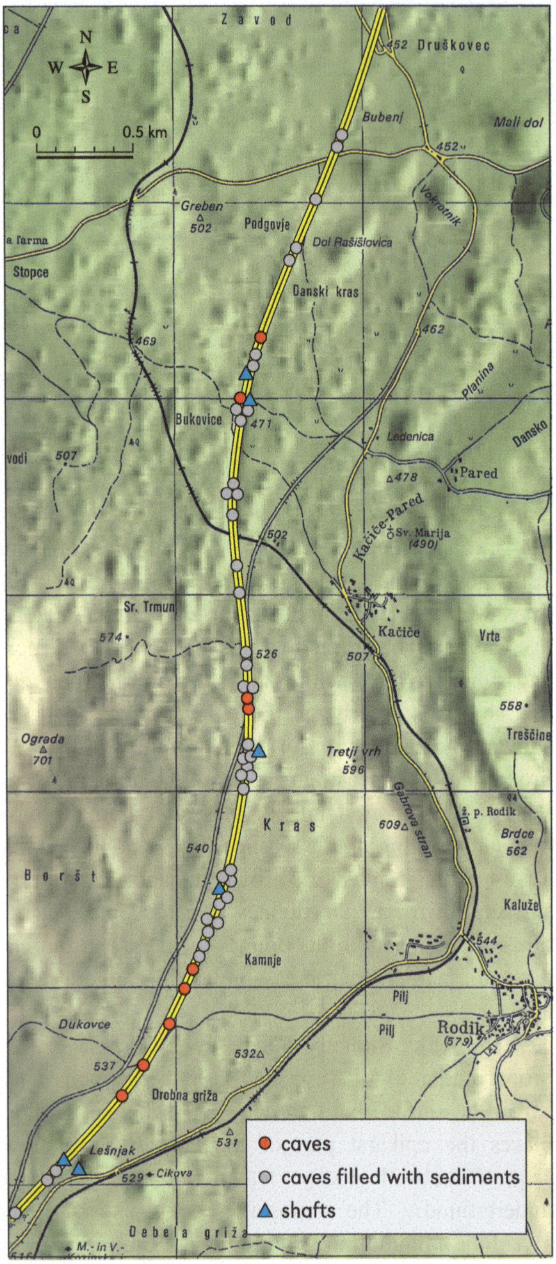

Fig. 4.1 Caves uncovered during construction works

4.2 Newly Discovered Caves

Caves occur in all types of limestone; their frequency can be mainly attributed to the local properties of bedded or fissured rocks and the aquifer development, but also to the methods used to reveal these caves in their respective setting.

Caves can be divided in old caves, mostly horizontal or inclined, and in shafts which are shaped by water percolating from the surface downwards. At one point, old caves were characterized by through-flowing water (Figs. 4.3, 4.4 and 4.5), however the drop of underground water meant that they would eventually dry out. There are also some old shafts that were either shaped by percolation water or that make up a part of bigger cave systems. The latter are depicted as old hollow caves on the map (Fig. 4.1) which illustrates the distribution of typical caves found along this particular motorway route.

This type of classification and the emphasis on their respective characteristics are the consequence, the predominance of caves that were filled up with either fine-grained clay or sand flysch sediments (Fig. 4.6). Moreover, this area was marked by the absence of flysch pebbles such as were found on the motorway routes around Divača. Speleothems and flowstone heaps were found under or on top of the alluvium in these passages, some of which had a diameter of 6 m, although the majority were smaller. There were 41 such caves and 9 empty caves. Two thirds of the caves filled with fine-grained alluvium already had a non-existent roof (Figs. 4.7 and 4.8). Drawing on our experience derived from monitoring the motorway construction in the Karst region, we were able to find two unroofed caves as early as the preliminary karstological study of this particular route (Šebela 1996). It is not uncommon that the road cuts off the same passage or cave system several times. Fine-grained alluvium completely filled all the cracks in caves, preventing the runoff to transport it away. However, due to the lowering of the karst surface for several tens of metres, these caves today occur on the karst surface as elongated, winding notches or as peculiar dolines. Two associates from the Institute of Geology, Academy of Sciences of the Czech Republic, helped us sample this alluvium to be subjected to various palaeomagnetic analyses.

In addition, a small coal mine was explored, measured and mapped out. The coal was sandwiched between two layers of Palaeogene limestone. Similar lenses of coal were also found elsewhere along the route.

Fig. 4.2 Doline-covered surface

Fig. 4.3 An old cave in the embankment of a roadcut

Fig. 4.4 The entrance into an old cave

Fig. 4.5 Collapsed roof in the old cave

Fig. 4.6 Alluvia in the unroofed cave at Divača

Fig. 4.7 An unroofed cave

Fig. 4.8 The cross-section of the old cave filled with fine-grained alluvium

4.3 Motorway Construction and Karst Features

For the motorway construction, several, even large-scale dolines had to be filled up as they were right on the motorway route, while some of the close-by dolines were used to dump excess rocks and soil. Before they were filled up, we had the soil and alluvim from the dolines removed.

The road passes over a horizontal passage which we found at the bottom of Škrinjarica cave. Lying 100 m under the route itself, the passage is not affected by the roadway above. The caves (most of them being small-scale) that had opened during the construction works were filled in and covered with concrete slabs. Concrete slabs were also put over narrow caves (Fig. 4.9) and shafts (Fig. 4.10). The route was surveyed by ground penetrating radar in order to identify any possible hitherto unknown caves under the road.

West of Divača, a fascinating 14 m deep old cave (Fig. 4.11) had opened. Its bottom is characterized by a small hall which is 12 m in length, 10 m in width and 5 m in height. Its roof displays several chimneys. The

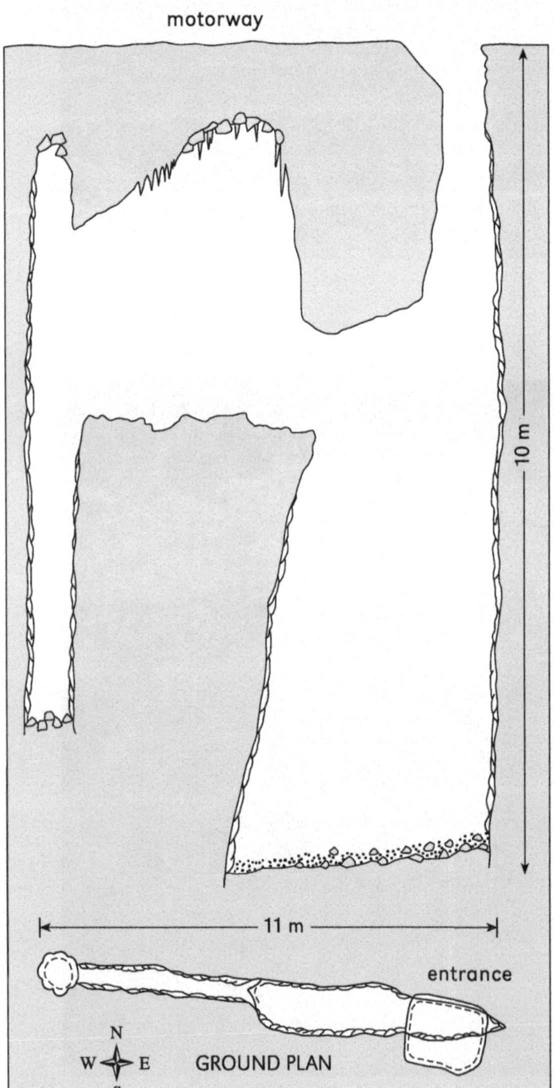

Fig. 4.9 A cave below the motorway route

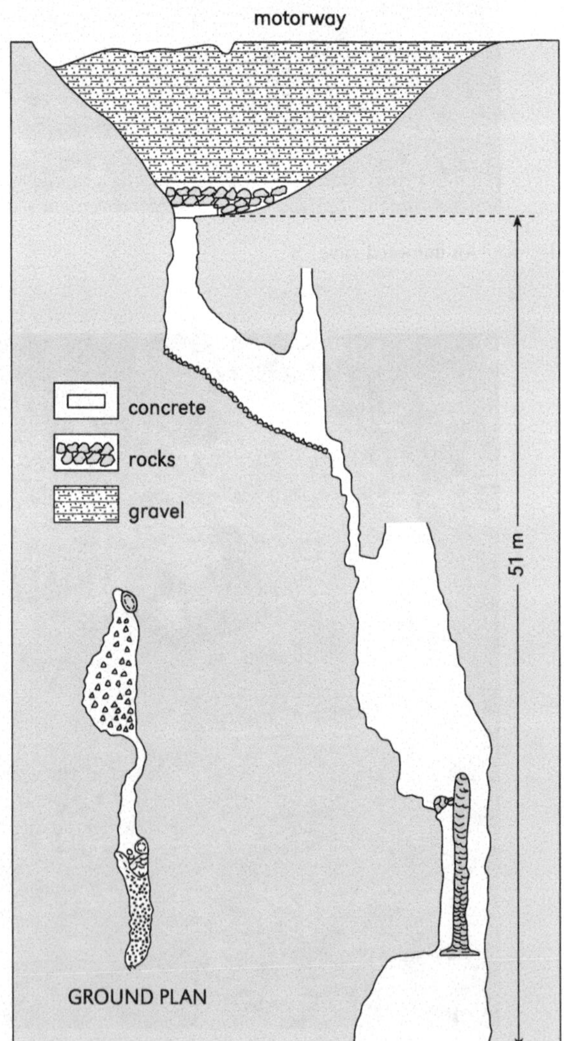

Fig. 4.10 A shaft at the bottom of a doline

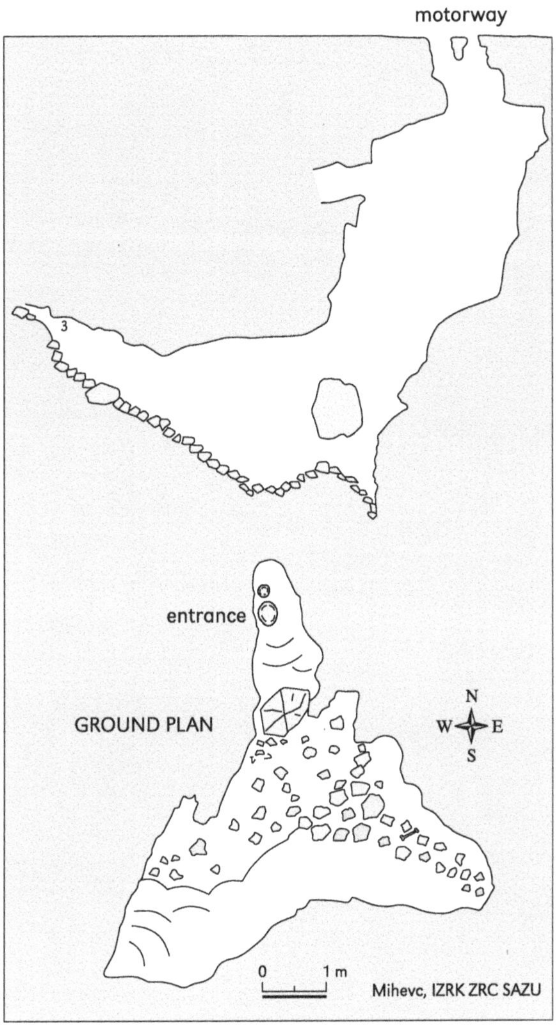

Fig. 4.11 A preserved old cave under the road

bottom is covered by rocks and loam. Fractured rock, the result of blasting, occurs only in the first metre of the cave below the surface. We proposed that this cave be preserved. It is accessed via a concrete tube that reaches to the lip of the underpass. It is now covered by a concrete lid.

The high permeability of the karst aquifer indicates how easily in fact underground water can be compromised. It can be compromised by the runoff from the road and potential hazardous substance spills, which may occur during construction or at a later point when the road is already used for traffic. For this reason, roads are designed to be impermeable, the runoff being collected in oil separators, allowing only treated water to drain onto the karst surface.

4.4 Conclusion

The construction of this particular section of the motorway revealed that the local karst features display very similar characteristics in terms of aquifer development to those observed previously on the route between Čebulovica and Fernetiči.

The old caves, which we described above, were once, in a time when the underground water table was still high enough, characterized by through-flowing water. At that time, the aquifer was encased in flysch high up and partly covered with it. The runoff from the flysch introduced loam and sand into the caves. After they had been dry for a while and featured flowstone deposits, flood waters filled the caves with fine-grained sediments in one of the last cave development phases. The alluvium sealed the cracks and withstood even the long periods of the aquifer development during which its surface lowered for several tens of metres. It thus appears that old unroofed caves are an important component of the karst surface, providing clues to the development of the aquifer. Seeing as today the underground water level is 200 m and more below the surface, the traces of ancient water flow through the aquifer are now preserved only in caves and their respective alluvium. There was no indication of surface water flows that could have reshaped the aquifer's surface (Melik 1960, 201; Radinja 1972, 13).

5 Great Cavernosity Between Dane and Fernetiči Points to Diverse Karst Formation

The motorway route between Dane and Fernetiči penetrates the karst ridge near Sežana, running over the karst plain to the border with Italy. The karst developed in Cretaceous limestone, which is intersected by smaller faults; this was best visible during the tunnel excavation under the Tabor Hill. The karst plain is studded with larger, mostly funnel-like dolines. The remaining relief, even though mostly covered by some type of flora, is covered by thin layers of soil. The surface is scarred by medium loam-filled notches. It was established that the depressions were in fact old caves which had lost their roof, and are now filled with fine-grained alluvium. Newly discovered karst features, caves in particular, provide us with a wealth of information with regard to the development of this part of the Karst region.

5.1 Caves

Caves can be divided into old caves, which are remnants of a former subsurface flow, and into shafts (Fig. 5.1).

Newly discovered shafts as part of old caves have a cross-section from 1.5 to 8 m in diameter. Most are filled up by fine-grained alluvium, with only the occasional smaller cave, as a rule with not more than 50 m^3 of volume, occurring void (Fig. 5.2). They often feature speleothems, the bottom and walls are adorned by heaps of flowstone. The roofs over such caves are thin, measuring 1–2 m at the most. The network of passages, their form and the rocky rims of these caves all indicate that they were formed as a result of slow water flow in the phreatic zone.

Many old caves which are filled up by fine-grained alluvium are already marked with an absence of roof (Figs. 5.3 and 5.4). The passages wind across the karst surface as shallow notches. It is therefore possible to make them out in the landscape even before commencing earthworks. In roadcuts and tunnels, the passages which are filled up by fine-grained sediments, may be accessed via their cross-sections. The caves are filled up by layers of yellow loam and sand, frequently consolidated; overlying are usually red loam and brown soil. The speleothems and flowstone heaps are covered with sediments, but the sediments are only rarely covered with flowstone. Small scallops were found which are relatively rare indications of faster water flows; in contrast, there were no pebble deposits found in these caves although they are a frequent occurrence in the old caves dotting the motorway routes around Divača. The passages of old caves are frequently cut off by shafts, and the sediments are, as a result, removed in a funnel shape. In addition, old caves may also be intersected by dolines. Percolation water cut out meandering grooves in the bottoms of some, usually smaller, passages. In some locations, patches of flysch remained on the surface for a longer period of time, acting as a layer from which runoff drained into the limestone.

Parts of newly discovered passages are filled up by coarse rubble (Fig. 5.5). Rubble, with a diameter ranging up to 5 cm, occurred as the result of the disintegration of the karst surface in the cold Pleistocene periods. It generally accumulated in such parts of caves, which were conducive to the transport of fine-grained sediments due to high vertical permeability of the aquifer, and where the thin roof caved in

M. Knez and T. Slabe (eds.), *Cave Exploration in Slovenia*,
Cave and Karst Systems of the World, DOI 10.1007/978-3-319-21203-6_5

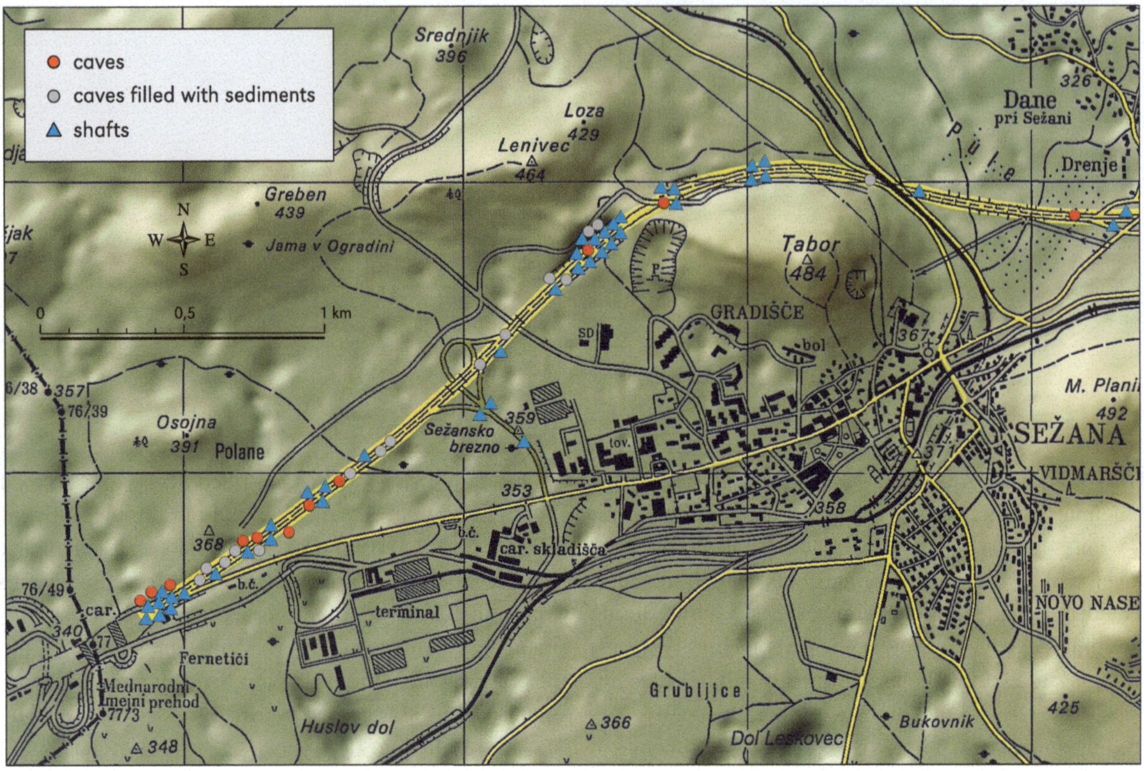

Fig. 5.1 Caves uncovered during construction

just before being filled by rubble. The passages are characterized by sections that are alternatingly filled by loam or rubble.

It appears that some of the newly discovered and relatively large old caves, either void or filled up by fine-grained sediments or rubble, which are located in the western part of the motorway route actually make up a large-scale lateral and vertical cave system which was gradually unearthed during the motorway route construction, seeing as the route cut right into the slightly undulating karst surface.

The surface runoff is drained into the underground by shafts and fissures. Along the route, two shafts were already known, the deeper one reaching 20 m deep. In the course of denuding the surface and during the earthworks, several shafts (even up to 30 m deep) were discovered in this relatively cavernous limestone. They may be divided into shafts with distinctive traces of percolation water, typically having circular cross-sections, and cracks of various dimensions that emerged along the fissures, their walls often covered with flowstone. Larger cracks developed along wrench-faults in fissured zones (Čar 1982). Spacious shafts with thick flowstone deposits also make up old cave systems that were apparently shaped in a phreatic zone. Shaft-like caverns emerged also among the breakdown boulders located in disintegrating old caves. Yet, the most common findings were newly discovered shafts through which the water infiltrates the aquifer. Most of them are located at the bottom and on the sides of larger, funnel-shaped dolines. Most of shafts had no visible natural entrances. After the soil was removed, there emerged narrow and impassable openings. Spacious shafts were mainly discovered and explored when excavating the roadcuts and the tunnel, i.e. only 10–30 m below the surface. Perhaps this is the result of the accumulation of water which diffusely seeps through the permeable surface and in effect turns the more spacious shafts—even those with a natural entrance, given the considerably lowered surface (Slabe 1996)—into water reservoirs. Klimchouk (1995) too established that there is diffuse water infiltration into epikarst going on, converging at the contact with the vadose zone into streams. The spaciousness

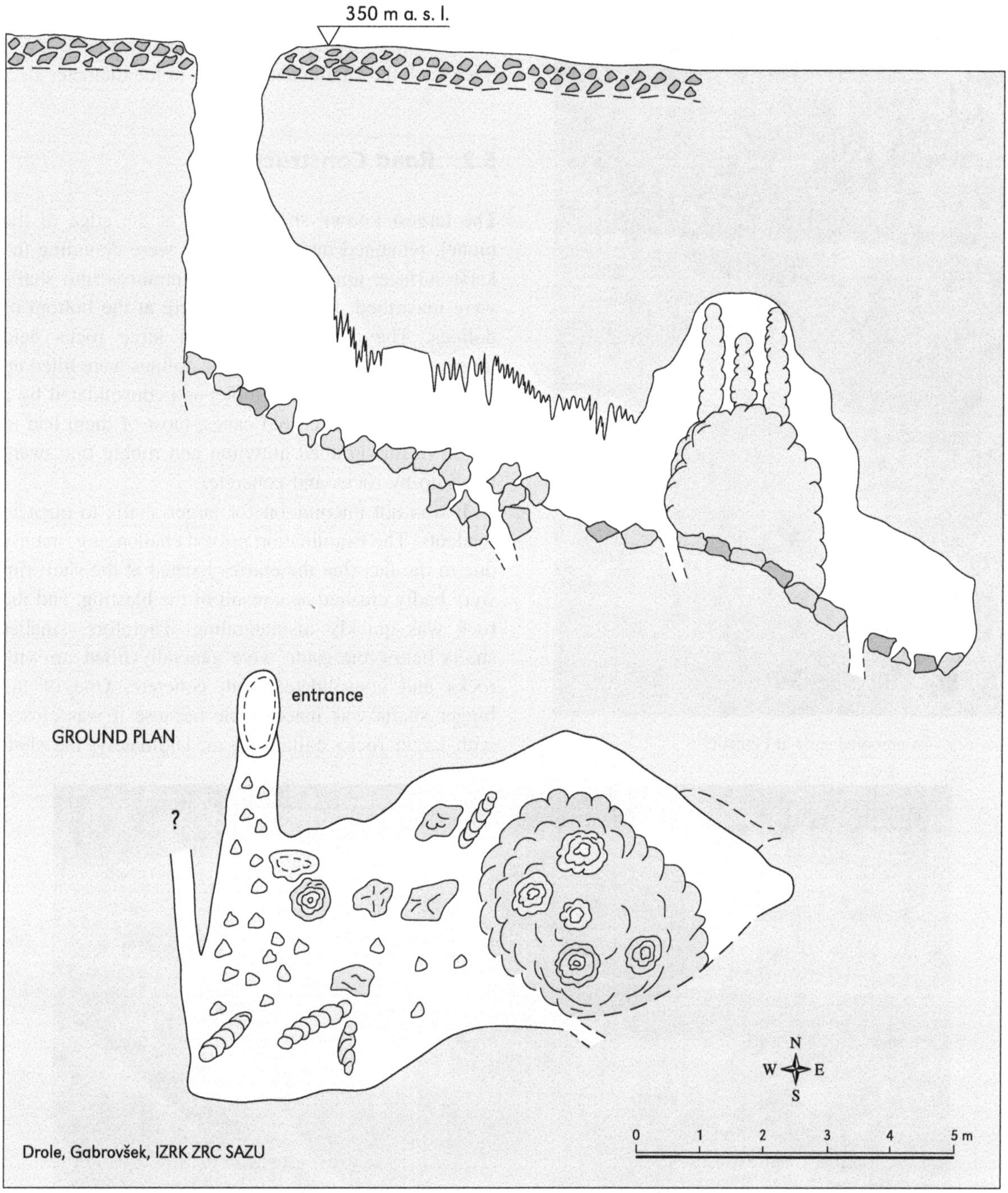

Fig. 5.2 A cave whose roof collapsed during gravel reinforcement works on the route

Fig. 5.3 An unroofed cave at Fernetiči

and shape of the shafts are also the result of the fissures along which they were formed. Shafts enable us to trace back the significant changes in the diameter size.

5.2 Road Construction

The largest known shaft, located at the edge of the tunnel, remained intact. While we were denuding the karst surface, numerous smaller entrances into shafts were unearthed. Most of them were at the bottom of dolines. They were closed with large rocks held together with concrete. Next, the dolines were filled up by layers of stones and rubble, and consolidated by a vibration roller. Even old caves, most of them had to be rid of fine-grained alluvium and rubble first, were filled up by rocks and concrete.

It was not uncommon for larger shafts to open in roadcuts. The examination proved challenging, mainly due to the fact that the entries located at the shaft rim were badly crushed as a result of the blasting, and the rock was quickly disintegrating. Therefore, smaller shafts below the route were generally filled up with rocks and consolidated with concrete. One of the bigger shafts was inaccessible because it was closed with larger rocks collapsing in. Ultimately, the shaft

Fig. 5.4 A winding passage of an unroofed cave near Sežana

Fig. 5.5 A rubble-filled cave

was buried though while we were deepening the route. We have proposed that the route running over it should be consolidated with concrete.

Fine-grained alluvium was washed out onto the route from old caves which were located in the embankments of roadcuts. For this reason, the caves had to be walled up (Fig. 5.6).

In addition, excavation works for the tunnel brought to light several old caves and shafts. They were small-scale and could be closed with the concrete rim of the tunnel. However, the caves in the ceiling of the tunnel could not be entered. The rock lining their rims was crushed due to blasting, with rocks and rubble pouring out the shafts, the rim was disintegrating.

5.3 Conclusion

It is quite remarkable the extent to which the surface of this part of the karst aquifer is dotted by old caves which are filled up by fine-grained alluvium and rubble and that had, at one point or another, lost their roofs. The higher parts of the aquifer have been subject to reshaping by long-term percolating water. Old caves reveal glimpses into a time when subsur-

Fig. 5.6 Caves in the embankment of a motorway roadcut that were closed with rock walls

Fig. 5.7 A subsidence in the motorway

face water was close to the actual surface of today. The continuous lowering of the karst regions led to the situation where today's surface is lower than the original karst surface was. The conditions we encountered on this route were similar to other motorway sections in the Karst region, only that the surface pitted with old caves was bigger in this locality. It appears that great cavernosity came about as the result of the high permeability of this part of the aquifer. The size of the passages suggests that they were passed by larger water flows which had originally completely filled the passages. It is assumed that a remnant of a flysch nappe was somewhere in the vicinity, feeding the aquifer with surface runoff. This is also suggested by the ponor properties of some caves. The cave bottoms display meandering patterns carved out by small-scale water flows. Even after the subsurface water table had dropped, flysch layers still lingered over the limestone. Despite the water drop, high waters occasionally still reached the caves, eventually filling them up with fine-grained sediments. In the cooler periods of the Pleistocene, some of the old caves were filled up by rubble. The great cavernosity of this part of the Karst region is highlighted by the numerous shafts and fissures that were discovered during earthworks. Similarly, this part of the Karst region showed no traces of surface water flows from the period when the limestone was still dammed. The same conclusion was made earlier while working on the route between Čebulovica and Dane. We discovered that all the alluvium is of cave origin.

One of the most reliable indicators of permeability, the great cavernosity characterizing this area, reminds us how delicately and carefully we must proceed with the construction and subsequent use of this motorway. The vibration roller used to consolidate the rubble on the roadway led to the emergence of subsidence (Fig. 5.7) and the collapse of roofs over smaller caves. However, considering the great cavernosity in this section, it is quite possible that other caves, even larger in scale, are hidden below the surface. It may be concluded that ground-penetrating radar surveys are an essential tool for road safety.

6 Unroofed Caves Discovered on the Route Between Čebulovica and Dane

Denudation of the karst surface and the corresponding earthworks necessary for road construction revealed a myriad karst features. These include old caves, which are either empty or filled with sediment; shafts, draining water from the surface into the karst interior; and various dolines. We undertook to explore, measure and map out the newly discovered caves. This provided us with new insights into the karst surface which is characterized by unroofed caves providing us, in turn, with clues as to the comprehensive development of the karst aquifer. In addition, we sampled some of the most typical flowstones and alluviums, which we found in old caves.

6.1 Karst Near Divača

The Čebulovica–Divača route runs transversely along the eastern edge of the karst ridge which extends from Štorje towards Čebulovica to the doline-dotted plain west of the Škocjanske jame Caves, whereas the section between Divača and Dane is laid out over the less karstified area of the Divača lowland around Divača (Melik 1960, 199).

It was established by the karstologists that the Karst region still holds well-preserved traces of the original surface water runoff towards the northwest. Melik based his (1960, 201) assumption on the elevations on the present landscape, and Radinja (1972, 13) on the abandoned valleys and sediment remains on the karst surface. It was believed that once, before the carbonate rocks became exposed, the limestone was confined and the subsurface water perched, thus recharging the surface flows. The former surface runoff from the Brkini hills over the Karst region formed a line-up of valleys between Divača and Brestovica (Habič 1974, 8). In his study of the development of the surface connecting the poljes of Postojna and Cerknica, Gams (1965a, 90) came to the conclusion that there must have been some morphological processes underway that accelerated karst dissection and the formation of doline-like shapes, which had been, all too often, labelled as the remnants of fluvial valleys. Relief characteristics of the Divača lowland are contingent on structure and lithology, seeing as different lithostratigraphic elements and the main faults are arranged from southeast to northwest (Habič 1974, 8).

Near Divača the limestone of the Brkini syncline comes in contact with Lower Cretaceous dark grey bituminous dolomites along a distinctive Dinaric-directed fault (Habič 1974, 4). The part of the route south of Divača runs across the dark grey bedded Cretaceous limestone alternating with rudist limestone. From Divača to Dane, i.e. along the Divača fault, which is reflected in the relief, the route runs over Palaeocene limestone—also over Cretaceous limestone for short periods—and over dark grey Cretaceous dolomite. Earthworks in the route also revealed tectonically fissured rock and numerous fault mirrors, indicating that the rock blocks had been moved in different directions. Along the fault area, the limestone has been crushed into rubble and mylonite.

The present hydrologic situation matches the lithological characteristics of this part of the Karst region. Today, subsurface flows run 200–300 m below the surface. South of Divača the Reka River works its way from the Škocjanske jame Caves in parallel to the route, crossing it in the vicinity of Divača and then

M. Knez and T. Slabe (eds.), *Cave Exploration in Slovenia*,
Cave and Karst Systems of the World, DOI 10.1007/978-3-319-21203-6_6

sinking into run beneath the Divača lowland towards northwest, tracing its way back to the springs of the Timavo river. Precipitation waters feed the springs through vertical percolation.

Near Dane, under a 0.5 m thick layer of brown soil, we found a hummocky rocky surface. The elongated hummocks measuring 2 m in width are rounded off and smooth, and alternate with fissures (Gams 1971). Individual smaller rock belts were crushed. They feature yellow rubble and red loam. Yellow rubble can be found in fractured zones of other route sections as well.

6.2 Dolines

Habič (1974, 6, 7) divided dolines, popping up in these sections of the motorway, based on their shape into funnel-shaped dolines from which loam was mostly washed out, bowl-shaped dolines with large patches of loam at the bottom, and shallow, infilled dolines with slightly permeable loam. The filled up dolines often retain water. Near Divača there are 11 dolines per kilometre of the route, and between Divača and Sežana there are only 5. The south section of the motorway Čebulovica-Divača is pitted with large funnel-shaped dolines which contain small amounts of loam. The limestone of the Divača lowland is predominantly scarred with bowl-shaped and filled up dolines. The latter are rather small, with 30–50 m in diameter and run 5–15 m deep. The greatest depth in a doline which was established by means of drilling, was 27.5 m (Habič 1974, 5). The dolines mostly feature red and brown karst loams which were down-washed from the doline slopes and accumulated on the rubble base. The loam, in turn, is the product of the weathering of the rocky bedrock. Some dolines in addition feature flysch sand and gravel as well as yellowish and layered silt, which is probably alluvium deposited by the former watercourses (Habič 1974, 6). Habič (1987) used the drilling method to study a particular doline near Sežana, which was in part already excavated. Small-scale filled up dolines are occasionally etched into the hummocky dolomite

Fig. 6.1 Removing soil and sediments from a filled up doline

surface, which is covered by a thin layer of soil. The upper layer of the doline infill is made up of brown soil, with red loam resting 1.5 m below the surface.

The motorway construction works required that we remove the alluvium from the filled up dolines (Fig. 6.1). The results obtained in the course of the preceding research, which included drilling into the doline alluvium, were thus confirmed. Frequently, we found considerable amounts of older, mostly flysch alluvium in the dolines. In one doline situated in the Divača–Dane section the layer of flysch gravel almost reached the surface. But it should be pointed out that there are many filled up dolines within alluvium-filled old caves that had lost their roofs at one point. Shafts were formed under passages which once had through-flowing water. This water flow deposited flysch loam and sand in the last distinctive stages of the development of the passages. Later, in the dry periods flowstone was deposited. These passages and shafts are permeated by infiltrating water which feeds the underground flows. Today, the underground flow runs deep below the surface. Funnel-shaped notches, filled with brown and red karst loam and rubble, were formed in the flysch alluvium rain-washed by the surface water.

Near the town of Povir, a pond (Fig. 6.2) was created in a sinkhole, which was surrounded by partly man-made walls, because water had collected at the bottom and could not drain off. The pond was located on an elevation, on top of the anticline. The elevation had to be cut through, and we were unable to preserve the pond. The excavation in the pond confirmed our assumption that the water was prevented from draining by an impermeable layer of loam, which was deposited in the old passage, the floor of which was covered by a large heap of flowstone (Fig. 6.3).

Shafts were a frequent occurrence in limestone and dolomite covering the bottoms or lining the edge of the dolines. Large dolines with openings (Šušteršič 1985, 94) featured numerous shafts giving rise to their particular shape. Near Žirje and Dane, artificial filled-up dolines were cut through because of the earthworks (Gams 1974, 177; 1991, 39). It was clear from the excavated dolines' cross-sections that they were filled in with rock and stone which had been

Fig. 6.2 A pond with water

collected while clearing out the doline and the adjacent karst surface (Fig. 6.4). The people transferred rocks and stones to the excavated doline and later covered them with soil. This method was used to obtain more flat arable land.

We approached the road construction by first removing the loam and rubble from the excavated dolines. The doline bottom, especially if cavernous, was then consolidated with rocks held together with concrete. Next, we put in several 30 cm layers of rocky rubble to fill the doline, consolidating each layer with a vibration roller. The soil and loam that were taken from the dolines were used as infill for several other dolines along the route.

Fig. 6.3 Flowstone and alluvia under the pond

Fig. 6.4 Uncovering of an artificial doline

6.3 Caves

The caves that we examined in the construction of this motorway route can be classified as old caves and as shafts (Fig. 6.5). Old caves are either empty or filled with alluvium. The Golobja jama Cave and Mošenjska (Srnja) jama Cave, both of which had to be blasted to make way for roadcuts, were known in advance. However, the new discoveries comprised of 76 caves, 24 empty old caves, 33 alluvium-filled old caves and 19 shafts.

6.3.1 Cavernosity of Different Rock Types

Hollow or alluvium-filled large-scale old caves emerge in all types of rock—Palaeocene and Cretaceous limestone and the Lower Cretaceous bituminous dolomites near Dane. The Palaeocene limestone south of the Čebulovica hill is scarce with old caves, in contrast to the Palaeocene limestone east of the village of Povir. A road stretch of a mere 50 m revealed 4 hollow caves and 3 alluvium-filled caves. A large alluvium-filled cave was also found in the Upper Cretaceous limestone south of Divača. Dolomites are dotted with smaller caves. Mainly their passages have a diameter of less than 1 m, most of them are filled with alluvium and contain flowstone deposits. Larger caves are a rather rare occurrence. Nevertheless, there were 2 alluvium-filled canyon passages found extending down to 5 and 6 m below ground.

Shafts are present in limestone and dolomite. The largest share of the big shafts was identified in the Cretaceous limestone close to the Škocjanske jame Caves. It is relatively rare for shafts to emerge in

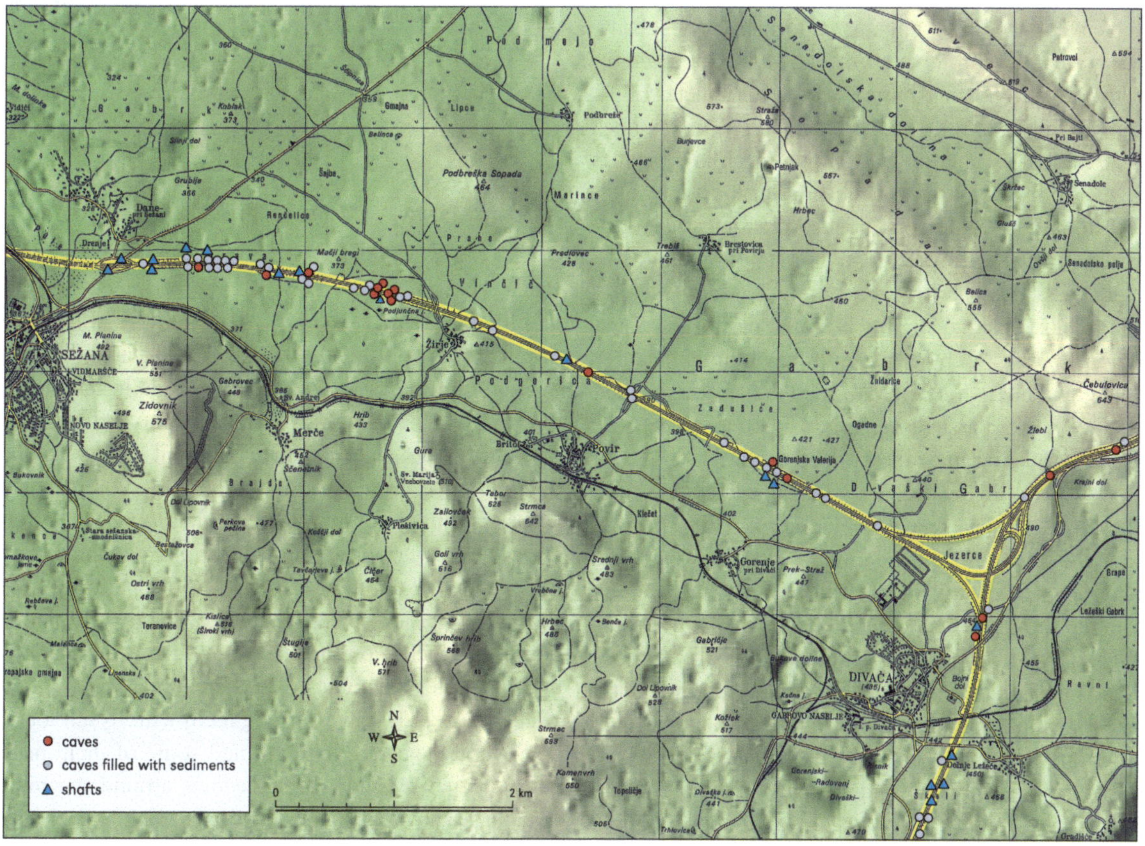

Fig. 6.5 Caves uncovered during motorway construction

Palaeocene limestone. Several shafts opened in the Lower Cretaceous dolomite. However, they are mostly small and cannot be entered. At the rim of a doline, some 10 m east off the route, a 30 m deep fissured shaft had opened, with a stepped shaft not far off.

Dular (1993a, b) established that Palaeocene and Eocene limestone and Cretaceous dolomite are less karstified and Upper Cretaceous highly karstified. Indeed, Palaeocene limestone close to the Čebulovica hill is scarce with old caves, and there were no shafts discovered. The poor cavernosity is also indicated by the banks of the large roadcut under Čebulovica hill, which featured two small, hollow old caves and one alluvium-filled cave. There are relatively many caves in the Palaeocene limestone lining the route from Divača to Dane. That section is pitted with old caves —hollow and alluvium-filled—and fissured shafts.

6.3.2 Old Caves

The caves, which are remains of former early subsurface water flows in this part of the Karst, are made up of networks of slating passages with a diameter of 1–5 m, and dotted with smaller chambers. The passages were created by a slow water flow, which is indicated by the complex system of meandering passages and, occasionally, by the rocky relief. Passage walls are partly covered with flowstone and stalactites and stalagmites. This type of caves, though relatively rare, occurs mostly in the slightly elevated part of the Karst region, below Čebulovica hill. Dolomite features smaller cavities which were formed in between pitched beds of rock. Passages emerged next to such beds which weathered into fine-grained sand.

Alluvium-filled caves are a far more common find. These types of caves have lost their roofs (Figs. 6.6 and 6.7) and are now meandering as furrows over the landscape. Alternatively, the filled in passages can be seen in the cross-sections in the banks of roadcuts (Figs. 6.8 and 6.9). Cave passages vary in size, at times exceeding a diameter of 5 m. Cave walls are frequently covered with flowstone, their bottoms displaying large flowstone heaps. Flysch and limestone gravel, flysch loam and sand are deposited in caves. In the biggest of caves, we found flysch beds over 5 m thick. Alluvium-filled caves can be tracked along the entire route Čebulovica–Divača–Dane. We sampled several sediments and applied X-ray diffraction, mineralogical, granulometric and pollen analyses to establish the origin and age of the alluvium. We were able to make out some of the most distinctive periods in the development of these caves. The meanders and passage shape provide clues to their original shape in the phreatic zone. Smaller scallops on the walls of some passages plus the gravel indicate that there was a relatively fast water flow for a short time, which appears to have carved out the passages. The dry periods in the formation period were characterized by flowstone deposits which have in places covered wall scallops and the water sediments elsewhere. Eventually, flood water reached the caves, filling them with flysch loam.

Fig. 6.6 Exploring an unroofed cave at Povir

Several caves are fissured, and some of them have been created by short-period eddies around the most notable fissures, as suggested by the cave shape.

Teeth (Fig. 6.10) and the skull remains of a Pleistocene horse, which we found while clearing the alluvium from the cave near Povir, allow us to

Fig. 6.7 Preserved walls of an unroofed cave

conclude that there was life in this part of the Karst region in that period.

6.3.3 Shafts

The motorway route revealed several small entry shafts and several deeper stepped shafts. The deepest shaft reached 109 m deep (Fig. 6.11). In this part of the Karst region, water flows are very deep below the surface—200 m and over. For a longer period, they have been infiltrated by precipitation water. Additionally, water carved its way along the fissures. The less prominent fissures, which allowed smaller drainage of water, meant that large caves could be formed only on more open sections. The walls in fissured caves are already partly covered with flowstone (Fig. 6.12).

Prominent fissures and a steady water influx were conducive to the formation of shafts with an often circular cross-section. Their walls are dissected into larger vertical channels or flutes, pointing to water percolation. Flowstone also covers walls in larger shafts.

The largest share of active shafts is at the bottom or at the edge of dolines. However, they also appear under old caves which were dissected by water that flushed away the old alluvium.

6.3.4 Road Construction and Caves

A growing number of new caves emerged as the construction of the motorway went on and the vegetation and soil from the karst surface had been removed. We found a rather large number of old caves with horizontal passages filled up with alluvium. Several shafts opened at the bottoms and slopes of dolines.

Other caves have been revealed during excavations of roadcuts either under the rocky surface or in the walls. These are hollow caves or old caves filled up by sediment and shafts. In the roadcut near the overpass of the main road, a larger old cave opened up when the roof over the passages collapsed (Fig. 6.13). That breakdown occurred at the vertical fault running parallel with the motorway route. The same fault is where the largest passage in the cave was formed (Fig. 6.14). The roof over the passage was 10 m thick. The cave bottom and its potential extension, which can be seen through impassable straits, were choked because of the breakdown. At the foot of the roadcut side, along the Divača–Dane route, a cave opened up along a fissure which was perpendicular to the motorway route. The cave is interfingering with the roadcut side, extending in part also under the route. In this part of the route, further caves with smaller widths (up to 4 m) and up to 20 m deep were revealed. Caves kept emerging during the excavation of the roadcut and later, when the rubble bottom infill was consolidated.

Blasting and roadcut excavation resulted in the fracturing of the rock in the surrounding caves. In the cave in the Čebulovica–Divača route, which opened following a roof collapse, the rock is fractured down to 12 m deep in pieces of a few cm^3. The fissures, which resulted from the blasting, are up to 1 cm wide and reach down to the cave bottom. However, most of the dripstones at the bottom of the cave remained attached to the walls. In another cave, which opened up in the

Fig. 6.8 Caves filled with alluvia in the embankment of a motorway roadcut

Fig. 6.9 Hollow and alluvium-filled caves uncovered during excavation works for the motorway roadcut

Fig. 6.10 Teeth of a Pleistocene horse

side of the roadcut on the same route, the rock in the entrance part was highly fractured. The rim has in part disintegrated and small pieces of rock fell from the roof of a small chamber within the cave. The entrance part of the fissured shaft, located on the Divača–Dane route, displays fractured rock. Hence, a part of the roof disintegrated, leading to the flowstone deposits to spall as well. Deeper down the shaft, we could detect no lasting impact of the blasting.

In the southern part, the Čebulovica–Divača route came close the canyon passage Hankejev kanal in the Škocjanske jame Caves within 400 m, which is as far as the western boundary of the Škocjanske jame Regional Park. Although it was presumed that the blasting in the route would not affect the Škocjanske jame Caves, it was decided, as a precaution, to keep monitoring the cave.

Granted, the cave is relatively far from the several blast sites, but the cave's final section harbours immense underground caverns, such as the 140 m high hall Martelova dvorana with a volume of 2,000,000 m^3. Besides, the precise location where the Reka River crosses the route has not been pinpointed yet. As a precautionary measure, we therefore continued to observe possible consequences of the blasting. Due to issued warnings, the blasting took place with adequate delays of the explosions in the boreholes. The roadcuts in this part of the route are relatively small, requiring smaller explosions at a lower depth. We could establish no immediate impact of the motorway construction on the Škocjanske jame Caves.

The objective was to preserve as many caves as possible, as far as this did not interfere with the construction of a solid road base and safe road use. The caves, which could not be preserved because of the technically challenging road construction, were at least examined. Cave examination proved to be a highly challenging task as we were excavating the roadcuts. The problem is that the rock gets blasted into relatively small rubble particles, not leaving much of the cave to be examined. The rock disintegrates into flowstone pieces and rubble. In this way the Golobja jama Cave, a relatively large cave under Čebulovica hill, was destroyed, because it was in the middle of the route. The caves that were preserved were located in the roadcut sides. Due to the blasting, the rock is often too fractured, making the caves unsuitable for entering.

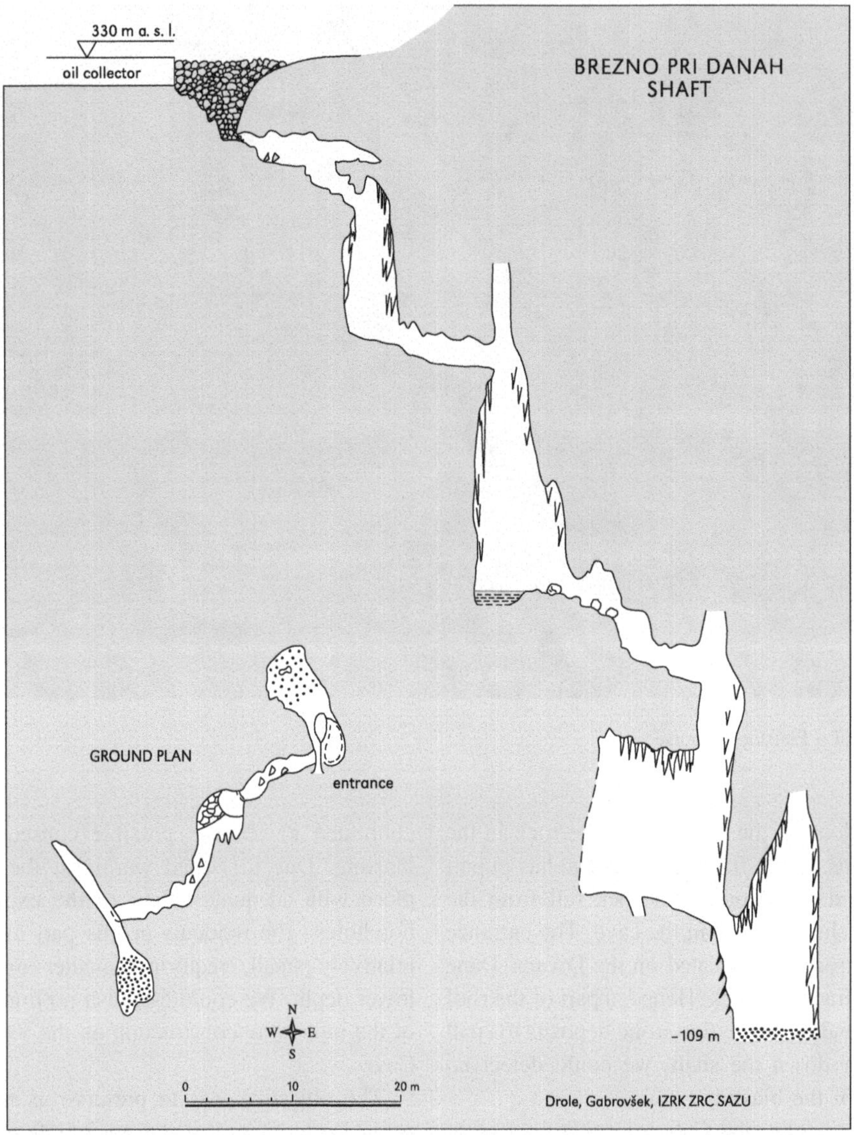

Fig. 6.11 The deepest shaft

In fact, some entrances had to be walled up with large rocks.

Minor and speleologically less important caves in the route had to be filled up, as well as dolines and old caves from which fine-grained deposits were cleared. Large excavation sites were filled in with several 30-cm thick layers of rubble, consolidating each with a vibration roller. Large caves and shafts with chimneys reaching up to the route had to be blasted and filled up in the way described above. Concrete covers were installed over deeper shafts with narrow entries. While we were investigating the caves and attempting to establish their impact on the construction of this road, we also came across some narrow sections. It was inferred that the shaft located on the Divača–Dane route has a large cavern beyond the narrow fissure, expanding below the route. A vibration hammer was used to widen the narrow part at first, but ultimately we had to resort to blasting. The solid rock in the sides of roadcuts ensures that the caves do not become compromised. Still, it is necessary to assess how blasting might affect the integrity of their rim, and

Fig. 6.12 Aragonite crystals on the walls of the shaft

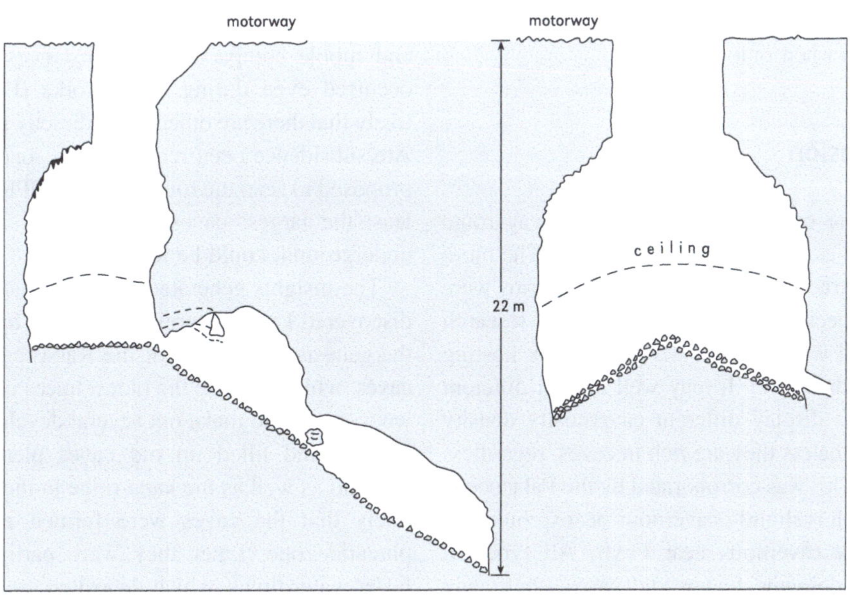

Fig. 6.13 A cave whose roof collapsed during excavation works for the motorway roadcut

Fig. 6.14 Cave roof which collapsed along the fault

clear away unstable and collapsed rock. The rock in the dolomite was especially compromised. The alluvium-filled caves located in the sides that can be entered via their cross-section had to be walled up because the alluvium on the surface is not consolidated and could be washed onto the road.

6.4 Conclusion

Caves that were opening up in the motorway route were examined as the construction went on. The more important or attractive caves in speleology terms were treated with special care. The fundamental research was augmented with the analysis of the rock hosting the caves and flowstone. It may well be that different carbonate rocks display different cavernosity density patterns, nevertheless they are rich in caves, regardless the rock type. This was corroborated by the Palaeocene limestone which is slightly cavernous near Čebulovica hill, but highly cavernous near Povir. All types of limestone and dolomite feature old caves, shafts and surface rock forms—solution pans (Fig. 6.15). The largest share of shafts is perforating Cretaceous limestone. We attempted to preserve as many caves as possible, but this proved rather challenging, since our top priority was to build a safe road. Moreover, caves opened up even in the course of the final clearing of the surface, right before we wanted to start the filling in and rubble compacting. But instances of subsidence occurred even during these works (Fig. 6.16). It is likely that there are other caves directly under the route. Are subsidence occurrences on the road possible? We proposed to scan the route with our GPR. In this way at least the largest caves, should there be any hidden underground, could be identified.

The insights generated from the study of the newly discovered karst features gave rise to hypotheses as to the genesis of this part of the Karst region. In the old caves, which provide the oldest traces of karstification, we were able to make out several development phases. Hollow and filled up old caves pierce the Divača lowland as well as the karst ridge to the northeast. It is likely that the caves were formed in the flooded, phreatic zone. Later they were partly modified by faster water flows, which deposited gravel and sand in the caves, in places hollowing out small scallops on the walls. Where did the watercourses spring from?

Fig. 6.15 Solution pans

Fig. 6.16 A road collapse

Perhaps in the Brkini flysch edge, which was closer to the caves. Was the alluvium brought over by the watercourses from the flysch which lined the Karst region in the north, or by the streams from the elevations where the flysch was preserved after the folding of the anticline (Gams 1974, 197)? In the Pliocene, the Karst region was lower than the flysch surface of the Vipava and Trieste syncline despite the anticline construction (Radinja 1972, 212). Since flysch pebbles were crumbled and ground, they were not carried far underground (Kranjc 1986, 114). Flowstone deposited in the caves in the intermediate dry climate periods, after the underground water level had dropped. In some caves the flysch loam infill remained in place, even though the caves were eventually reached by flood waters. The roofs spanning over the old alluvium-filled caves, which are located at higher levels, were already carried away. We can say that caves dissect the karst surface in the form of furrows. Concluding by the flowstone in the caves we can infer that the roofs were several metres thick. Cucchi and associates (Cucchi et al. 1994, 61) conducted measurements, establishing that the karst terrain, when exposed to weather conditions, tends to lower by 0.02 mm a year. Gams (1965, 86) argues that in the Quaternary the surface above the Postojna Cave had lowered by about 40 m. The surface has been radically transformed. The alluvium transported here by water has been mostly preserved but only in caves. Dolines often came about as the consequence of the opening of shafts under old passages. Created by the weathering of the surface layer, rubble and karst loam can frequently be found together with older flysch alluvium.

Part II

Classical Karst: Newly Discovered Significant Karst Features

7 Unroofed Caves Provide Important Clues to the Karst Development

Unroofed caves (Fig. 7.1) are old caves that were revealed on account of the lowering of the karst surface. They are preserved by their infill—mostly alluvium and flowstone. We were able to identify their most typical incarnations and their significance for the research of the karst aquifer cavernosity, the epikarst and topography.

It became clear during the motorway construction undertaking in Slovenia that unroofed caves constitute a relatively common karst landform. In fact, more common than we had imagined before the karst surface was uncovered through earthworks. The various types of notches occurring on the surface have long been interpreted as types of dolines or as the result of the lithological properties of rock and its fracturing. 75 km long and, on average, 25 m wide stretch of the motorway gave up 350 caves, of which 90 are unroofed caves. Some of them make up the same cave system. New findings prompted us to become more aware of these unique surface karst forms. In the process, we discovered numerous unroofed caves filled with all types of alluvium (Mihevc 2001; Šušteršič 1978). There were several attempts at typification of the characteristic shapes of unroofed caves (Mihevc et al. 1998; Knez and Slabe 1999b) and to design partial models to explain their typical formation processes (Šušteršič 1998; Mihevc 1999a; Knez and Slabe 2002a).

This paper sums up our years of experience gained in the study of this fascinating karst feature, while it was also supported by the latest insights. We believe that this, even though a familiar karst feature, deserves more spotlight than it has received in the past. For the purpose of this paper we shall provide examples from the Classical Karst.

The surface and subsoil dissolution of carbonate rock and its disintegration from back in the Ice Age, brought about the lowering of the karst surface. Old caves, which were formed by erstwhile water flows and are partly intersected by shafts which drain water from the permeable karst surface, pop up as either empty or filled with alluvium. The caves were formed as a part of a system of cavities in a period when impermeable rocks had enclosed the aquifer higher up, causing the underground water in the aquifer to be at a higher level. The hypothesis was that the karst topography and its remarkable systems of valleys can be traced back to former surface water throughflow. However, revealing the surface did not provide us with the evidence to support the hypothesis, instead we identified obvious signs of former water throughflow in carbonate rock—manifested as open and cut through old caves.

7.1 Identifying Unroofed Caves on the Karst Surface

Drawing on our years of experience, which we have obtained during planning and construction of motorways across the Slovenian Karst region, we were able to pool the insights on the type of the unroofed cave on the karst surface. This proved to be of great help for our study of the karst surface. The distinctive shape of the unroofed cave which is set in the karst surface is the result of the type and form of the alluvium-filled transformed cave, and of the development of the karst surface. Surface development is in turn dictated by the rock structure and fracture as well as point surface

M. Knez and T. Slabe (eds.), *Cave Exploration in Slovenia*,
Cave and Karst Systems of the World, DOI 10.1007/978-3-319-21203-6_7

Fig. 7.1 An unroofed cave

permeability, the geomorphologic embeddedness of the aquifer and its development in specific climate conditions. The distinctiveness of the unroofed cave surface form is associated with the washout velocity of the cave alluvium.

The dissected karst surface is dotted with caves of different shapes. We can single out two of the most obvious examples which dictate the shape of unroofed caves. In the first instance, the passage which had opened up in the surface runs parallel to said surface (Fig. 7.2), in the second the passage cuts through the surface (Figs. 7.3 and 7.4). A very specific version of an unroofed cave emerges when the surface opens up in several higher parts of a winding passage or, alternatively, when dolines are formed within one passage (Fig. 7.5). In the first instance, unroofed caves occur as elongated notches, individual dolines or dolines lining up in the open passage. Unroofed caves that are born from passages which had been laterally intersected by the surface come about as doline-like forms. The repeatedly pierced winding passage therefore appears as a string of dolines and notches.

To a large part, the degree of recognizability of an unroofed cave on the karst surface is the result of the velocity with which the sediments were washed out of the cave. The most prominent shapes occurred in instances where the velocity of the sediment washout from the caves exceeded the downcutting rate of the adjacent carbonate terrain. Obviously, empty caves merged with the surface much easier.

When we set out to study the karst features that had been discovered in the southeastern part of the Slovenian Karst region in the framework of the motorway construction, we concluded that the Cretaceous limestone terrain is dissected with sharp features such as very developed, partly subsoil karren and dolines. Yet it is necessary to distinguish between karren grooves and the notches that originated in unroofed caves.

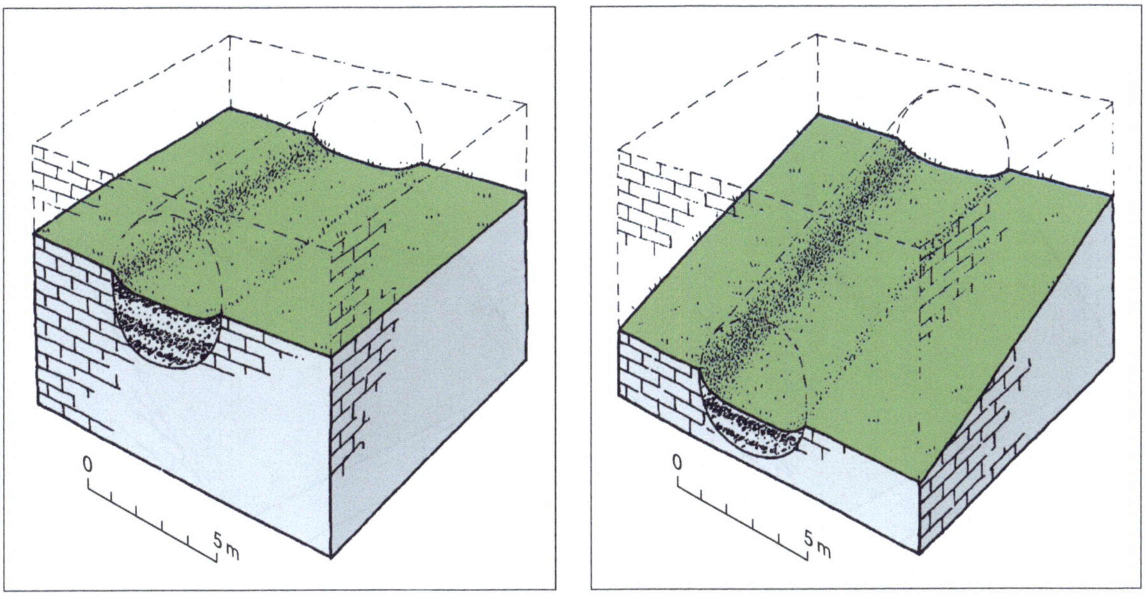

Fig. 7.2 Two unroofed caves and an elongated indentation on a horizontal and inclined karst surface (the legend provided in Fig. 7.6)

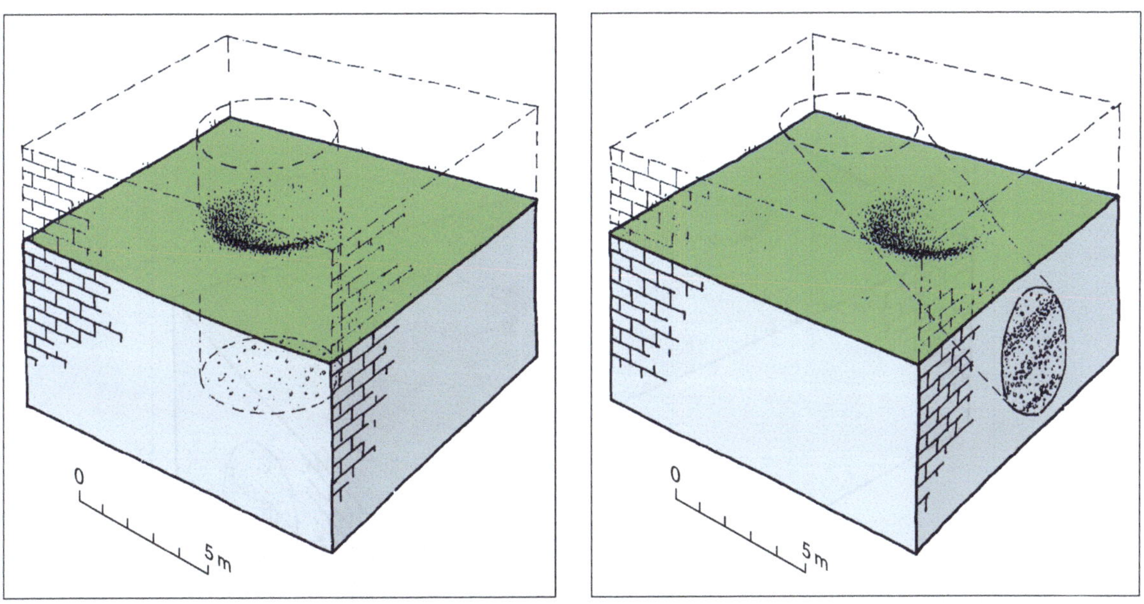

Fig. 7.3 Two unroofed caves, doline-like forms (the legend provided in Fig. 7.6)

Palaeogene limestone terrain, overlain by coarse rubble from the same period, is characterized by greater smoothness, occasional dolines and less visible notches. Flowstone patches may be recognizable after it is denuded.

The unroofed caves, which are the subject of our research, were filled with several metre thick pebble layers and fine-grained sediments, and displayed traces of fluvial action and flooding. The infill in some consisted of coarse rubble emerging in response to

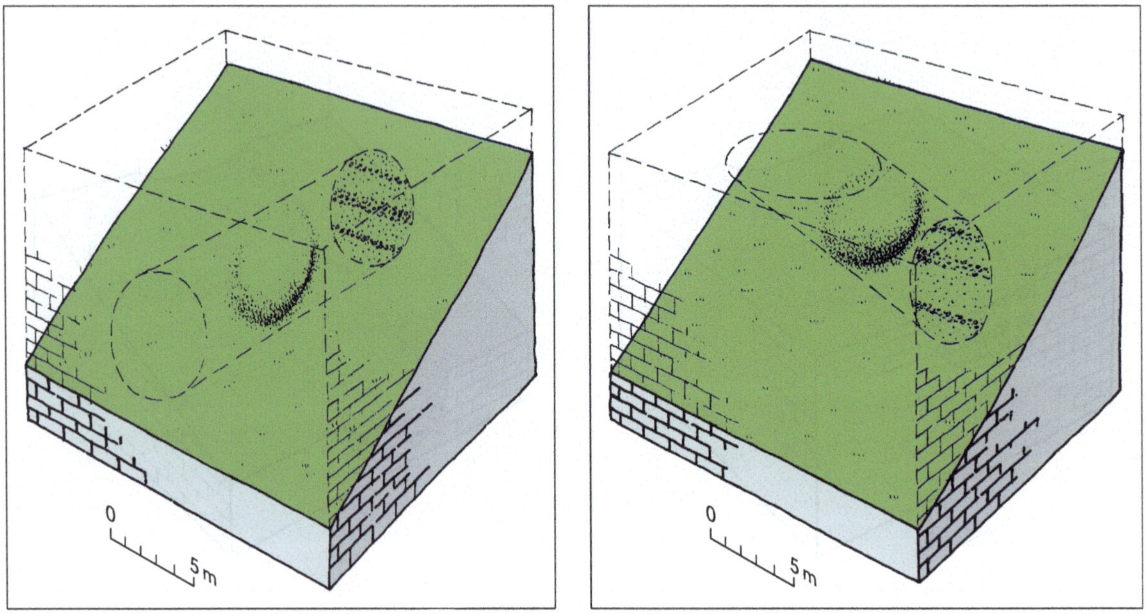

Fig. 7.4 Two unroofed caves, doline-like forms on an inclined surface (the legend provided in Fig. 7.6)

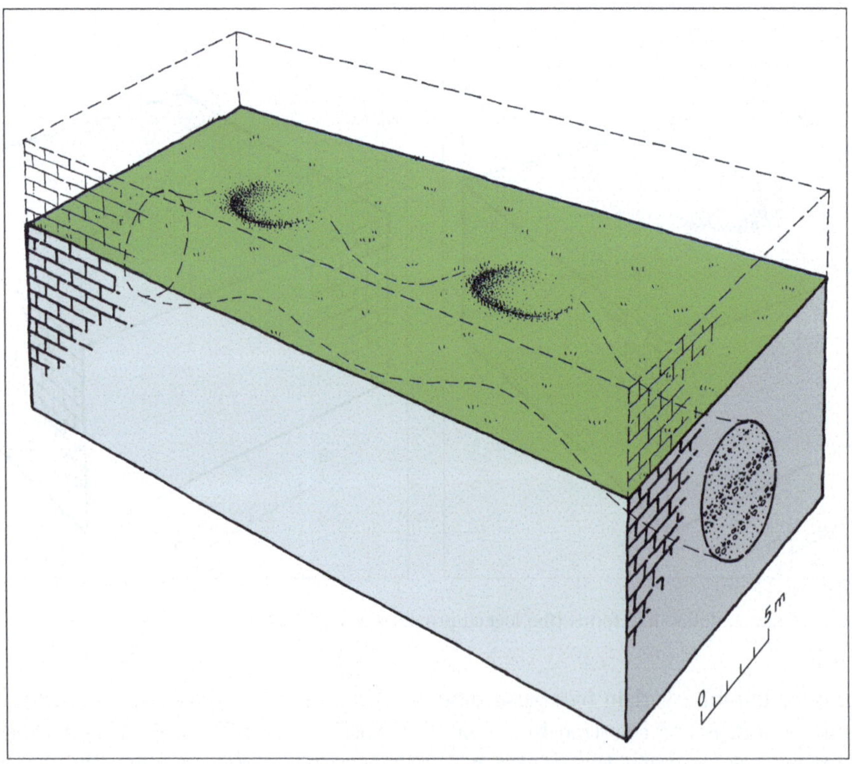

Fig. 7.5 A winding passage cut by the karst surface (the legend provided in Fig. 7.6)

rock weathering in cold Pleistocene periods. Almost all caves featured masses of flowstone and stalagmites.

The flowstone from unroofed caves included specimens with remarkably large calcite crystals and flowstone with tiny sugar-like crystals. The colour scale of the flowstone spanned from pure white to yellow, red and brown to black. In the process of sedimentation and later recrystallization and weathering, iron and manganese cations present in the solution formed calcite crystals. After the cave chambers were filled in, silt, clay and organic matter were included.

Due to the recrystallization process, flowstone is often made up of large calcite crystals and only rarely do we encounter large crystals that may be the reason for the clear, oozing and saturated water from the primary sedimentation environment. The share of finely crystalline flowstone was much smaller. In this case, flowstone was either white or yellow-white. Many flowstone samples contain clay or its own weathered material, layered in individual colour-distinct bands. On few occasions, three chronologically different generations of flowstone from different events have come in contact at one point. Most of the discovered flowstone is consolidated, although small sections do disintegrate into calcite fragments in contact with the surface. The large and tiny crystals from unroofed caves are likely the result of long-term weathering in non-carbonate alluvium. The varied inventory of flowstone occurring in a myriad of morphological and genetic variations deserves far greater attention than was given up to date, and not only in terms of dating.

In many instances the cave rock relief was preserved overlain by alluvium and flowstone. Of course, the velocity of alluvium washout from the caves also depends on the type of alluvium, its disintegration and dissolution rates.

We took sediment samples for palaeomagnetic, pollen and mineralogical research (Mihevc and Zupan Hajna 1996) and for dating. Palaeomagnetic reversal in the alluvium suggests that the caves are older than originally hypothesised by karstologists attempting to explain their evolution, particularly throughout Pleistocene periods. In one of the higher-lying caves, right below the surface, we were able to detect a reversal somewhere around 1.6–1.8 million years, but it can also be older, ranging from 3.8 to 5 million years, but there are no additional indications to support this dating result (Bosák et al. 1998b).

7.1.1 Types of Unroofed Caves

Unroofed caves, which are characterized by slow washout of alluvium, may be recognized based on the following signs:

- patches of karst terrain which are covered by unique soil and vegetation,
- flowstone and cave alluvium lying open on the karst surface.

In contrast, unroofed caves characterized by a fast washout rate of alluvium comprise:

- dolines and semidoline-like shapes (Figs. 7.3 and 7.4),
- strings of dolines (Fig. 7.5) and
- notches spanning over 100 m or even several kilometres (Fig. 7.2).

7.1.2 Patches of Karst Terrain Which Are Covered by Unique Soil and Vegetation

One of the most striking features of unroofed caves, which we had the chance to observe in the field, is that the patches of karst terrain are marked with a typical pedological horizon and vegetation cover.

The most obvious signs of an unroofed cave in the karst terrain are smaller areas of grass or perhaps areas with lush vegetation set between woods or shrubbery.

Unlike the adjacent terrain, grassy areas also do not have larger rocks strewn around. If there are any rocks, they are covered with a relatively thick layer of soil. This is especially the case with Cretaceous rock which appears in the region of the Karst as distinctive and jagged karren which can be up to several metres high, and is difficult to pass. Palaeogene limestone in contact with Cretaceous limestone is mechanically significantly less stable and resistant against weathering on account of its different lithostratigraphic properties. In contact with Cretaceous limestone, Palaeogene limestone tends to fuse with the surface almost seamlessly over time, obscuring the traces of unroofed caves.

Another indicator of an unroofed cave, in addition to the typical surface form described above, is the presence of a cultivated area. Beside fields where soil is on top, the meadows too signalize a more favourable pedological horizon, which may indicate the contact between a filled-in intersected underground passage with the surface.

We had been aware of these forms before initiating earthworks and had marked them on the prognostic map. Once the terrain was uncovered, it turned out that our assumptions were correct.

7.1.3 Flowstone and Cave Alluvium on the Karst Surface

Unroofed caves which we identify on the karst surface by recurring pieces and flowstone blocks of sizes up to several 10 dm^3 can be described as longitudinal areas, tens to hundreds of metres long and several metres wide or mere patches.

In this case, the surface karst morphology reflects the lithostratigraphic foundation. This particular karst surface is characterized by the prevalence of Palaeogene limestone, and is relatively levelled, with slight waves, and marked by the absence of jagged karst forms indicative of surface erosion. On the other hand, the part of the karst surface exposing Upper Cretaceous rock has been intensively eroded, demonstrating numerous karren and various surface notches. Bands of weathered flowstone are much more obvious on Palaeogene terrain than odd Cretaceous rock, although, statistically speaking, Cretaceous rock in the Karst region displays twice the amount of karst forms.

Seeing as Palaeogene limestone in this part of the Karst region occurs in thin beds which are tectonically badly fractured and hence susceptible to weathering, they have disintegrated into coarse rubble, particularly under the influence of cold Pleistocene periods.

Coarse rubble, which allows us to trace flowstone bands, can occasionally cover areas stretching several 100 m^2, and act as cave infill elsewhere. It was commonly found to be overlying old alluvial deposits or filling the space below the abris. The Pleistocene material displacement caused the coarse rubble to become combined with loam, flysch alluvium, fine sand and flowstone pieces that were sometimes found in their original manifestation (stalagmites).

7.1.4 Doline-like Shapes and Strings of Dolines

Unroofed caves can appear similar to dolines (Figs. 7.3 and 7.4) when the karst surface is pierced by an old shaft (Mihevc et al. 1998, 169), for instance a phreatic jump in an old cave, which was filled with deposits and flowstone. Unroofed caves (Fig. 7.4) which are formed when an oblique or horizontal

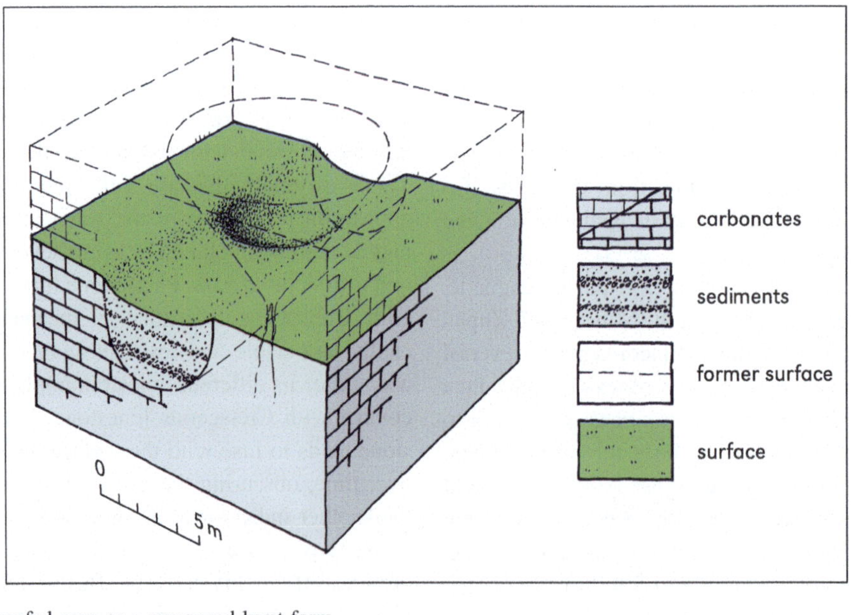

Fig. 7.6 An unroofed cave as a composed karst form

passage cuts through a steep side appear as semi-dolines with an elliptical cross-section (Knez and Slabe 1999b). In addition, single doline-like unroofed caves can emerge within old passages. In this case, only a part of the passage opens up or the doline is formed amidst an old unroofed passage (Fig. 7.6). Dolines can also form in old cave alluvium which was washed away in a funnel shape by the percolating water. Later, the alluvium is covered with layers of rubble and soil in various thicknesses. The so-described dolines can have a diameter of several tens of metres. Also, dolines can cut through old passages or merely encroach on new ones.

Strings of dolines occur when a passage gets opened in several places or, alternatively, when several dolines line up in a passage. We can often identify the direction and size of larger caves based on strings of dolines.

7.1.5 Notches

Notches are created from old unroofed passages which run parallel to the karst surface (Fig. 7.2). They may extend several hundred metres or even kilometres. They may occur as uninterrupted formations, or interrupted with preserved roofs or pitted with dolines. They may run several metres wide and several metres deep. They feature alluvium-filled bottoms, overlain by soil; their rims are often lined with flowstone. Notches may occur on horizontal or sloped sides either transversely or longitudinally.

7.1.6 Variegated Shapes of Cave Systems

A system consisting of various shapes (described in the preceding sections of this paper) was unearthed as the result of denuding a larger and complexly formed cave.

The largest cave system (400 m long) that had been discovered at the start of the route near Kozina was made up of empty passages, one of which was already recognized before the earthworks, and alluvium-filled passages with thin roofs, as well as passages with no roof at all. On the surface, the cave was recognizable as a system of more or less distinctive notches, the most prominent ones dotting the slopes of the dolines and acting as connection between dolines, and of smaller, relatively shallow dolines. Most of the cave was filled with fine-grained alluvium, in some parts there were even layers of flysch pebble. The bottom of the southwestern part was almost completely covered with heaps of flowstone, stalactites and stalagmites. Fine-grained alluvium was deposited over flowstone. Coarse rubble was covering the surface or acted as cave infill in some locations; it was commonly overlying old alluvial deposits or filling the space below abris, which had formed in the beginning sections of the passages. The rubble came about as the result of rock disintegration during the cold Pleistocene periods. The bottoms of the two larger dolines that had formed at the perimeter of the old cave system were covered with layers of brown and red soil a few metres deep. Their rim featured rock formations suggesting notable water percolation. Also at their bottoms were entrances into narrow vertical shafts.

It was observed that the most distinct surface forms of unroofed caves were in fact notches that had formed on the edge of the dolines where high water velocity facilitated sediment down-washing. Shallow dolines were formed in old alluvium within passages. The unroofed passage which had a well preserved alluvium cross-section was used to take alluvium samples for palaeomagnetic and pollen analysis.

7.2 Conclusion

Unroofed caves have recently come to the fore as a surface karst form that is increasingly distinguishable. Unroofed cave shapes depend on the type and shape of the original cave and on the development of the karst surface. They also reflect the evolution of the aquifer with its typical geological, geomorphologic and hydrological properties, as well as the climate conditions. Doline-like shapes are formed when the lowering surface is intersected in its cross-section by an old passage, filled with alluvium and flowstone. Alternatively, they can be formed if the surface encroaches on the passage at a specific point or when dolines are formed in the cave alluvium choking a large denuded passage. A line-up of these forms can often provide information about the shape and size of the transforming cave. Notches emerge from old unroofed passages which run parallel to the surface. The degree of recognizability of an unroofed cave on the karst surface comes about as the result of the difference between the surface lowering rate and the rate at which the sediments were washed out of the cave. The connections between the varied karst surface forms often suggest the presence of cave systems which are subjected to modern transformation of the karst surface and the epikarst.

Other types of karst also boast their own distinctive and peculiar unroofed caves. In front of many entrances into old caves, which hollow out the heaps of flowstone in the fenglin karst in Puzhehei, Yunnan Province, China, there are bands of flowstone and dripstone which can extend 10 m and more. These are remains of former passage extension.

It may be summarized that unroofed caves are an important feature on the karst surface and on the epikarst, providing us with remarkable clues to the development of aquifers.

Ultimately though, the wealth of information we have assembled thus far about the various shapes of unroofed caves is highly conducive to the effective planning involving human interventions in the karst.

8 The Large Unroofed Cave Near Povir

The largest unroofed cave in the entire motorway route was the 230 m long unroofed cave named Brezstropa jama which was unearthed near the village of Povir. It contained flowstone, stalactites, stalagmites and various sediments. Several short unroofed passages dotted the motorway route near Divača, Dolnje Ležeče and the section from Sežana to the state border.

Although a cave passage without a roof no longer qualifies as a cave, but rather a ditch, it still deserves scientific attention. At a time when a part of the speleological community focuses its efforts on the study of the processes that lead up to the formation of caves, such a passage calls attention to the transformation of a cave into a surface karst relief form or, possibly, the disappearance of the cave altogether. It also contributes information to help us explain the flowstones which we found on the surface, the discoveries of silica gravel, sand and loam, and to understand the lowering of the karst surface.

8.1 Speleological Characteristics of the Brezstropa Jama Unroofed Cave

8.1.1 Discovery and Exploration of the Cave

During the preparation for the construction of the motorway on the Divača–Dane section, a 40 m long and not more than 10 m wide area was detected east of Povir, on the Zadušice fallow, along the axis of the roadway. This stretch displayed less favourable geomechanical properties, and ended in a shallow doline. Boreholes showed that this was a depression filled with karst loam. The next step was to excavate the loam down to the supporting bedrock (Fig. 8.1). After the sod, the top layer of soil and of red loam, and the remaining sediments were stripped down by the construction crew, it became clear that this was an old cave passage. Outside of the doline, the passage, now slightly narrower, continued along the route for about 100 m, and then left the route (Fig. 8.2). The form and contents of this cave and its location at an elevation of about 400 m are important for our understanding of the development of this part of the Karst region. We therefore proceeded by measuring the cave and taking several sediment samples for further analyses.

Before the motorway construction started, this particular stretch was barely perceivable as a shallow sag. However, the aerial photographs (Fig. 8.3) made in the infrared part of the spectrum in black-and-white (cyclic filming of Slovenia in 1980) show a well expressed band of brighter, i.e. warmer ground, while a part of the passage is marked by a band of shrubs on thick soil. Similar traces of cave passages can be found on deserted pastures nearby, and some off the route (Fig. 8.4).

8.1.2 Shape of the Excavated Cave

The excavation revealed a 320 m long and up to 5 m deep passage. The dip direction of the cave was 305°. The two passage extremes were at the following coordinates:

- northwestern extreme: X = 5418233, Y = 5062587, Z = 395 m and
- southeastern extreme: X = 5418484, Y = 5062402, Z = 400 m.

M. Knez and T. Slabe (eds.), *Cave Exploration in Slovenia*,
Cave and Karst Systems of the World, DOI 10.1007/978-3-319-21203-6_8

Fig. 8.1 Unroofed cave Brezstropa jama near Povir after the excavation of cave sediments. Remnants of flowstone and stalactites are still visible on the walls of the passage

The southeastern extreme of the cave is located somewhere around the motorway overpass bridging the road to the Gabrk hill. Three segments of the passage were excavated from an undoubtedly larger cave. In the southeastern part, where the cave encroached on the motorway route, the passage is 6 m wide and was excavated to a depth of 5 m. Gradually widening, the passage meandered slightly towards the northwest, gradually tapering in. After approximately 140 m, it ended above the bottom of a smaller doline, and resumed its way on the other end. This section of the passage was narrower, only some 3 m wide and 45 m long. The excavated section of the passage ended with massive flowstone, canopied with a metre-thick rock roof. The passage excavation was resumed after 35 m. This section was highly compromised and resembled a sort of chamber. Only its northern wall was left intact and as such boasted preserved scallops. From there the passage continued towards the north and off of the motorway route. We were able to trace it based on the flowstone on the surface for another 10 m, where it ultimately ended in a large shallow doline (Fig. 8.5).

The larger part of the passage was up to 6 m wide, whereas at its narrowest point, east of the doline, it was only 2–3 m wide. For the most part, the excavation depth was around 3–4 m and at no point touched bedrock. Only in the central part of the passage, just before the doline, did the walls close together to such a degree that the bedrock of the passage was close, probably at an elevation of around 395 m. The two boreholes in this part of the passage (Dular 1993b) reached limestone at an elevation of 396 m, but not necessarily limestone at the lowest or representative point in the passage.

8.1.3 Rock Relief on the Passage Walls

The bedrock roof of the passage has been preserved only in that part of the passage that leads from the doline towards the northwest. The roof was only around 1 m thick. The part of the passage, where the roof was missing, and the excavated sediment

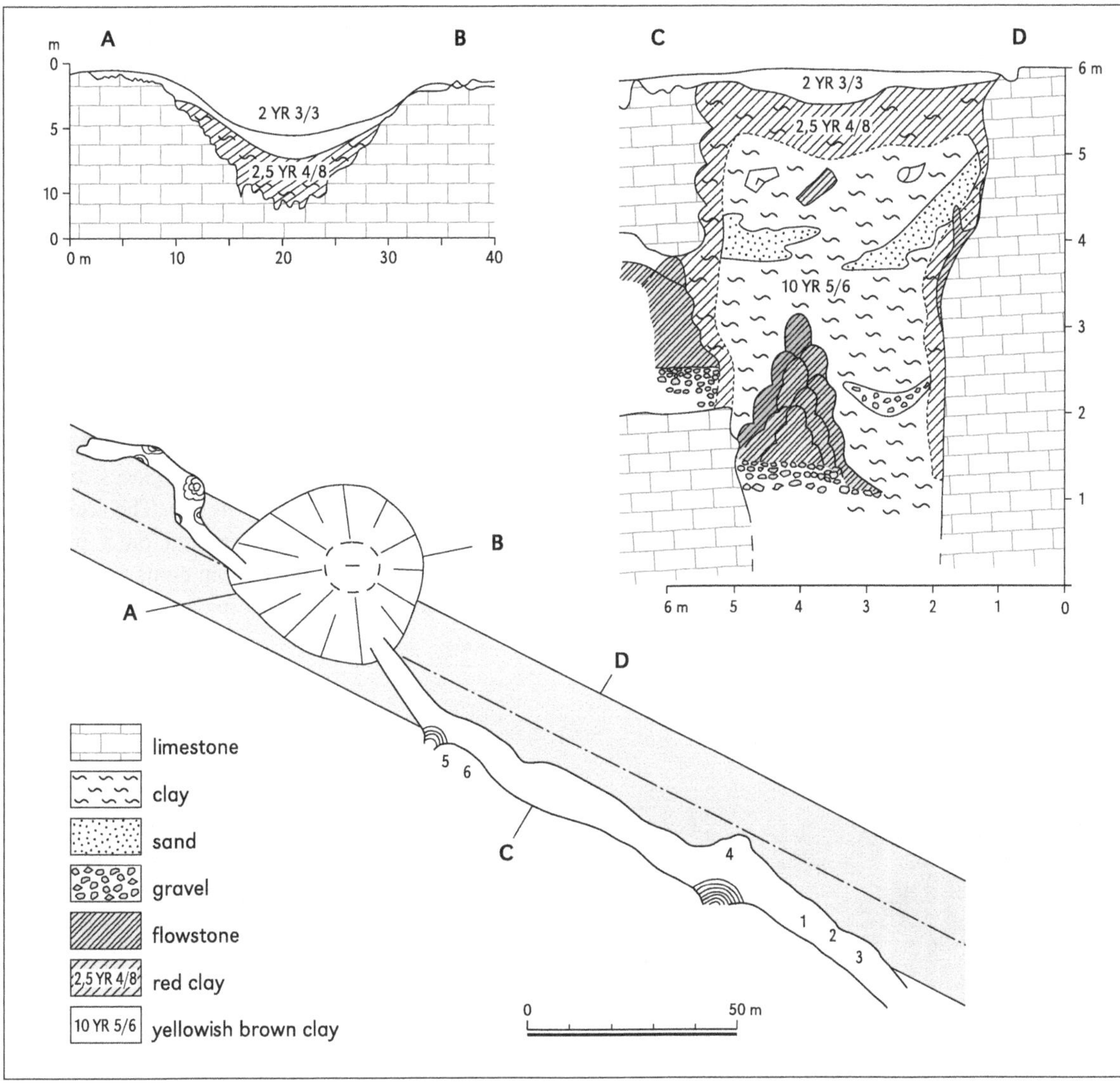

Fig. 8.2 Unroofed cave Brezstropa jama near Povir. Layout and schematic cross-section of the passage and of the doline

displayed a few rocks floating in the sediment, but we were unable to determine whether these rocks belong to the cave roof or if they are rocks that had broken off from the top part of the walls. Some stalactites, up to 0.5 m long, were found set in the sediment.

The walls of the passage were vertical, occasionally overhanging. The walls' feature preserved traces of formation in the cave environment, traces of corrosion formation along contact areas with sediment, and traces of mechanical and corrosion transformation in the subsoil zone of karst.

The most notable of all the cave forms were the wall notches, which appear as relatively horizontal semicircular channels, along the water level line, and along the line of sediment deposits in the cave (Fig. 8.6). Similar notches, slanting in the direction of the water flow, can be found in sink caves transporting gravel.

Inside the cave, notches were created and preserved in several places. Usually, they occur two or three in parallel alignment (Fig. 8.7). They are 20–60 cm high, 12–20 cm deep, and up to 7 m long. Sometimes, they are lined with loam, and other times with sand and gravel. The dip of the notches was measured in seven places. Five notches were striking NW–SE. With two notches we were unable to

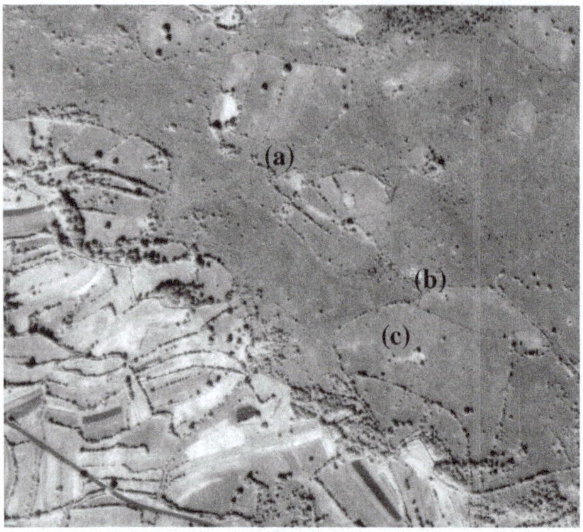

Fig. 8.3 On aerial photographs in an infra-red technique the unroofed cave **a–b** is visible as a zone of brighter, i.e. warmer ground. The section of the cave that has been preserved **c** can be seen in Fig. 8.4

measure the dip—they appear to have been created during a period of a still water table or close to sediments. The incline of the notches ranged from 1.5 to 3.9 %, which indicates a drop of 10 cm within a distance of 5 m. Disregarding the possibility of tectonic inclination of the terrain from the time the notches were created, the water flow direction indicates a southeast flow. Several parallel notches indicate that the cave was filled with sediment or that the infill level has changed several times.

Scallops are shallow niches, up to a few cm long, that are shaped by eddies. Clusters of scallops can reliably indicate the direction of the water flow, while their average size points to the velocity of the water flow. Scallops were found in several places, but they have been well-preserved only in two. The scallops in the wall notch have been preserved on a patch of around 0.5 m^2, but they have been compromised over time. The scallops in the northern part of the cave were

Fig. 8.4 A flat surface at an altitude of 398–400 m a.s.l. A winding unroofed cave lies in front of it. It is separated from the rest of the surface by various vegetation. This unroofed cave was not damaged during motorway construction and was thus preserved

Fig. 8.5 A small corrosion doline which formed in the Brezstropa jama unroofed cave

overlain by flowstone, which preserved their shape in full. We were able to peel off the flowstone from the wall, thus revealing the scallops. The average scallop measured 2–3 cm. They indicated that the flow direction was toward SE. Such scallops are created if the water flow velocity is around 1 m/s.

8.1.4 Forms Created Due to the Filling of the Passage with Sediments and Soil

Where soil or sediment come in contact with limestone, the rock is evenly infiltrated with aggressive water. This causes sheet corrosion, which compromises the previous forms that were created under different conditions. This type of transformation also affects individual rocks floating in the sediment.

In some places in the passage the rock is covered by flowstone and therefore remained fully intact, whereas in adjacent areas corrosion has worn away a more or less thick layer of rock. From these areas we were able to evaluate the extent of subsoil corrosion, all the way back from when the cave was filled up by sediments. Subsoil dissolution wore down no more than 5–20 mm of the film deposited on the rock. The passage demonstrated several centres of strong vertical percolation and the associated down-washing, which were expressed as penetrations of brown surface soil into the cave sediment. Vertical and shallow channels were made on the rock, but the rock failed to demonstrate the white bloom, characteristic of subsoil corrosion.

There was noticeable damage to the walls in the passage, which had been caused by the lowering of the earth's surface and the coming nearer of the surface and the cave. Inside the passage, these features are expressed as highly dense fissured areas and fissures filled with red loam and widened by corrosion in the top part of the passage walls, up to about a metre and a half beneath the surface.

8.1.5 Cave Sediments

The cave passage was completely filled with sediments and soil. On the surface, there was an

Fig. 8.6 Nicely visible in the wall of the excavated Brezstropa jama unroofed cave are wall flutes and the characteristic colour of the sediment, which next to the wall changes into a *reddish colour*. Traces of epikarst decomposition of the walls are only visible in the *top part* of the cave walls

approximately 10 cm thick layer of brown soil. It was underlain by a layer of terra rossa (0.5–1 m thick), which transitioned into clastic sediments, loam, sand and gravel, with its predominant and typical yellow-brown colour. The sediment was composed of fragments ranging from the size of silt to 25 cm pebbles. Samples were analysed with the x-ray diffraction method, and we conducted microscopic thin sections of certain sediments (Mihevc and Zupan Hajna 1996). The colour of the loam and sand was identified based on the Munsell colour charts.

It was established that non-carbonate flysch sandstone pebbles were predominant. The largest pebble demonstrated a longer, 25 cm axis, and the majority of the pebbles measured around 5 cm. The pebbles were well rounded. After they dried up, they cracked and disintegrated into smaller pieces. There were a few patinated carbonate pebbles in between. The pebbles, including the largest ones, were set in quartz sand, which had probably been created as a result of the disintegration of some pebbles (Fig. 8.8).

A distinct feature among the pebbles are the highly weathered, porous siliceous stones, cherts, which are up to 20 cm long. These rocks are rounded off, but they still feature several equally rounded niches, whose roundness could not have come about as the result of fluvial transport. The rock surface is smooth and covered with a thin shiny black coating.

The conglomerate was found accumulated all in one place. It was in fact the base for a metre-thick flowstone heap (Fig. 8.9). This is a true cave conglomerate, which emerged where the flowstone-forming water had dripped down, indicating that gravel deposits and the flowstone growth took place in the same period. The conglomerate was composed of flysch sandstone pebbles, up to 15 cm big, and mixed with several limestone pebbles. The pebble share in this instance exceeds the one in the unconsolidated gravel, pointing at their corrosion.

Sands appeared in the form of lenticular bodies or lamellae in loam, between gravel or independently. The predominant colour of sands was yellow-brown.

Fig. 8.7 View of the Brezstropa jama unroofed cave after its bottom was already filled in and fortified. Wall flutes are visible on the walls of the passage and in the *left part* of the photograph, in front of the car, also a remnant of an eroded flowstone heap

In the upper part of the sediment profile, upon contact with terra rossa, their colour transitioned into the same colour. In the same vein, we could observe pure sand of yellow-brown colour (10YR 5/6), which changed colour close to the passage wall turning into a reddish brown (2.5YR 4/8), even though this was undisputedly a sediment of the same layer.

Mineralogical analysis of the sand has shown a relatively similar composition; the majority is formed by siliceous grains (97 % or over), while the remaining minerals are only found in traces. The grains are poorly rounded and their roundness does not set them apart from the grains of sand in the flysch pebbles. The different types of sand are distinguished by colour which comes about as a reaction to the surface—yellow colour is provided by goethite and red by hematite, two minerals that were present in traces (Mihevc and Zupan Hajna 1996).

Sand derives from flysch sandstone and was created with the disintegration of pebbles, which could have occurred as early as during river transport or later on in the cave. The varied sand pigmentation can be traced back to the conditions in the current environment. An eolian origin of the grains is not likely, since the sedimentation environment, the gravel, flowstone and loam do not indicate a dry climate.

In the cave, loam (silty clay with admixture) was combined with layers or lenticular bodies of sand and gravel. A peculiarity of loam is the highly folded varved sedimentation structure, which points to a strong kneading of plastic sediments after they were deposited. The predominant colour of the loam was yellowish brown in the bottom part of the profile (10YR 5/8), which transitioned into the red colour of terra rossa in the upper part; the loam likewise changed colour along the walls of the passage, with the yellowish brown loam transitioning into a band of red loam next to the walls. In terms of mineralogical composition, there were no significant differences between both types of loam. It is dominated by silica (over 90 %) and illite.

Inside the cave, flowstone has been preserved in the form of crusts, massive flowstone heaps, and free-standing stalagmites, but we also found some broken off stalactites. In addition to this old flowstone,

Fig. 8.8 The typical sediment in the cave are flysch sandstone gravel and carbonate gravel, in which a reduction occurred

flowstone was deposited in certain areas along the walls of the cave within subsoil fissures which were widened by corrosion. But this flowstone was much different from the old one.

Most of the old flowstone was located in the northwestern part of the cave. It appeared as though this part of the passage was close to its former roof. We found that some massive heaps, a few curtains and even a column without a roof, have been preserved in this particular part.

A large flowstone heap emerged in the wider part of the passage in the northeastern section. The heap's base was created simultaneously with the deposits of gravel. North of this heap, a wall notch had corroded into a layer of flowstone which was more than a metre thick. This leads us to believe that this flowstone is older than the last stage of the formation of the cave walls. A flowstone sample from this site was dated with the ^{234}Th/^{235}U method, which showed that the age of the flowstone exceeds 350 thousand years, which is the dating limit of this method.

8.2 Comparison of Sediments from Different Caves Along the Motorway Route

The findings of quartz sand, gravel and loam in the Karst region have been described repeatedly. The most similar to the Brezstropa jama unroofed cave appears to be the cave or remnant of a cave in Lipove doline (Pleničar 1954). In this unroofed cave quartz sand was excavated and a large flowstone heap was left on the surface. Quartz sand is also reported by D'Ambrosi and Legnani (1965), Habič (1992), Radinja (1972), and others.

Similar sediments to those in the Brezstropa jama unroofed cave were also discovered on several other locations during the motorway construction. As such they were found on the fallow of the Grintavca between the valleys Dol Češnjevec and Dol Rebidnik, in the Bojni dol Valley, in several caves piercing the Divaški hrib Hill, in several smaller caves between the

Fig. 8.9 The conglomerate, created beneath the flowstone heap, has also preserved carbonate pebbles in the sediments due to the carbonate-saturated water

villages of Povir and Žirje, and in three large unroofed caves lined up between the tunnel under the Tabor hill (484 m) and the state border.

The sediments, all of them cave sediments, displayed a few striking common characteristics, but also some differences attributable to the autonomous development of each cave.

A common feature of the sediments is their typical colour. The larger sediment bodies have a yellow-brown colour, which goes over into red along the edges, along the contact with the wall, directly below the surface and in zones of vertical down-washing. In these places, the mineral composition also slightly changes, attesting to the migration of substances from the surface or to the oxidation of certain minerals.

The different types of loam are similar in terms of minerals, whereby most are displaying banded, but strongly kneaded sedimentation structures. In view of mineral composition, the sand and clay contained predominantly quartz which was obviously derived from flysch rocks. Flysch sandstone pebbles and cherts were found only in the Brezstropa jama unroofed cave and approximately 1 km northwest. In Bojni dol and at Grintavca, we identified chert pebbles which were not present in the Brezstropa jama unroofed cave. This difference in composition seems relevant, prompting us to undertake further detailed studies.

We have observed notable differences regarding the occurrence and absence of flowstone, collapse rocks or coarse rubble on allochthonous sediments, which is an indication of the method of opening up of individual caves on the surface.

8.3 The Brezstropa Jama Unroofed Cave in Time and Space

It is possible to reconstruct a part of the development of the Brezstropa jama unroofed cave based on the shape of the walls and the sediments. The cave is a remnant of a larger cave system which drained sinking stream waters from flysch. The preserved section of the passage was located deep beneath the surface. The cave had a through-flowing sinking stream, carrying large pebbles. The size of the pebbles and the large share of flysch sandstone pebbles (Kranjc 1986, 1989) allow us the conclusion that ponors were not far off. The flow rate fluctuated from a few tens of L/s to several m^3/s. The flow direction in the cave was running towards the southeast.

Today the closest flysch is found above the valley of the Raša River, around 5 km away, while the Brkini flysch, located in the southeast, is even farther, 7 km away. Apparently though, at that time, flysch must have been somewhere closer to the cave, possibly at the Gabrk hill near Divača, which is today composed of Palaeogene limestone and is only a few km away. In that particular period it was probably covered by flysch, which is only some 100 m higher in the stratigraphic column.

There were no relevant barriers for the water flow on its course from the ponors to the cave. The formation of stalactites/stalagmites and flowstone heaps does not reflect external influences, although the growth of the stalactites and stalagmites was interrupted several times by stages of either erosion or sedimentation. One of these cave erosion stages also left its mark on the flowstone which is over 350 thousand years old. Next, the cave was filled with fluvial sediments. The infill prevented the further filling with flowstone, the collapse of the roof and the transformation of the walls by corrosion. Thus, the cave transformation came to an end, but not that of its surroundings.

There were several horizontal caves identified in the vicinity of the Brezstropa jama unroofed cave, at a similar elevation. One of these, the Trhlovca cave, is located immediately above the southwestern end of the Divaška jama Cave, at an elevation between 404 and 420 m. The cave is but a short meandering passage, mostly sediment-filled. The sediments are mainly varved yellow-brown loams with sand laminae, overlain in places with thick flowstone. Scallops and wall notches have been preserved on the passage walls. In terms of the dimensions of the cave's passage, particularly its width, the type of sediments and the elevation, this cave has a striking similarity to the Brezstropa jama unroofed cave.

A much larger cave is the Divaška jama. The entrance is only 2.5 km south of the Brezstropa jama unroofed cave, at an altitude of 426 m. At the entrance, one can already notice the massive flowstone accumulations, today extending right up to the surface. At an elevation ranging 356–390 m a.s.l., there are yellow-brown banded loam deposits, overlain by red loam (Gospodarič 1985), and several generations of massive flowstone. A large passage of 600 m runs towards the southwest from the entrance. According to Gospodarič, the yellow-brown loam had been deposited during the Mindel Glacial Stage, and the red loam is supposed to be terra rossa which had been down-washed from the surface. It was washed into the cave during the warm Mindel-Riss Interglacial.

The extension of the Divaška jama Cave, extending from the entrance towards the northwest, is filled in. In that direction, 250 m from the entrance, the side of Gorenjski Radvanj collapse doline cuts through the passage which is filled up with sediments, yellow-brown loam and flowstone, and located at an altitude ranging from 390 to 415 m. It is quite possible that the extension of the Divaška jama Cave is of similar size to the cave itself. Chert pebbles were also found among the loam.

Based on their similar sediments and identical elevation, both caves and the filled-in passage in Radvanj can be set in the same time window as the Brezstropa jama unroofed cave. This time window also includes the remnant of the cave in Lipove doline, which can be traced along elongated dolines, in which flowstone and yellow loam may occasionally occur, extending several hundreds of metres.

For the time being, we are still unable to date the precise age of the Brezstropa jama unroofed cave. The only indirect clue is the rate with which the surface is lowering. It was argued that rainwater and percolating water infiltrating through the soil lose most of their dissolution capacity after the first few metres under the surface (Gams 1962b). This leads to even surface lowering. The cave, which is now free of water flow and completely filled up with sediments, therefore no longer undergoes active transformation. Eventually the surface will reach the cave and the cave will ultimately disappear. Based on the measurements of the limestone dissolved in rivers and the river output, Gams calculated that the surface of the river basins of the rivers Ljubljanica, Soča and Krka is lowering at the rate of 1 m in 16.6, 12.0 and 17.2 thousand years, respectively, and that the surface in the river basins of the mentioned rivers had lowered in the past million years by 60, 83 and 58 m, respectively. It took

750 thousand to 1.5 million years for the Brezstropa jama unroofed cave's 50–100 m-thick roof to be corroded down, while the cave itself or its infillings are probably even older.

The age of the Brezstropa jama unroofed cave is defined by the time in which the water level of free-flowing rivers, such as the one flowing through the cave, dropped from 400 to 180 m a.s.l. The Reka River, today sinking into the Škocjanske jame Caves at an altitude of 317 m, runs through the Kačna jama cave at an elevation ranging from 156 to 180 m. Its extreme northwestern end, the chamber named Cimermanova dvorana, 1200 m away from the Brezstropa jama unroofed cave, is situated at an altitude of 180 m, plunging 220 m lower.

Thus, the surroundings of the Brezstropa jama unroofed cave can be regarded as a 220 m deep vadose zone, i.e. a zone marked with predominantly vertical percolation, shaft growth and sediment and soil down-washing; it appears though that these occurrences are not evenly distributed, since the cave also features all of the sediments. Two dolines emerged in the vicinity of the cave, one of which adjoined the cave passage. Their impact on the cave was minimal. In the portion where the doline cuts into the older passage, it appears as though the sediment slid down to the doline bottom; however the impact of the doline did not extend beyond this point, suggesting slight lateral sediment displacement. It seems that this movement was slight even in the cold Pleistocene climates, provided that the doline is not in fact younger. Considering, in addition, the thick red soil at the bottom of the doline, it can be established that the washing away of soil through the bottom of the doline was also lees intensive.

Caves, which are similar to the Brezstropa jama unroofed cave, appear frequently in the Karst region. Some of them were used to harvest the flowstone as a decorative stone, quartz sand or loam. Nevertheless, such findings, particularly quartz sand and chert pebbles derived from flysch, were attributed mostly to the surface Reka River, from when it had supposedly ran across the surface in the pre-karst phase. It was believed that its sediments, carried over from the flysch edge, were washed down into the caves at a later point.

There is no doubt that the Brezstropa jama unroofed cave features fluvial sediments which had been deposited in the cave. Large-scale work on the motorway, hitherto unparalleled in the Karst region, revealed that there are many such caves here and that, possibly, the majority of the surface non-carbonate sediments in the Karst region is of cave origin, pushing the age of the pre-karst surface flow of the Reka River even further into the past. The surface of the Karst region has been transformed so profusely up to today that it is impossible to find any remnants of a pre-karst relief.

9 Origin and Mineral Composition of Clastic Sediments on the Karst Surface Around Divača

During the motorway construction around Divača, construction works had unearthed a large number of dolines and caves filled with mechanic sediments. The clastic sediments in some of them have been subjected to in-depth analyses (Fig. 9.1). A sediment analysis has shown that some of these depressions were in fact old caves, filled with fluvial allochthonous sediments which the lowering of the surface has opened up on top, transforming them into surface depressions. Sediment analysis helped us to differentiate the dolines from eroded cave passages near Divača. Cave sediments found on the surface of the Karst region differ from other surface sediments and may provide us with important signals based on which we can distinguish between corrosion dolines and underground caverns that have been transformed into dolines. Sediments similar to those that we found along the route, have been already observed in the Karst region and described multiple times. They were usually attributed to surface deposits which were the remains of a once much more extensive bar of the former Reka River at a time when it had supposedly crossed the Karst (D'Ambrosi and Legnani 1965; Melik 1961; Radinja 1972). Accordingly, these sediments had supposedly only been preserved as resediments at the lowest points of the Karst surface, in the bottoms of dolines. Less frequently and for the most part restricted to a narrower stretch, some authors attributed them to cave sedimentation (Pleničar 1954; Habič 1992). What is striking about these findings is their undisputedly cave sedimentation environment, large surface distribution and their large quantity. This opens up a possible explanation to the origin of numerous sediments, now found at the surface, as being the remains of the cave sediments which had appeared as a result of the surface lowering. Besides providing us with a possible explanation as to the development of individual shapes, their quantity and the sites of fluvial sediments indicate that it is likely that most of the allochthonous sediments found hitherto on the surface of the Karst region have a cave origin, rather than being remnants of superficially flowing streams, i.e. the Reka River.

The qualitative mineral composition of the samples was analysed by powder x-ray diffraction. The samples were analysed with a Philips diffractometer at the Department of Geology at the Faculty of Natural Sciences and Engineering in Ljubljana, Slovenia. The recording conditions were a $Cu_{K\alpha}$ anode, with the voltage of 40 kV and a current of 30 mA; a Ni filter and proportional metre and an automatic divergence valve were used. The recording was continuous, at the rate of 2° 2Θ/min in the range of 2Θ from 2° to 70°. Some samples were also examined in the thin sections. The amount of minerals in the samples is provided in percentages, which do not represent the absolute quantity of an individual mineral in the sample, but a calculated share of the mineral in the sample with regard to the height of its main reflection. Therefore, the percentages are purely of an informative nature and for the sake of comparison. The difference between the concentration of individual minerals in the thin section and in the samples, examined by the

M. Knez and T. Slabe (eds.), *Cave Exploration in Slovenia*,
Cave and Karst Systems of the World, DOI 10.1007/978-3-319-21203-6_9

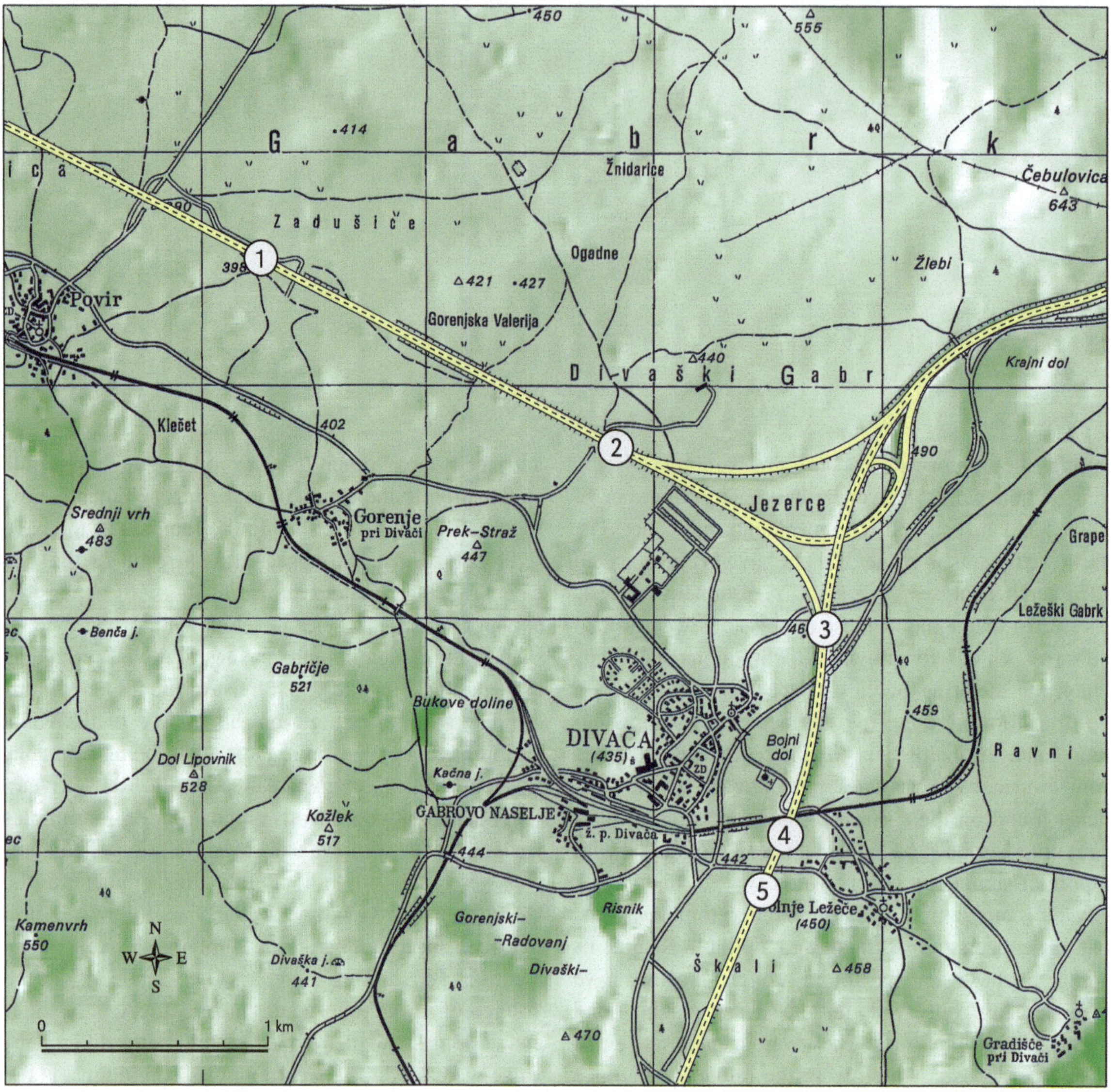

Fig. 9.1 Site of the described sediments. Legend: *1* Brezstropa jama unroofed cave near Povir, *2* Profile 650, *3* Divaški hrib, *4* Bojni dol, *5* Grintavca

powder x-ray diffraction method, can be accounted for by the fact that the minerals that make up less than 3 % are not detected, while goethite and hematite, which covered other mineral grains, were only noticeable in the thin section. Plagioclases are not defined in the samples, because their amount was too minute and their reflections were covered by the reflections of other minerals represented in greater numbers. The same applied to all other minerals that were present in traces, and the background itself was rather high.

9.1 Sites and Description of Clastic Sediments in Karst Depressions Around Divača

9.1.1 The Brezstropa Jama Unroofed Cave Near Povir

At the section of the motorway north of Divača, an elongated area with less favourable geomechanical properties was detected in the axis of the motorway.

The boreholes (Dular 1993a) showed a depression filled up by mechanically unstable sediments. After the sod, the top layer of soil and of red loam, and the remaining sediments were stripped down by the construction crew, it became clear that this was an old cave passage which was interrupted in the middle with a shallow doline (see Fig. 8.2). The passage featured well-preserved flowstone, speleothems, channels and scallops. Before the motorway construction started, the passage was barely visible in the surface relief. On aerial photographs made in the infrared spectrum in black and white (cyclic filming of Slovenia 1980) it stands out as a belt of lighter, hence warmer soil (see Fig. 8.3). Nearby, but off of the motorway route, we observed traces of other similar caves. The whole passage, which we excavated in a length of 320 m and to a depth of 5 m, i.e. a volume of 6900 m^3, was filled up by allochthonous fluvial cave sediments. In the passage we found yellow-brown loams, quartz sand and gravel which measured up to 25 cm. For the most part, the excavation depth was around 3–4 m and at no point touched bedrock. Cave features are preserved in several places on the walls. In the northern part of the cave the scallops were overlain by flowstone, which preserved their shape in full. The scallops averaged from 2 to 3 cm in size. There are places in the passage where the rock was covered by flowstone, preserving it perfectly, yet elsewhere, close by, corrosion has worn away a more or less thick layer of the rock. At such places we were able to assess the extent of subsoil corrosion since the period the cave had been filled up by sediments. Sheet subsoil corrosion wore down between 5 and 20 mm of the film deposited on the rock. The passage demonstrated several points of strong vertical percolation and the associated down-washing, which were expressed as penetrations of brown surface soil into the cave sediment. Vertical, shallow channels occurred on the rock. In the sediment filling the passage, there were no remains or traces of a collapse. Among the excavated sediments there were a few rocks floating on the sediment but it was impossible to establish whether these rocks belong to the cave roof or if they are rocks that had broken off from the top part of the walls. Some stalactites, up to 0.50 m long, were found set in the sediment. On top, there was an approximately 10 cm thick layer of dark red-brown soil (2YR 3/3). It was underlain by a 0.5–1 m thick layer of terra rossa (2.5YR 4/8), which transitioned into clastic sediments, loam, sand and gravel, in a predominantly yellow-brown colour.

Most sediments are from the adjacent non-carbonate area. The sedimentation conditions in the passage changed in short-distance increments. The lenses of sand and gravel disappeared and transitioned into thinner fractions. Occasionally, the gravel was cemented in a base of massive flowstone heaps. The deposit layers were heavily folded, indicating that the sediment moved after it had been deposited in the cave. The most characteristic types of sediments were sampled. Figure 8.2 demonstrates the ground plan of the cave and the sites of the individual analysed samples. We were mostly interested in the origin of clastic sediments and how they would compare to the sediments from other sites from the route of the motorway. A reconstruction of the sedimentation conditions in the passage was not possible because of the rapid changes in granulometry and the damage caused due to machine excavation (Fig. 9.2).

9.1.1.1 Gravel

Well rounded non-carbonate pebbles of flysch sandstone, about 5 cm in size, were predominant in the cave. The largest pebble had a longer axis that was 25 cm long. Some of them had dried, cracked and ultimately disintegrated into smaller pieces. There were also some patinated carbonate pebbles in the mix. The pebbles, including the largest ones, were set in quartz sand. The conglomerate that formed on the spot where the flowstone-depositing water infiltrated through a gravel bar, suggests gravel deposition and flowstone growth in the same period. The conglomerate was composed of flysch sandstone pebbles, up to 15 cm big, and mixed with several limestone pebbles. The share of limestone pebbles was higher than in unconsolidated gravel, indicating the corrosion within.

Chert nodules are a peculiarity among the gravels. They were found throughout the passage, mixed between the gravels but also among the quartz sands in the southeastern part of the cave. The sample of chert nodules (Fig. 9.3) from the northwestern part of the cave demonstrated a predominance of quartz, making up 98 % of the sample, whereas only 2 % were calcite, and plagioclase only in traces. A thin section was made from the pebble, which revealed the silification of biomicritic limestone.

Fig. 9.2 Uncovered cave Brezstropa jama near Povir. View of the cave after the excavation of cave sediments

9.1.1.2 Sands and Loams

Sands appeared in the form of lenticular bodies or lamellae within loam, or gravel or independently. The predominant colour of the sand was yellow-brown. In the upper part of the sediment profile, in contact with terra rossa, its colour transitioned into red. The same was observed in pure sands of yellowish brown colour (10YR 5/6) which changed over to red (2.5YR 4/8) along the passage wall, although the sediment was undoubtedly from the same layer. In between the sands, there were remains of non-disintegrated pebbles of flysch sandstone in some places. The loams in the

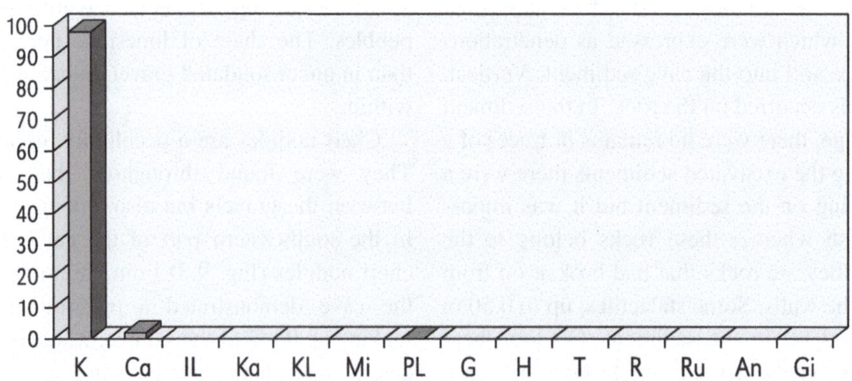

Fig. 9.3 Sample 1: quartz gravel stone from the northwest part of the Brezstropa jame unroofed cave. X-ray: *K* Quartz 98 %, *Ca* Calcite 2 %, in traces: *PL* Plagioclase; *IL* Illite, *Ka* Kaolinite, *KL* Chlorite, *Mi* Microcline, *G* Goethite, *H* Hematite, *T* Tourmaline, *R* Hornblende, *Ru* Rutile, *An* Anatase, *Gi* Gibbsite

cave were mixed with layers or lenses of sand and also gravel. A peculiarity of the loams is their very folded varved sedimentation structure, implying strong kneading of plastic sediments after they were deposited. The predominant colour of the loam in the lower part of the profile was yellowish brown (10YR 5/8), abruptly transitioning to red (2.5YR 4/8) in the upper part of the profile. Similarly, a colour change of the loam occurred along the walls of the passage where yellowish brown loam passed into a belt of red loam.

Yellow-brown loam (Fig. 9.4) was located in the southeastern part of the cave. The yellow-brown loam featured sand lenses that passed into red loam on the top of the profile. Yellow-brown loam was composed of 90 % quartz, 8 % illite, 1 % calcite and 1 % chlorite. Tourmaline, kaolinite, microcline, hornblende and rutile occurred in traces. This yellow-brown loam is a cave loam deriving from weathered remains of Eocene flysch. The share of clay minerals in this sample was higher than in Sample 3.

Sandy lamina (Fig. 9.5) in the yellow-brown loam is composed of 97 % quartz, 2 % illite, 1 % microcline, whereas kaolinite, chlorite and goethite are in traces. The sand is well washed, which is the result of the predominance of quartz grains. The presence of other minerals suggests that the sand originates in silicate sandstone, which is a component of Eocene flysch.

The sample shows the predominance of red sand (Fig. 9.6) from the contact between the sediment and the cave wall. It consists of 98 % quartz, 1 % illite, 1 % kaolinite and hematite in traces. Hematite

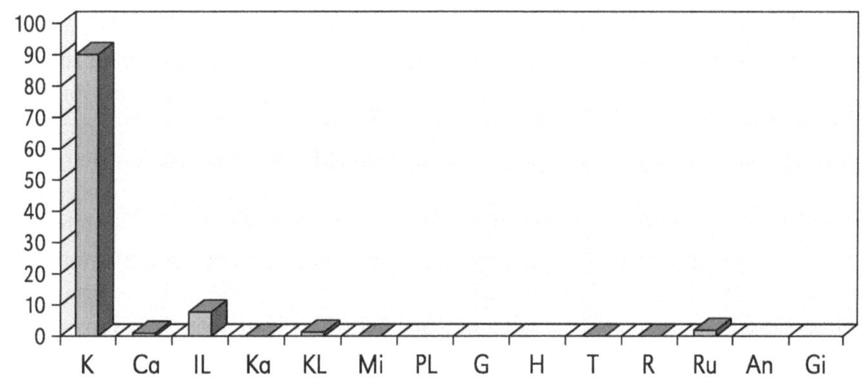

Fig. 9.4 Sample 2: yellow-brown loam from the Brezstropa jame unroofed cave. X-ray: *K* Quartz 90 %, *IL* Illite 8 %, *Ca* Calcite 1 %, *KL* Chlorite 1 %, in traces: *T* Tourmaline, *Ka* Kaolinite, *Mi* Microcline, *R* Hornblende, and *Ru* Rutile; *PL* Plagioclase, *G* Goethite, *H* Hematite, *An* Anatase, *Gi* Gibbsite

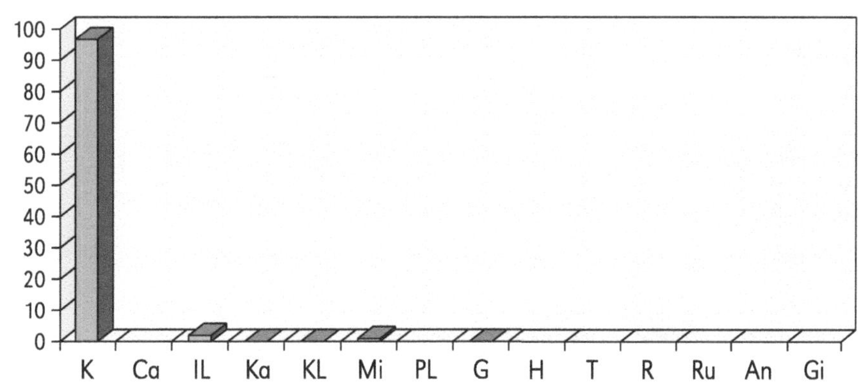

Fig. 9.5 Sample 3: sandy lamina from the yellow-brown loam from the Brezstropa jame unroofed cave. X-ray: *K* Quartz 97 %, *IL* Illite 2 %, *Mi* Microcline 1 %, in traces: *Ka* Kaolinite, *KL* Chlorite, and *G* Goethite; *Ca* Calcite, *PL* Plagioclase, *H* Hematite, *T* Tourmaline, *R* Hornblende, *Ru* Rutile, *An* Anatase, *Gi* Gibbsite

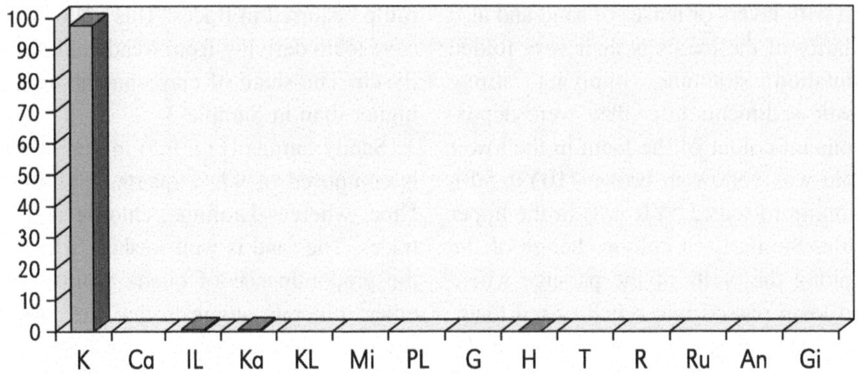

Fig. 9.6 Sample 4: red sand from the contact point with the cave wall in the Brezstropa jama unroofed cave. X-ray: *K* Quartz 98 %, *IL* Illite 1 %, *Ka* Kaolinite 1 %, in traces: *H* Hematite; *Ca* Calcite, *KL* Chlorite, *Mi* Microcline, *PL* Plagioclase, *G* Goethite, *T* Tourmaline, *R* Hornblende, *Ru* Rutile, *An* Anatase, *Gi* Gibbsite

represents the pigment over the grains of other minerals and came about with the oxidation (dehydration) of goethite. The process of goethite transitioning into hematite marked the cave sediment which is in contact with the cave wall. This is where the sediment was infiltrated by oxygen-rich precipitation water. This sand was also deposited in the cave and has its origin in weathered flysch sandstone.

The yellow loam sand (Fig. 9.7) overlain by a layer of flowstone is comprised of quartz (93 %), illite (6 %) and chlorite (1 %). Goethite, microcline and plagioclase occurred in traces. This sand was also deposited in the cave and traces back to weathered remains of Eocene flysch, based on its mineral composition.

In the yellow washed sand (Fig. 9.8) quartz dominates with 99 %, there was 1 % of chlorite, and traces of tourmaline, muscovite and zircon. A thin film of goethite coated the grains of some other minerals. Sand is a cave sediment which was well washed, since it is made up almost exclusively of quartz grains.

Sand and loam in the Brezstropa jama unroofed cave near Povir derive, based on their mineral composition, from flysch sandstone that disintegrated to varying degrees in the process of weathering. They were deposited in the cave during different hydrological periods. In terms of mineralogical composition, there were no essential differences between the two types of loam—red and yellowish brown. Both display a prevailing share of quartz (more than 90 %) and illite. An eolian origin of the grains is not likely, as neither gravel nor loam and flowstone indicate a dry climate. They were deposited at a time when the cave was crossed by a river. The sedimentation pattern suggests fluvial transport and sedimentation, and the shape and composition of grains originate from Eocene flysch.

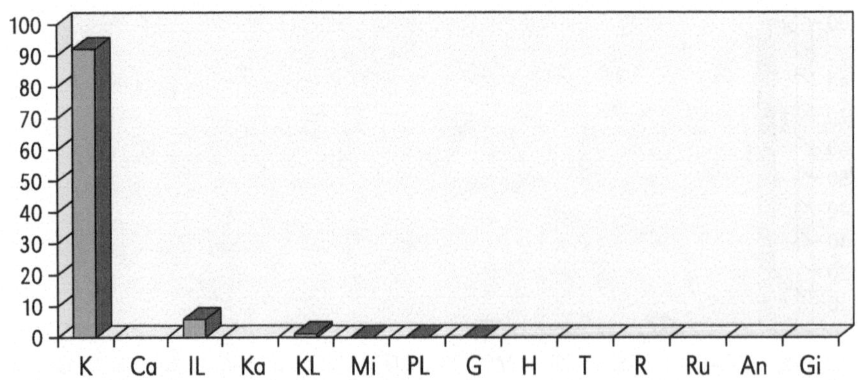

Fig. 9.7 Sample 5: yellow loam sand from underneath of the flowstone heap in the Brezstropa jama unroofed cave. X-ray: *K* Quartz 93 %, *IL* Illite 6 %, *KL* Chlorite 1 %, in traces: *G* Goethite, *Mi* Microcline, and *PL* Plagioclase; *Ca* Calcite, *Ka* Kaolinite, *H* Hematite, *T* Tourmaline, *R* Hornblende, *Ru* Rutile, *An* Anatase, *Gi* Gibbsite

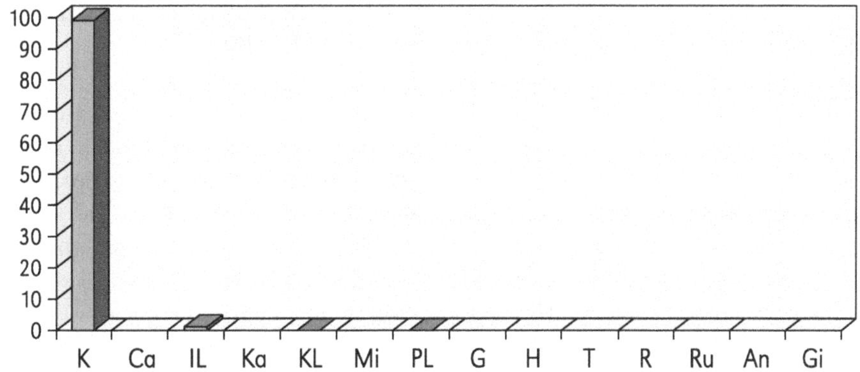

Fig. 9.8 Sample 6: yellow washed-out sand opposite the flowstone heap in the Brezstropa jama unroofed cave. X-ray: *K* Quartz 99 %, *IL* Illite 1 %, in traces: *KL* Chlorite and *PL* Plagioclase; *Ca* Calcite, *Ka* Kaolinite, *Mi* Microcline, *G* Goethite, *H* Hematite, *T* Tourmaline, *R* Hornblende, *Ru* Rutile, *An* Anatase, *Gi* Gibbsite

9.1.1.3 Infill from the Surface

Red loam (Fig. 9.9) was brought into the cave from the surface down a fissure and deposited next to the yellow sand and loam. The loam in the colour of 2.5YR 4/8 is mostly comprised of quartz (84 %) and illite (12 %). The chlorite, kaolinite, plagioclase and hematite made up 1 % of the sample each, while goethite and calcite were in traces. The specific site of this sediment, but mostly its composition and the high share of clay minerals, indicate the cave bottom is resedimented.

Unfortunately, the doline (profile a–b in Fig. 8.2) that intersected the cave in the centre was compromised by the construction work and we were unable to establish conclusively the relation between the cave and the doline. Apparently it was filled up by red loam up to 2 m thick, without yellow layers which are typical of cave sediments. Although the passage was open into the doline, the sediment in the passage did not slide in. The centre of the doline lay slightly out of the axis of the passage. It seems as though the doline and the passage came about completely independently of each other, influenced by various original structures, and in vastly different conditions. The impact of the doline on the cave was insignificant.

9.1.2 Sediments from Two Filled-in Caves South of Povir

In the profile that lies at an elevation of 408 m, about 1 km southeast from the Brezstropa jama unroofed cave, we intersected two caves, filled up by sediments. One of the caves was filled up by rubble and the other

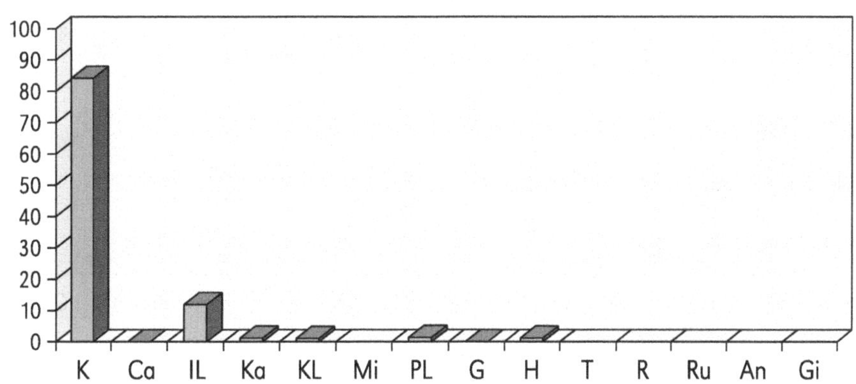

Fig. 9.9 Sample 7: red surface infill in the Brezstropa jama unroofed cave. X-ray: *K* Quartz 84 %, *IL* Illite 12 %, *KL* Chlorite 1 %, *Ka* Kaolinite 1 %, *PL* Plagioclase 1 %, *H* Hematite 1 %, in traces: *G* Goethite and *Ca* Calcite; *Mi* Microcline, *T* Tourmaline, *R* Hornblende, *Ru* Rutile, *An* Anatase, *Gi* Gibbsite

by loamy and sandy allochthonous sediments. We were intrigued by the similarity of these sediments to those from the Brezstropa jama unroofed cave, so we took samples to be analysed.

9.1.2.1 Loams

The profile of the first passage was filled up by laminated loam sediment with highly folded layers. In the lower part of the profile the yellowish brown (10YR 5/8) loam with thin grey layers prevailed. The upper part of the infill was the same colour as the subsurface red loam (2.5YR 4/8) in the Brezstropa jama unroofed cave.

Laminated yellow-brown loam (Fig. 9.10) was made up of 62 % quartz, 26 % calcite, 7 % illite, 2 % anatase, 1 % kaolinite, 1 % goethite, and microcline and plagioclase in traces. Again, we deal with a cave sediment originating from the weathered remains of Eocene flysch, whereas calcite is probably derived from weathered cave walls eroded by running water; calcite grains have accumulated in the cave sediment interspersed with fluvial deposits.

9.1.2.2 Rubble

In the same roadcut, but a few metres apart, there was a karst cave filled by coarse rubble, in which the particles averaged about 10 cm in size. Unconsolidated rubble was mixed with red loam. A few pieces of stalactite, about 10 cm across, were also in the mix. Apparently, this was an older, void cave which was filled up by pleistocene climatic rubble.

9.1.3 Filled-in Cave from the Divaški Hrib Hill

in a roadcut in the Divaški hrib Hill there was a large cavern which was cut through at an elevation of 450 m. Inside the roadcut, it appeared as a 4 m high and 10 m wide infill of red and yellow-brown loam (Fig. 9.11). The cavern featured a preserved roof about 2 m thick. The cavern was filled up by laminated, strongly kneaded yellow loam with lenses and layers of sand, and macroscopically similar loam of red colour. Layers, lenses and laminas were kneaded and folded. It was overlain by more rubble which has been intensively dyed by red loam. The roadcut cut through the cavern, leaving us unable to establish its shape and size. It is fairly obvious though that the sediments have been significantly displaced. In one part, the cave sediment came into contact with the loam through vertical percolation, down-washing and transport of rubble into the fossil infill, becoming red in response. Two samples of the infill were analysed.

Yellow loam (Fig. 9.12) with laminas of fine sand (10YR 6/6) was made up of poorly rounded grains varying from 0.1 to 0.025 mm. The sample was made up of 98 % quartz, 1 % illite, and 1 % chlorite, while kaolinite, goethite, and hornblende appeared in traces. It was a cave sediment originating from weathered rock of Eocene flysch.

The red loam (Fig. 9.13) with the colour 2.5YR 4/8, overlying the yellow sand, is made up of 85 % quartz and large portions of illite (5 %), gibbsite (4 %),

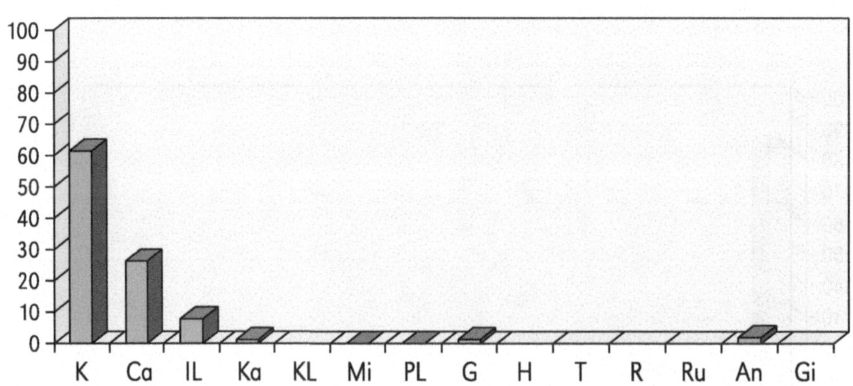

Fig. 9.10 Sample 8: grey-yellow loam from the cave on the motorway in the 650 + 10 profile. X-ray: *K* Quartz 62 %, *Ca* Calcite 26 %, *IL* Illite 7 %, *An* Anatase 2 %, *Ka* Kaolinite 1 %, *G* Goethite 1 %, in traces: *Mi* Microcline and *PL* Plagioclase; *KL* Chlorite, *H* Hematite, *T* Tourmaline, *R* Hornblende, *Ru* Rutile, *Gi* Gibbsite

Fig. 9.11 View of the profile of the big passage filled with loam sediment in the Divaški hrib

kaolinite and hematite (2 % each), chlorite and tourmaline (1 % each) and plagioclase in traces. Based on its mineral composition, especially gibbsite which is a bauxite material, we may infer that this is the onetime floor which was resedimented through the fissure down into the cave which already featured fluvial deposits of yellow loam.

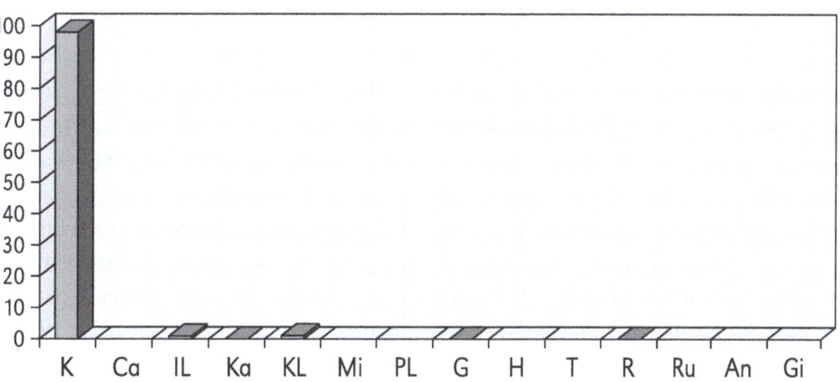

Fig. 9.12 Sample 9: yellow sand and loam from the cave under Divaški hrib. X-ray: *K* Quartz 98 %, *IL* Illite 1 %, *KL* Chlorite 1 %, in traces: *Ka* Kaolinite, *G* Goethite, and *R* Hornblende; *Ca* Calcite, *Mi* Microcline, *PL* Plagioclase, *H* Hematite, *T* Tourmaline, *Ru* Rutile, *An* Anatase, *Gi* Gibbsite

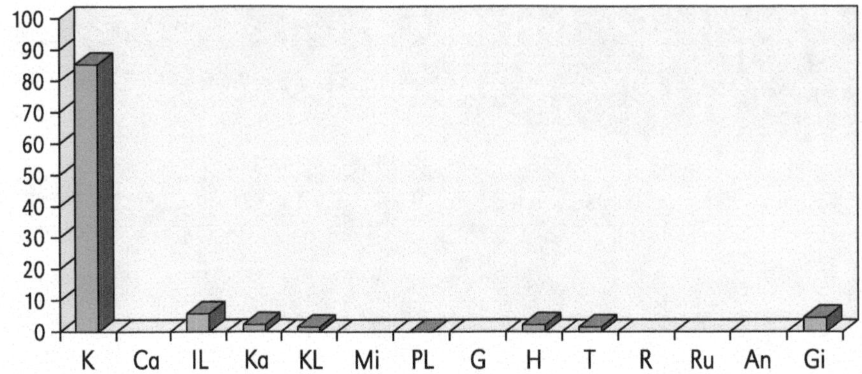

Fig. 9.13 Sample 10: red loam, washed out from the surface in the cave under the Divaški hrib. X-ray: *K* Quartz 85 %, *IL* Illite 5 %, *Gi* Gibbsite 4 %, *Ka* Kaolinite 2 %, *H* Hematite 2 %, *KL* Chlorite 1 %, *T* Tourmaline 1 %, in traces: *PL* Plagioclase; *Ca* Calcite, *Mi* Microcline, *G* Goethite, *R* Hornblende, *Ru* Rutile, *An* Anatase

9.1.4 The Unroofed Cave at Grintavca

Between the two larger dolines, Dol Češnjevec located to the west of the route, and Dol Rebidnik, the motorway route was traversed by a series of connected shallow dolines. The surface is located at the altitude of approximately 449 m, while the bottom of the string of dolines can be found at approximately 446 m. The larger valleys have a wide, flat bottom at the altitudes of 433 and 431 m, respectively. The aerial photos of the infrared part of the spectrum in black and white (cyclic recording of Slovenia 1980) show the series of dolines as a strip of lighter, i.e. warmer ground. The bottoms of both collapse dolines exhibit the same shade. It was revealed in the course of excavating the ground that dolines or shallow relief depressions formed above the old, up to 6 m wide cave passage without a ceiling. Since a relatively small part has been excavated, it was not possible to examine it to a satisfactory degree. The excavation to the depth of around 5 m (at the altitude of 441 m) did not reach the bedrock of the passage. In the excavation area, the passage was interrupted in two places where the ceiling was preserved. The ceiling of the passage was preserved at the altitude of around 448 m. A major part of the passage was filled with allochtonous fluvial sediments.

9.1.4.1 Gravel

In the lowest part, there were deposits of multi-coloured chert gravel mixed with quartz sand. The pebbles ranged from 1 to 3 cm in size and contained no flysch sandstone pebbles. Sedimentation structures were not visible in the lower part.

9.1.4.2 Sand and Loam

Upwards, the gravel bar transitioned, without any sharp boundary, into yellow sand (2.5Y 6/6) mixed with loam lamellae of the same colour. The bar was about 3 m thick. In the upper part and certain spots along the walls, the sediment colour turned red (5YR 4/4). In one spot, this sedimentation sequence was covered by a layer of flowstone that was up to 1 m thick and extended over several square metres. A 0.5 m thick layer of red loam was deposited atop the flowstone. Within the loam, either a cave or even subsequent surface sediment, individual smaller rocks of Cretaceous limestone floated in places. Only up to 10 cm of humus horizon was positioned on top of it.

Only red loam (Fig. 9.14) from the upper part of the profile underwent more thorough analyses. A sample of this red loam contained 66 % of quartz, 30 % of illite, 2 % of kaolinite, 1 % of hematite and 1 % of chlorite as well as traces of microcline and plagioclase. In view of the mineral composition, especially the high content of clay minerals and the content of hematite, it would be possible to assume that the sediment, perhaps at one time a cave sediment, has been on the surface for a longer period of time and has passed over the stage of further weathering or was already heavily weathered prior to being deposited in the cave.

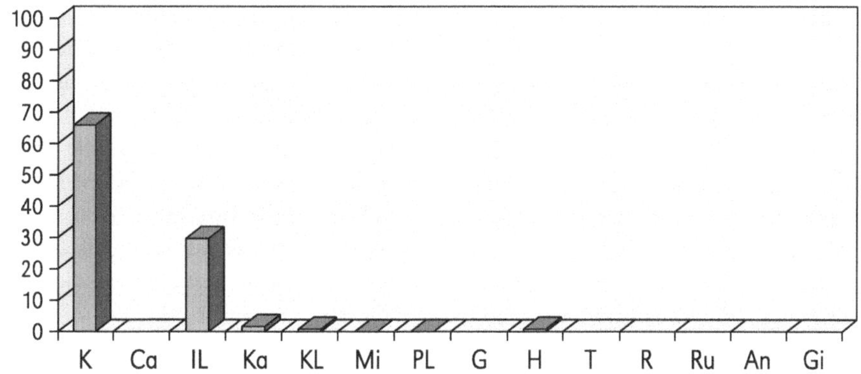

Fig. 9.14 Sample 11: red loam from an open cave at Grintavca. X-ray: *K* Quartz 66 %, *IL* Illite 30 %, *Ka* Kaolinite 2 %, *H* Hematite 1 %, *KL* Chlorite 1 %, in traces: *Mi* Microcline and *PL* Plagioclase; *Ca* Calcite, *G* Goethite, *T* Tourmaline, *R* Hornblende, *Ru* Rutile, *An* Anatase, *Gi* Gibbsite

9.1.5 Sediments from the Unroofed Cave in Bojni Dol

In the southern part of Bojni dol, a larger shallow depression with the bottom at the altitude of 423 m, a larger doline filled with sediments was emptied. Autochthon rubble material prevailed, mixed with larger individual rocks and red loam, while the lower part of the doline featured yellow-brownish sands and loams as well as individual conglomerate blocks.

9.1.5.1 Conglomerate

The conglomerate consisted of up to 2 cm large pebbles of colourful chert and individual limestone pebbles. It was deposited in layers and was well-sorted. In the course of excavation, it was not possible to identify the form of the sediment body; it was obviously old cave sediment intersected by a doline.

9.1.5.2 Sand with Slope Rubble

The yellow sand (Fig. 9.15) from the bottom of the doline near Bojni dol was mixed with slope rubble. The sample contained 62 % of quartz, 37 % of calcite and 1 % of illite, while chlorite and plagioclase were present in traces. In view of the mineral composition, this is cave sediment which was later mixed with slope material.

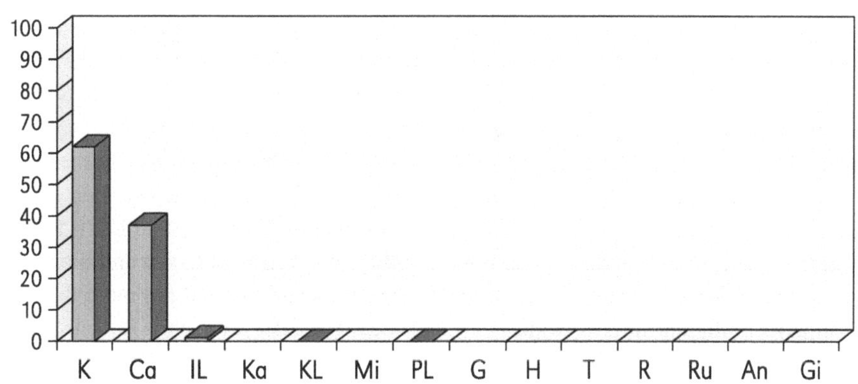

Fig. 9.15 Sample 12: yellow sand mixed with talus debris in the doline of Bojni dol. X-ray: *K* Quartz 62 %, *Ca* Calcite 37 %, *IL* Illite 1 %, in traces: *KL* Chlorite and *PL* Plagioclase; *Ka* Kaolinite, *Mi* Microcline, *G* Goethite, *H* Hematite, *T* Tourmaline, *R* Hornblende, *Ru* Rutile, *An* Anatase, *Gi* Gibbsite

9.2 Conclusion

All cases described undoubtedly reveal cave sediments formed in different periods or which differ due to a different course of development in the caves. According to their mineral composition, the samples of gravel, sand and loam from the Brezstropa jama unroofed cave, the bottom part of the cave in the Divaški hrib Hill, and the unroofed caves at Grintavca and Bojni dol originate from weathered remains of flysch rocks. When caves open up to the surface or are severed by an emerging doline, the cave fluvial sediments reach the surface and become exposed to the surface conditions, no longer protected against weathering. When the sediments are in contact with the surface, due to the percolating rainwater and thus exposed to oxidation conditions, goethite becomes dehydrated once the brown-coloured sediment changes colour as the grains of hematite turn it red. That this is cave sediment, which merely changed colour due to the leaking of water from the surface, is evident from the fact that it contains less clay minerals than the sediments that have already survived a partial pedogenesis, i.e. soil formation. These were only carried into caves after having undergone substantial weathering and may also contain bauxite minerals (Zupan Hajna 1999).

Similar sediments to those in the Brezstropa jama unroofed cave were also discovered on several other locations during the construction of the motorway. The sediments, all of them being cave sediments, showed a few distinct common features, as well as a few differences, which resulted from the individual development of each cave. The distinct differences include the appearance of flowstone, collapse rocks or coarse rubble on allochtonous sediments, which points to the way individual caves opened up to the surface. Sediments of loam, silt and sand were found in places, as well as gravel, while only one of the fractions appeared elsewhere. These are caves that were formed at different altitudes, were part of various water flows or which were formed in different time periods.

Denuded cave sediments represent an important source of soil in the Karst. Their location can be either primary or they could have undergone partial or full resedimentation. In several cases cave sediments were found filling up dolines or other depressions on the route, e.g. on the fallow of Grintavca between Dol Češnjevec and Dol Rebidnik, in the doline at Bojni dol, in several dolines between Povir and Žirje, which attests to the widespread range of fluvial cave sediments on the karst surface.

Infiltration of material from the surface, which was exposed to pedogenesis, is evident from the last sample from the Brezstropa jama unroofed cave, the upper part of the Divaški hrib Hill cave and the cave at Grintavca. The mineral composition of the red infiltrated soil, however, differs from one sample to another. The types of red soil differ in terms of age and thus the pedogenesis stage. The weathered remains of Eocene flysch served as the original material in cases listed here, including in the formation of red soil. At Bojni dol, the mineral composition of the sample points to the cave sediment being mixed with slope rubble.

Sediments and cave passages also enable us to partially depict the time in which these caves were formed. The Brezstropa jama unroofed cave is a remnant of a larger cave system, which conveyed the water of the sinking streams from the flysch. The external influences were not reflected in the formation of dripstones and flowstone heaps, and the amount of deposited flowstone points to a ceiling that was at least several ten metres thick. The growth of dripstones was interrupted by erosion at least once. A sinking stream flowed through the cave, carrying large pebbles. From their size and the large share of flysch sandstone pebbles, which are very fragile and could not survive a lengthier river transport (Kranjc 1986, 1989), it can be deduced that ponors were not far away. Today, similar pebbles can only be found in the caves of the Brkini sinking streams which are, at the most, 1 km away from the contact area of flysch with limestone. The flow of water varied from a few tens of L/s to several m^3/s. It can be assumed that the cave was completely filled with fluvial sediments when the surface above it was still at the altitude of around 450 m and the ponors at the contact area of flysch and limestone were at least at the same altitude but not more than 1 or 2 km away. Today, the closest flysch is the one above the valley of the Raša River, which is around 5 km away, while the contact area between flysch and limestone is at the altitude of around 500 m. The Brkini flysch, located in the southeast, is even farther, 7 km away. Since both flysch areas are far away, we assume that the source of the flysch sandstone was located closer, perhaps in the area of the Gabrk Hill near Divača, 1 km to the northeast of the cave. There is no flysch there today. Since Gabrk Hill consists of younger Palaeocene and Eocene alveolinid and nummulitid limestone, above which there is only about 120 m to the Eocene flysch layers, this area seems to be the most likely source of the sediments. If this area really is the source of these pebbles, the Reka River had to be sinking already at that time, but in separated ponors. Other caves discovered were formed in similar conditions, but it is more difficult to depict the circumstances of their formation as they are smaller and less well preserved. They were probably formed by different sinking streams. Today's speleohydrological conditions are different. The most important sinking stream is the Reka River that sinks into the Škocjanske jame Caves at the altitude of 317 m; it is located at the altitudes ranging from 156 to 180 m in the area of the described sites of cave sediments, and fluctuates by app. 100 m (Mihevc 1984). Thus, the surroundings of the Brezstropa jama unroofed cave can be thought to contain a 220 m deep vadose zone, i.e. a zone in which vertical percolation prevails, along with the expansion of shafts, and the washing away of sediments and soil downwards. In caves, the water level of the free-flowing rivers and the level of karst water dropped from the altitude of 400–180 m, while the surface was lowered by at least 50 m. According to the calculations, the surface is said to have lowered by app. 60 m over the last million years (Gams 1962b). With the presumed thickness of the ceiling ranging from 50 to 200 m, the Brezstropa jama unroofed cave was filled with sediments around 0.8–3 million years ago.

During that time, dolines could also have been formed on the surface, since all the conditions were met, i.e. a vertical gradient and a satisfactory draining of water through the karst. Despite the existing vertical gradient, the passages remained filled up with sediments. A great deal of sediments was resedimented and originated from caves already completely removed by erosion, and can today be found on the surface, sometimes even in dolines, in which the former cave sediments could also have been partially altered.

Composition of Sediments in Dolines 10

Mechanical sediments in the Karst have always stirred great interest, with researchers mainly interested in the red soil, its composition and origin (Hrovat 1953; Gregorič 1967, 1969) as well as their accumulation in dolines (Gams 1971; Habič 1974). The formation of red soil was mostly defined on the basis of non-soluble limestone remains and Eolian deposits (Gregorič 1967; Šušteršič 1978). Quartz pebbles and sands were found on the karst surface, which have always been attributed to the fluvial transport of the weathered remains of flysch rocks across the Karst in the "pre-karst phase" (Melik 1961; Radinja 1972; Habič 1992). It is interesting that Habič (1992) defines the colour of unconsolidated clastic sediments in the Karst on the basis of climate; he believes the yellow colour reflects their depositing in a colder climate, while the red colour is the result of a tropical climate. Through research of the soil and surface clastic sediments in the Karst (Mihevc and Zupan Hajna 1996; Zupan Hajna 1998; Mihevc 2001), we ascertained that they, in many cases, originate from clastic cave sediments which surfaced due to the denudation of limestone above the cave (Mihevc 1999b).

Between 1994 and 2002, archaeological test probes (Bavdek 1998, 2003a, b; Bavdek and Križman 2000) were excavated in a great number of dolines along the motorway route between Divača and Kozina, thus opening up the possibility of examining the mineral composition of the clastic sediment layers found in them. Sample analyses were conducted within the project entitled Archaeological Research of Dolines Along the Divača–Kozina Motorway Route (Arheološke raziskave vrtač na trasi avtoceste Divača–Kozina) which was carried out by the Notranjski muzej Postojna, headed by Alma Bavdek. In dolines, red and yellow sediments prevailed, occurring in several shades and transitioning from red to yellow and then to brown. The same layer of clastic sediment could transition upwards to darker, browner shades, which showed the sediment's contact with the surface. Such spots reflect both the soil formation processes and the impact of human intervention. If the upper layer sediments of the doline were in direct contact with the surface for a longer period of time, they were exposed to pedogenesis. The upper layers began forming soil which people probably cultivated, as evidenced by the small fragments of prehistoric ceramics discovered in the course of archaeological excavations. Later on, they were buried beneath new layers of clastic sediments which originated from the weathered remains of the Eocene flysch. Whenever the new layers of sediments were in lengthier contact with the surface, the pedogenetic processes took place again, causing another colour change of the upper layer. That this is soil formation is evidenced by the continuous transitions from the fresh, lighter-coloured sediment to darker shades, which also points to the increased presence of organic substances in the sediment.

10.1 Sediments from Dolines

Clastic sediments in dolines were separated according to their graininess and colour within individual layers. Samples were taken from these layers, some of them to be analysed by means of the powder x-ray diffraction method.

The positions of dolines examined are presented in Fig. 10.1. Four of the examined dolines (dolines No. V1, V2, V3, and V4), a minor red sediment filling

M. Knez and T. Slabe (eds.), *Cave Exploration in Slovenia*,
Cave and Karst Systems of the World, DOI 10.1007/978-3-319-21203-6_10

in the slope of doline No. V4 and a cave featuring yellow fluvial sediments were located on and along the Divača–Kozina motorway route, on the western side of the Bubenj Hill, i.e. by the Matavun–Lokev road. Doline No. V5 was located on the route south of the Jama pod Škrinjarco Cave, while doline No. V6 was on the motorway route west of Rodik.

Our research was mainly focused on the origin of sediments found at the bottom of dolines, and whether they could provide answers to the following questions:

- Do the described cases involve surface or cave sediments?
- Could their origin be attributed to the weathered remains of flysch rocks or the already formed red soil?
- What is the mineral composition of the red soil in the cases described, and where does it originate from?

The qualitative mineral composition of the samples was determined by the powder x-ray diffraction method described in Chap. 9. The amount of minerals in the samples is expressed in percentages which do not represent the absolute quantity of an individual mineral in the sample; rather, they express a calculated share of the mineral in the sample with regard to the height of its main reflection. Two samples taken from the bottom of each doline were analysed using this method.

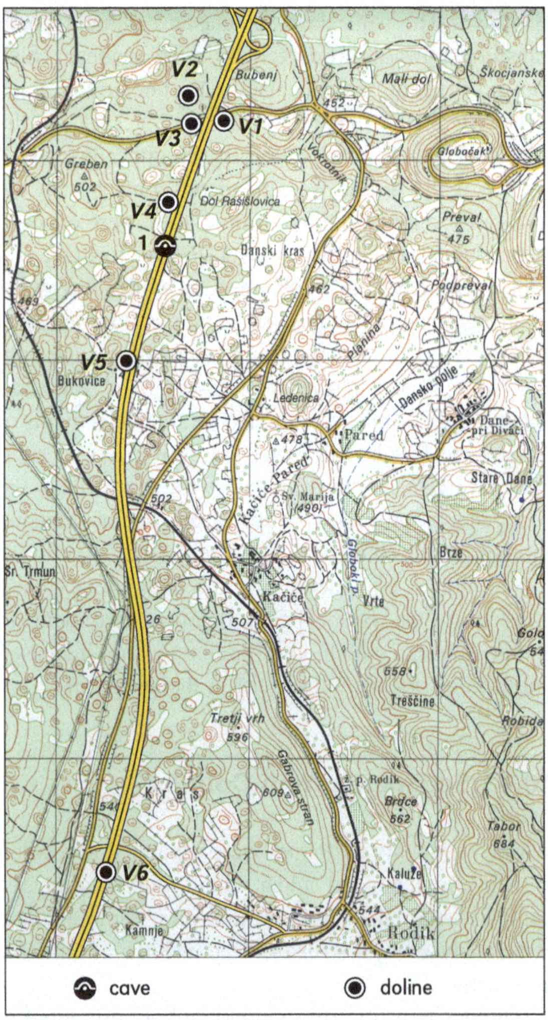

Fig. 10.1 Map with the locations of dolines with the described sediments

10.1.1 Doline No. V1

This doline was overgrown, measuring around 60 m from east to west, and was 6 m deep. The bottom was flat and filled with sediments, while the eastern and western slopes were steep, with individual pieces of limestone jutting out to the surface. The doline was formed in the black limestone of the Liburnia Formation, i.e. in the limestone of the Lower Palaeocene (Pc1). The dip of the layers ranged from 170/20–30 to 180/30, with two fissured zones in the doline walls, the first striking in the direction of 210°–30°, and the second in the direction of 290°–110°.

The probe was excavated in the direction of 250°–70°, and was 2.5 m deep in the middle. In terms of colour, the excavated profile displayed 5 different sediment layers (Fig. 10.2). The bottom of the profile contained a layer of yellow loam (7.5YR 4/4) followed by a thin layer of grey-yellow loam (7.5YR 5/4), after which came the red-brownish loam (5YR 4/4) with individual fragments of limestone, and a layer of brown-yellow soil (7.5YR 4/4) above it. The western section of the profile clearly exhibited two thinner layers of red loam (5YR 3/4). The grey-yellow loam was probably merely pedogenetically altered yellow loam from the bottom layer, since the transition was gradual. The colour and the shape of the layer indicated the former bottom of the doline, which was then filled with slope rubble and younger sediments which again underwent the pedogenetic processes in the upper part.

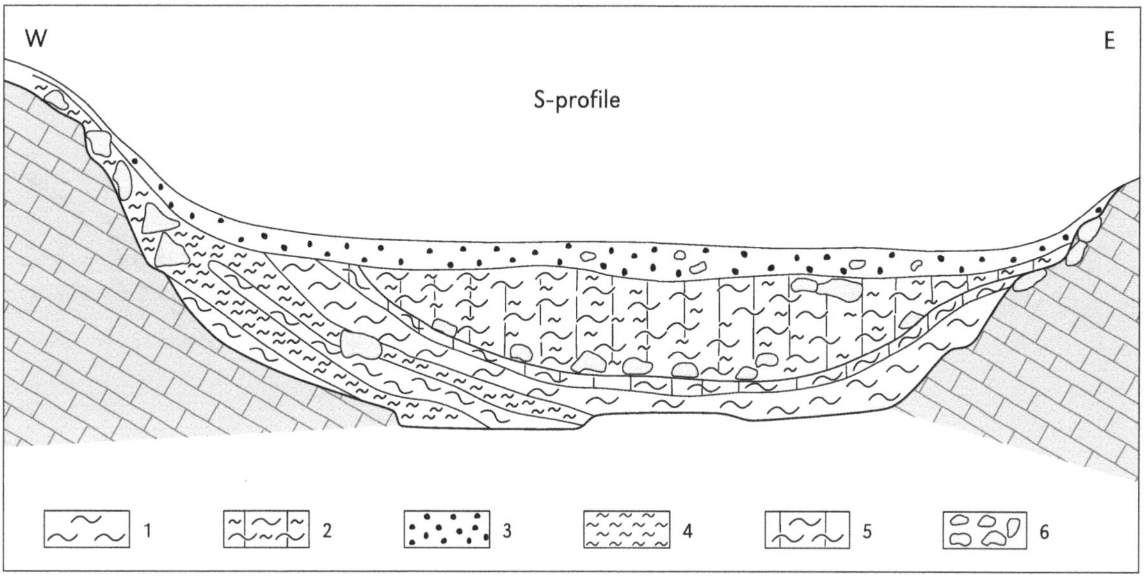

Fig. 10.2 Sketch of the doline V1 with diameter of about 60 m, excavation depth 2.5 m. *Legend 0* source rock, *1* yellow loam (7.5YR 4/4), *2* reddish-brown loam (5YR 4/4), *3* brown-yellow soil (7.5YR 4/4), *4* red loam (5YR 3/4), *5* grey-yellow loam (7.5YR 5/4)

Since we were primarily interested in the origin of the initial sediments from the bottom of the doline, we applied the x-ray diffraction method to analyse the V1/1 yellow loam sample and the V1/4 red loam sample. The V1/1 sample, yellow loam with 7.5YR 4/4 colour, contains 71 % of quartz, 11 % of muscovite/illite, 9 % of chlorite, 5 % of plagioclase and 4 % of microcline. The V1/4 sample, red loam with 5YR 3/4 colour, comprises 64 % of quartz, 19 % of chlorite, 10 % of muscovite/illite, 5 % plagioclase and 1 % of hematite.

Both samples contained the same minerals which point to their origin in the weathered remains of the Eocene flysch. The only difference can be seen in the content of the hematite mineral that gives the sediment its red colour. We attribute the presence of hematite to the transformation of goethite into hematite, a process that necessitates suitable conditions, i.e. contact of the sediment with rainwater. In view of the mineral composition and colour of both layers, we gather that this is the same sediment which, upon contact with the surface, merely oxidized down to a certain depth and turned red. Later on, new, fresh material was deposited and covered both layers described.

10.1.2 Doline No. V2

The doline was overgrown, measuring 50 m in diameter and featuring steep eastern and southern slopes. Prior to the excavation of the probe, the doline was 5 m deep, with soil accumulated at the bottom. The doline was formed in the limestone of the Liburnia Formation, i.e. in the black limestone of the Lower Palaeocene (Pc1). The layers dipped at 160/20. They were intersected by a fissured zone striking in the direction of 310°–130°, with fissures following one another at 10 cm intervals at first, and then at 1 m intervals.

The probe was excavated from the sediment in the direction of 310°–130°, and up to 2.5 m deep in the middle. In terms of colour, the northern profile exhibited 5 different types of sediments (Fig. 10.3). The bottom displayed red loam (2.5YR 3/4), followed by a layer of yellow loam (7.5YR 4/6). Above it was brown-grey-yellow loam (7.5YR 3/4) with individual limestone fragments, which was followed by a layer of red-yellow loam (5YR 3/4), and a layer of dark brown-yellow soil (7.5YR 3/4) with rubble in between.

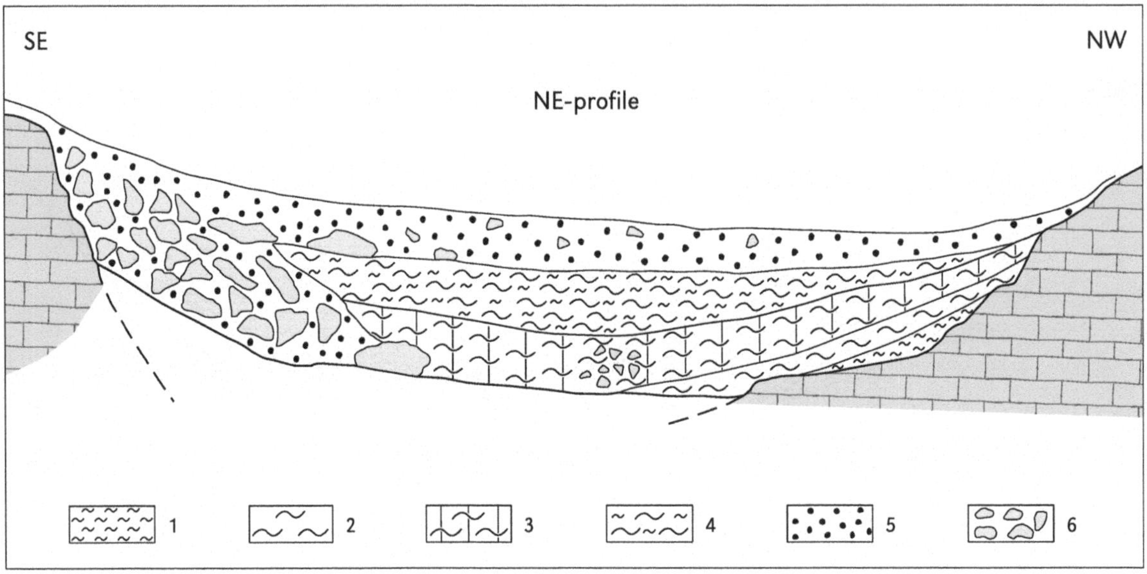

Fig. 10.3 Sketch of the doline V2, with diameter of about 50 m, excavation depth 2.5 m. *Legend 0* source rock, *1* red loam (2.5YR 3/4), *2* yellow loam (7.5YR 4/6), *3* brown-grey-yellow loam (7.5YR 3/4), *4* red-yellow loam (5YR 3/4), *5* brown-yellow soil (7.5YR 3/4) *6* talus debris

Applying the method of x-ray diffraction, we analysed the V2/1 red loam sample and the V2/2 yellow loam sample (Fig. 10.4). The V2/1 sample, red loam featuring 2.5YR 3/4 colour, comprises 76 % of quartz, 12 % of chlorite, 5 % of muscovite/illite, 3 % of kaolinite, 3 % of plagioclase and 2 % of hematite. The V2/2 sample, yellow loam with 7.5YR 4/6 colour, comprises 73 % of quartz, 13 % of chlorite, 7 % of plagioclase, 4 % of muscovite/illite and 3 % of microcline.

Since the bottom of the doline was not reached in the process of excavation, we don't know whether the layer of red loam was the oldest sediment in the doline. In view of the mineral composition, this is also probably weathered and resedimented flysch sediment which turned red under the influence of rainwater, namely due to the changing of goethite into hematite.

The layer of brown-grey-yellow loam used to be yellow loam from another layer, which underwent pedogenetic alterations and represented the former bottom of the doline. Yellow-red loam was then deposited over this layer, with a layer of brown-yellow soil on the surface. The southeastern part of the profile is missing loam layers, but contains a lot of slope rubble and brown-yellow soil, probably indicating subsequent subsidence of the doline bottom, and thus the subsidence of the loam layer.

10.1.3 Doline No. V3

The doline was overgrown, measuring 40 m in diameter, featuring steep and rocky southern and eastern slopes, and a distinctly flat northern side. It was 5 m deep, with soil covering the bottom. The doline was formed in the limestone of the Liburnia Formation, namely in the black limestone of the Lower Palaeocene (Pc1). On the southern, nearly vertical slope of the doline, the layers dipped at 160/25. This wall was caused by a tectonically broken zone striking in the direction of 230°–50°, while the limestone layers were also cracked by a fissured zone striking in the direction of 330°–150°.

The probe was excavated in the direction of 350°–170°, up to 1.8 m deep in the middle. In terms of colour, the western profile exhibited 5 different types of unconsolidated sediments (Fig. 10.5). The bottom displayed yellow, slightly reddish loam (7.5YR 4/6), followed by a layer of grey-brown-red loam (5YR 4/3). Above it was a smaller wedge of red-yellow loam (5YR 4/4), followed by a layer of grey-brown-red loam (5YR 4/6), with a top layer of grey-brown soil (5YR 3/3) featuring rubble in between. There was quite a lot of slope rubble in the southern part of the profile, while the top of the profile featured a thin layer of humus.

Fig. 10.4 Profile in the doline V2. Visible in the *bottom* are the layers of *red* and *yellow* loam from which two samples were taken for mineralogical analyses

Using the x-ray diffraction method, we analysed the V3/1 yellow loam and the V3/3 red-yellow loam samples. The V3/1 sample, yellow loam with 7.5YR 4/6 colour, comprises 77 % of quartz, 12 % of chlorite, 6 % of muscovite/illite and 4 % of plagioclase. The V3/3 sample, red-yellow loam with 5YR 4/4 colour, comprises 76 % of quartz, 13 % of chlorite, 4 % of muscovite/illite and plagioclase and 2 % of dolomite.

In view of the mineral composition, this sediment also originates from the weathered remains of the Eocene flysch, while the upper part of the sediment, influenced by the rainwater, oxidized and turned red due to hematite. Dolomite originates from the doline slopes where part of the limestone has obviously undergone dolomitization. The grey-brown loam from the second layer and the grey-brown-red loam from the fourth layer represented the former doline floor. In both instances, the loamy sediment was deposited in contact with the surface for so long as to undergo alteration through the pedogenetic processes. In the southern part of the profile, a lot of slope rubble and intermediate grey-brown soil was accumulated, indicating subsidence of the doline bottom after the deposit of loamy sediment layers.

10.1.4 Doline No. V4

The double-bottom doline measured 80 m from north to south, and 130 m from east to west. On the eastern side, the bottom was 13 m deep, and this is where the archaeological probe was excavated. Both the southern and the eastern slopes of the doline were very steep. The doline was formed in the dark limestone of the Liburnia Formation, i.e. in the limestone of the Lower Palaeocene (Pc1). On top of the southern slope edge, light milliolid limestone was found, which, stratigraphically speaking, belongs to the Upper Palaeocene (Pc2). The dip of layers ranged from 180/25 to 170/30. From the tectonic aspect, the limestone was rather fractured, with the fissured zone striking in the direction from north to south being especially distinct, much like the broken zone striking in the direction of 340°–160°, along which the doline slopes formed as well.

The probe was excavated in the direction from north to south, up to 2 m deep in the middle. In terms of colour, the eastern profile exhibited 4 different types of loamy sediment layers (Fig. 10.6). The bottom of the southern part of the probe featured fine rubble, perhaps of climate-based origin, over which yellow loam was deposited later on. In the middle of the doline, several limestone rocks were accumulated along with slope rubble, indicating the subsidence of the bottom in this part. Above it was a layer of red, slightly yellowish loam (7.5YR 4/6), followed by a layer of yellow-red-brown loam (5YR 4/6) and a layer of yellowish-brown soil (7.5YR 4/4) on top.

We analysed the V4/1 yellow loam and the V4/2 red-yellow loam samples. The V4/1 sample, yellow loam with 7.5YR 5/6 colour, comprises 73 % of quartz, 12 % of chlorite, 7 % of muscovite/illite, 5 % of plagioclase, 3 % of microcline and 1 % of dolomite. The V4/2 sample, yellow loam with 7.5YR 4/6 colour, comprises 74 % of quartz, 10 % of chlorite, 6 % of plagioclase, 5 % of muscovite/illite, 5 % of microcline and 1 % of hematite.

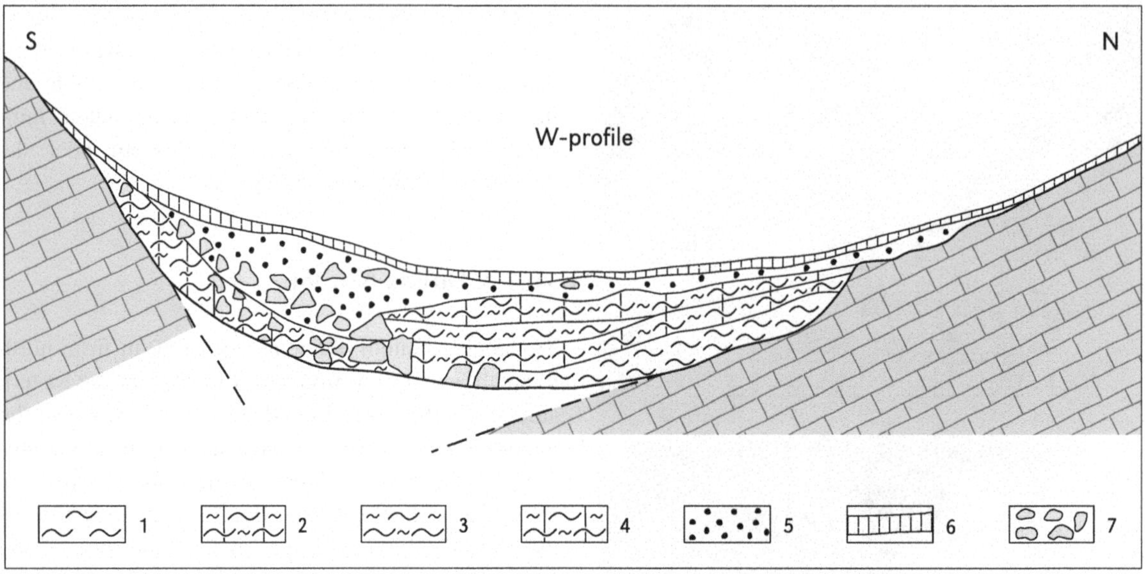

Fig. 10.5 Sketch of the doline V3, with diameter of about 40 m, excavation depth 1.8 m. *Legend 0* source rock, *1* yellow loam (7.5YR 4/6), *2* grey-brown-read loam (5YR 4/3), *3* red-yellow loam (5YR 4/4), *4* grey-red loam (5YR 4/6), *5* grey-brown soil (5YR 3/3), *6* humus, *7* talus debris

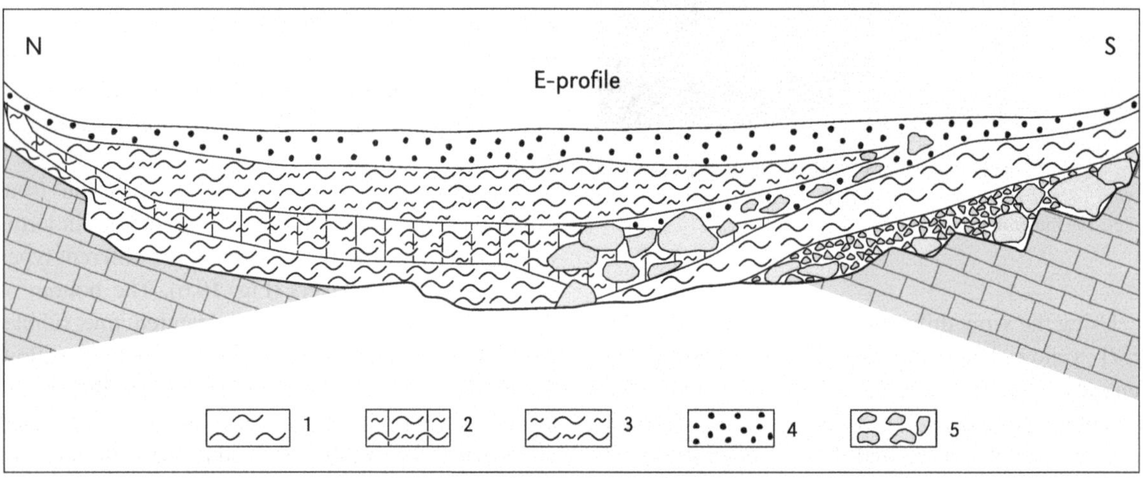

Fig. 10.6 Sketch of the doline V4, with diameter of about 80 m, excavation depth 2 m. *Legend 0* source rock, *1* yellow loam (7.5YR 5/6), *2* red, somewhat yellowish loam (7.5YR 4/6), *3* yellow-red-brown loam (5YR 4/6), *4* yellowish-brown soil (7.5YR 4/4), *5* talus debris

The yellow-red-brown loam from the third layer presented the former doline floor, where pedogenetic changes were noticeable in the loamy sediment. In this spot, sedimentation took place continuously, with lengthier interruption after the previously described third layer was deposited. Judging from the mineral composition, the yellow loam at the bottom of the southern part, deposited over the rubble, also originated from the weathered rocks of the Eocene flysch.

On the southern edge of doline No. V4, where the terrain is already flattened, there was a red loam-filled widening found between the limestone layers next to the fissures striking in the direction from northeast to southwest. From it, the V4SR sample was taken. The red loam with the 2.5YR 4/6 colour featured the following mineral composition: 77 % of quartz, 12 % of chlorite, 3 % of muscovite/illite, 3 % of plagioclase, 2 % of kaolinite and 2 % of hematite. In view of the mineral composition, the sediment originated from weathered flysch which was on the surface for a longer period of time, enabling goethite to alter into hematite, giving the sediment its intense red colour.

10.1.5 Infilled Cave with Stalactites and Stalagmites

One of the caves was completely filled with yellow sandy sediments (Fig. 10.7), among which there was a large heap of disintegrating white flowstone. The sample JSED, yellow sand with the 10YR 6/6 colour, comprises 57 % of quartz, 16 % of muscovite/illite, 15 % of chlorite, 4 % of calcite, 4 % of microcline and 4 % of plagioclase.

Fig. 10.7 A cave filled with a yellow clastic sediment in the motorway route Divača–Kozina, from which a JSED sample was taken for analysis

In view of the mineral composition of the cave sediment, it could be deduced that it originated from the weathered remains of the Eocene flysch which were resedimented to the cave. The calcite grains probably originate from the disintegrating flowstone.

10.1.6 Doline No. V5

The shallow doline located south of the Jama pod Škrinjarco Cave was formed in the grey, coarse-grained Eocene limestone with fossil fragments. The Eocene limestone layers dipped at 100/20. From the tectonic aspect, the limestone was highly fractured, with a well-visible fissured zone located on the northern slope, striking in the direction of 210°–30° (dip at 128/80), in which individual fissures followed one another at intervals ranging from 1 to 10 cm. The doline was 3 m deep and filled with soil at the bottom, featuring gentle slopes and overgrown with turf grass. It measured around 40 m in diameter from north to south, and 30 m from east to west.

The fissures located between individual limestone blocks measuring 5–30 cm contained embedded brown soil in the upper part, while the bottom of the profile featured yellow loam mixed with slope rubble. The probe was excavated in the north–south direction, and ended up in the middle of the doline (Fig. 10.8). The excavation was up to 1.8 m deep but did not reach the rocky bottom. On the northern slope of the doline, the probe was shallow, with up to 2 cm of brown soil on the surface and containing individual pieces of limestone. At the rocky bottom of the northeastern part of the profile, the limestone exhibited a well-visible fissured zone which transitioned into a fractured zone. Brown loamy soil was seen between individual blocks of limestone. The fissures located between individual limestone blocks measuring 5–30 cm contained embedded brown soil (10YR 4/3) in the upper part, while the bottom of the profile featured yellow loam (10YR 5/6) mixed with slope rubble. This was visible especially in the western wall of the profile, while the eastern wall contained more fine-grained rubble and red loam (5YR 3/3). The yellow loam is also visible at the bottom of the profile, in the middle of the doline. It surrounded larger pieces of limestone in rather advanced stages of weathering, with red (5YR 4/6), greasy loam above it, extending to the top of the profile.

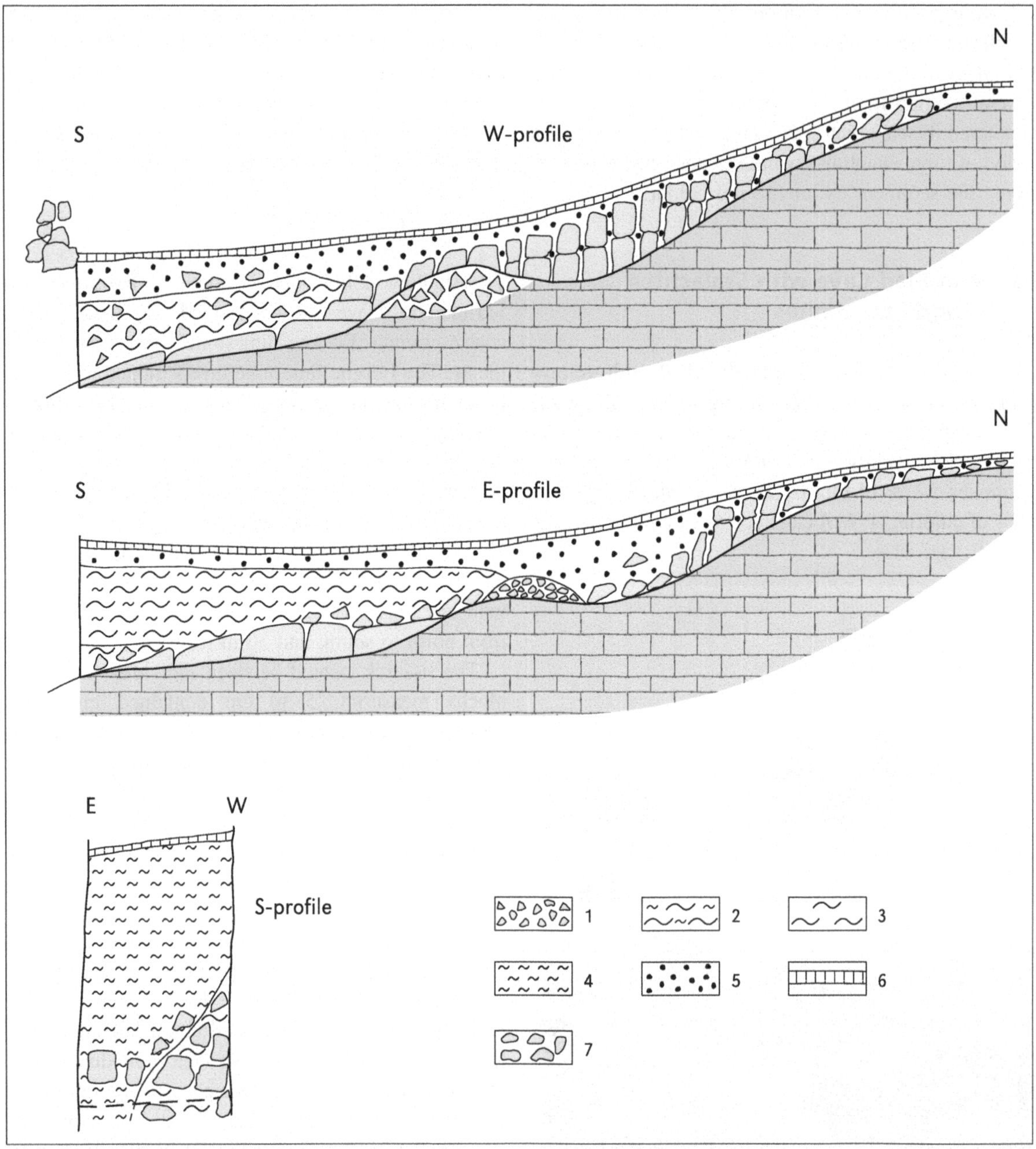

Fig. 10.8 Sketch of the doline V5, with radius of about 20 m, excavation depth 1.8 m. *Legend 0* source rock, coarse-grained limestone, *1* fine rubble with a layer of red-brown loam in between (5YR 3/3), *2* yellow-red loam (5YR 3/3), a few somewhat more brown parts (7.5YR 5/6) from the eastern section of the profile, *3* yellow loam (10YR 5/6) and rubble from the lower and western section of the profile in between, *4* red loam (5YR 4/6), individual parts were dyed (2.5YR 4/6) from the front of the profile, *5* brown loam soil, *6* humus, 7 talus debris

The V5/2 sample, yellow-red loam featuring 5YR 3/3 colour, comprises 75 % of quartz, 13 % of chlorite, 5 % of muscovite/illite, 5 % of plagioclase and 2 % of microcline. The V5/4 sample, red loam featuring 5YR 4/6 colour, comprises 75 % of quartz, 12 % of chlorite, 8 % of muscovite/illite, 5 % of plagioclase and 1 % of hematite.

This sediment also originates from the weathered remains of flysch rocks, as the red loam has the same mineral composition as yellow loam, the only difference being the hematite content.

10.1.7 Doline No. V6

The rocky doline located west of Rodik measured 40 m in diameter and featured a rather steep slope in the east. It was 4 m deep, with soil accumulated at the bottom. The doline was formed in the dark limestone of the Liburnia Formation, i.e. in the limestone of the Lower Palaeocene (Pc1). This limestone exhibited changes typical of the soil formation processes, i.e. darker areas measuring a few millimetres to a centimetre that initially give a breccia-like appearance. The dip of layers on the northern edge of the doline ranged from 250/10 to 260/5. The limestone was highly tectonically fractured, but most fissures were calcitised. The fissures striking in the direction of 160°–340° were most prominent, with the eastern edge of the doline formed along them. The direction of the fissured zone running across the middle of the doline is the same, with the open fissures following one another at intervals ranging from 1 to 1.5 m. The probe was excavated across the doline in the direction of 280°–100°, while the excavation went 5 m deep at the bottom that was closer to the steeper eastern edge of the doline. The southern profile revealed 5 different types of sediments in terms of colour and graininess, while the northern profile featured 3 types (Fig. 10.9).

The bottom of the southern profile contained a layer of very fine rubble with yellow loam (10YR 5/6) in between, and a thin layer of yellow-red loam (7.5YR 5/6) above it, followed by a layer of yellow loam and a layer of red loam (7.5YR 4/6). At the top and across the entire eastern part of the profile, there was brown soil (7.5YR 4/4) with thick slope rubble which abounds on the eastern edge of the profile. The present-day soil is brown and mixed with slope rubble which is especially abundant on the eastern slope of the doline where there were no loam layers. The bottom subsided in this part of the doline. The northern probe profile featured only yellow loam and brown soil with slope rubble in between.

The V6/2 sample, brown soil featuring the 7.5YR 4/4 colour, comprises 72 % of quartz, 10 % of chlorite, 6 % of plagioclase, 4 % of muscovite/illite, 3 % of kaolinite, 2 % of microcline, 1 % of dolomite and 1 % of hematite.

In view of the mineral composition, this sediment also originated from the weathered remains of flysch rocks. After the deposit of fine rubble, probably of climate-based origin, grey-yellow loam of the eighth layer was deposited and, for some time, represented the top deposit in the doline, featuring a darker loam colour due to the formation of humus in it. After the deposit of the yellow-red loam from the third layer and the red-yellow-brown loam from the fifth layer, the latter was in contact with the surface for a while, as indicated by a thin lense of black-brown loamy soil that, judging by its colour, contained a lot of organic matter. The present-day soil is brown and mixed with slope rubble which is especially abundant on the eastern slope of the doline where the above-referenced loamy layers were not present. This part of the doline underwent a recent subsidence of the bottom.

10.2 Conclusion

The examined dolines contained unconsolidated mechanical sediments which differed in terms of graininess; however, there was a prevalence of loams in various colours. The bottoms of all the selected dolines contained yellow sediments that were slightly reddish in some places, with intensely red-coloured loam (2.5YR 4/6) found next to them. In such cases, the colour reflected the sediment's contact with the surface and human interventions on the surface. Whenever the sediments in the doline were exposed to the surface for a longer period of time, they underwent the processes of diagenesis and pedogenesis. The upper layers began forming soil which people probably cultivated/turned, as evidenced by the small fragments of prehistoric ceramics discovered in the course of archaeological excavations. Later on, they were washed off the slope and covered with new layers, or the bottoms of dolines were levelled with freshly excavated sediments. Whenever the newly deposited sediments again came into lengthier contact with the surface, they underwent the processes of diagenesis and pedogenesis. At the same time, the top layers could be worked by people and so the upper layer again changed its colour. That this is soil formation is evidenced by the undefined transitions from the fresh, lighter-coloured sediment to darker shades featuring increased content of organic substances.

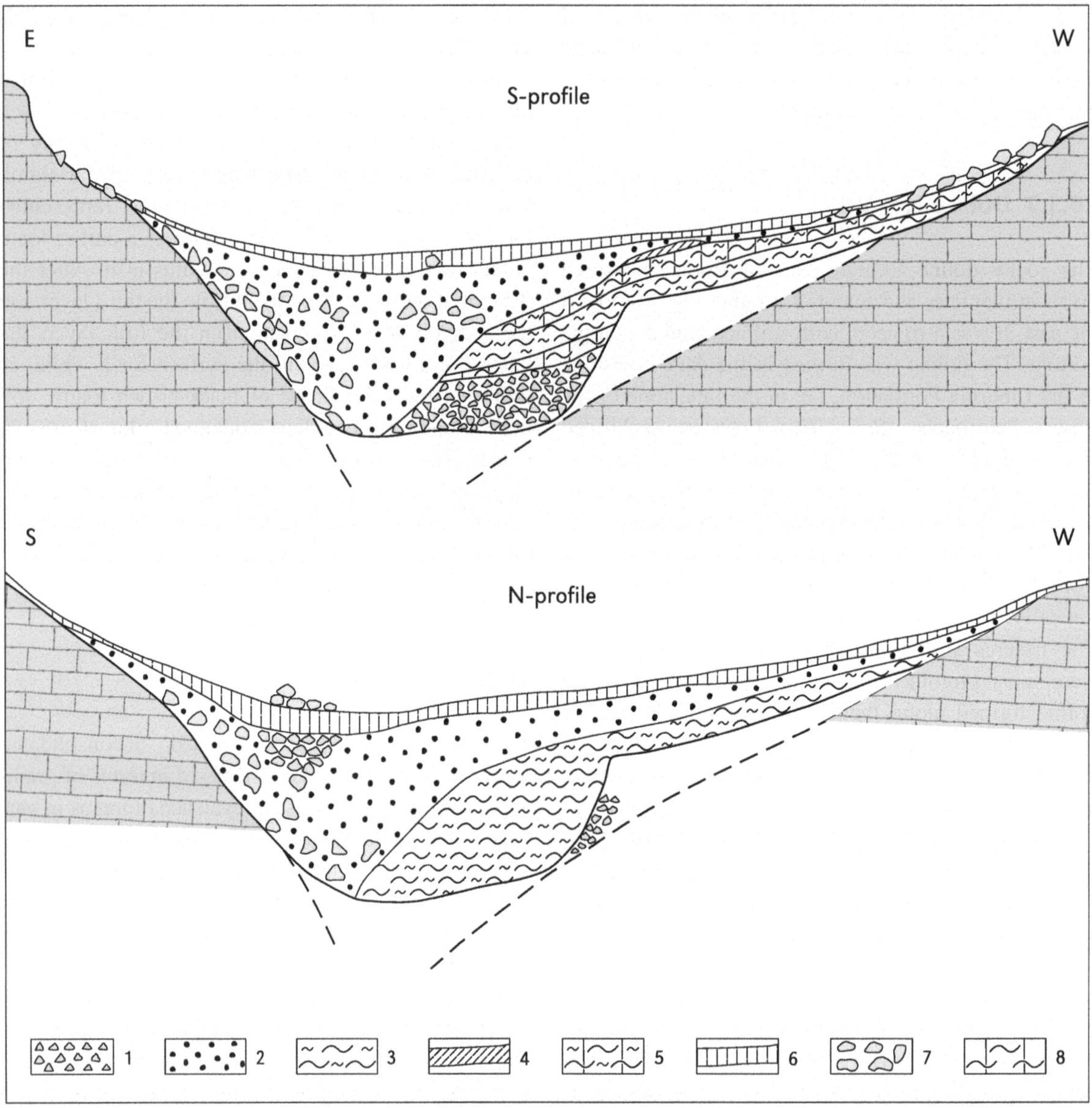

Fig. 10.9 Sketch of the doline V6, with diameter of about 40 m, excavation depth 5 m. *Legend 0* source rock, *1* fine rubble with a layer of yellow loam in between (10YR 5/6), *2* brown soil (7.5YR 4/4), *3* yellow-red loam (7.5YR 5/6), *4* black-brown loam soil (5YR 3/3), *5* red-yellow-brown loam (7.5YR 4/6), *6* humus, *7* talus debris, 8 grey-yellow loam

The different colours of sediments indicate their mineral composition which reflects their origin and conditions in the environment into which they have been resedimented. In this case, the colour is mainly related to the content of iron minerals, which include goethite, lepidocrocite, magnetite, maghemite and hematite, all of which we can expect to find in samples judging by the colour of the sediment.

Iron minerals are common in clastic sediments and soils in all environments in which they appear during the weathering of various rocks rich in iron. Iron is the fourth most common element in the Earth's crust where it is represented in igneous rocks in the Fe^{2+} form, silicates as well as sulphides such as pyrite. During weathering, these rocks disintegrate through processes of oxidation and hydrolysis. The resulting

Fe^{3+} oxides exhibit very low solubility. In the weathering zone, in the soil, the Fe^{3+} oxides can be remobilized under anaerobic conditions by means of microorganisms. A new increase in the concentration of the Fe^{2+} ions causes a decline in the pH value in the environment and a migration of the free ion in other environments, where the iron oxide can undergo precipitation again. The iron oxides created through weathering exhibit three main characteristics: they give colour to the material whose admixture they are serving as, they are present in very small crystals (5–100 nm) and represent an extensive area in terms of their form and presence.

The most frequently occurring minerals in sediments and soil are iron oxides and hydroxides such as goethite, lepidocrocite, hematite and maghemite, while the iron sulphide, pyrite, is also frequent. Limonite is a mixture of goethite ($\alpha FeOOH$) and lepidocrocite ($\gamma FeOOH$), which can also contain admixtures of other minerals. Larger quantities are generated especially during the weathering process of pyrite (FeS_2) and siderite or as a precipitate in marshes where the increased amounts of bituminous substances create a reduction environment. It is also common in laterites. Goethite is a very frequent mineral created in the oxidation of various iron minerals, but can also be created as a precipitate in marshes and springs. Goethite and lepidocrocite are not magnetic. Another frequently occurring mineral is hematite (αFe_2O_3) that colours the sediments and the soil red, and is either slightly magnetic or non-magnetic. Maghemite (γFe_2O_3) is created in the oxidation process from magnetite and goethite, and is magnetic. Magnetite (Fe_3O_4) occurs in the soil in the form of lithic grains, and is magnetic. Pyrite (FeS_2) is the most frequent sulphide created at high and low temperatures. Weathering first down into iron sulphates and then into goethite, it is almost completely non-magnetic. Since it is frequent in Slovenian flysch rocks, it can be found in sediments and soil originating from said rocks.

Precipitation, dissolution and re-precipitation of iron oxides in the environment depends on the water pH, Eh, temperature and activities, which is why they can serve as indicators of the environment in which they were created.

From the thermodynamical aspect, goethite and hematite are the most stable iron oxides under aerobic conditions, and therefore occur most frequently in sediments and soil. The yellow-brown goethite occurs in nearly all types of soil, in lakes and streams, while hematite is typical of the soil formed in the tropical and subtropical climate, colouring the soil red. The ratio between goethite and hematite changes in line with the local conditions, which is why this ratio serves as an environment indicator.

Goethite precipitates directly from solutions, while hematite requires the presence of ferrihydrite ($Fe_5HO_8.4H_2O$) as a predecessor—it is namely formed with its dehydration and rearrangement of atoms. There is no direct transition of goethite into hematite, except under conditions involving increased pressure, where water is squeezed from a goethite molecule and hematite is formed. In the soil, goethite and hematite are usually of irregular shape, even though goethite can be fibrous, and hematite can be found in the shape of hexagonal plates.

Lepidocrocite is a less frequent mineral but typical of environments with Fe^{2+} from which lepidocrocite is formed through oxidation. It gradually transitions to a more stable goethite.

Ferrihydrite is restricted to areas where Fe^{2+} oxidizes very quickly and where the crystallization inhibitors (organic substances, phosphates, silicates) are present. The crystals are small, and these inhibitors stabilize ferrihydrite, preventing it from transforming into more long-lasting minerals. It is typically formed in environments such as springs, drainage lines, lake oxides, groundwater, water in sediments, river sediments, oceans etc. It is an important predecessor of hematite formation, but can also produce goethite, depending on the pH. With neutral pH, the transformation is very slow (taking anywhere from a few months to a few years), but can accelerate considerably in the presence of organisms (taking up to a few days). It is thus safe to say that ferrihydrite is a young iron oxide.

Maghemite is a mineral linked with the pedogenetic processes. It can be found in the soil of tropical and subtropical zones, but also in the temperate zone where it is bound to the increased temperatures (near forest fires or fireplaces) and the presence of organic substances. It is formed through the transformation of other iron oxides such as goethite, for example.

The yellow loams and all loams featuring shades of yellow and brown get their colour from goethite, while hematite colours the red loams and all loams with red shades. The black shade in brown and grey-yellow loams can be attributed to the presence of organic substances, which also indicates the formation of humus in former soil that was once exposed to the surface.

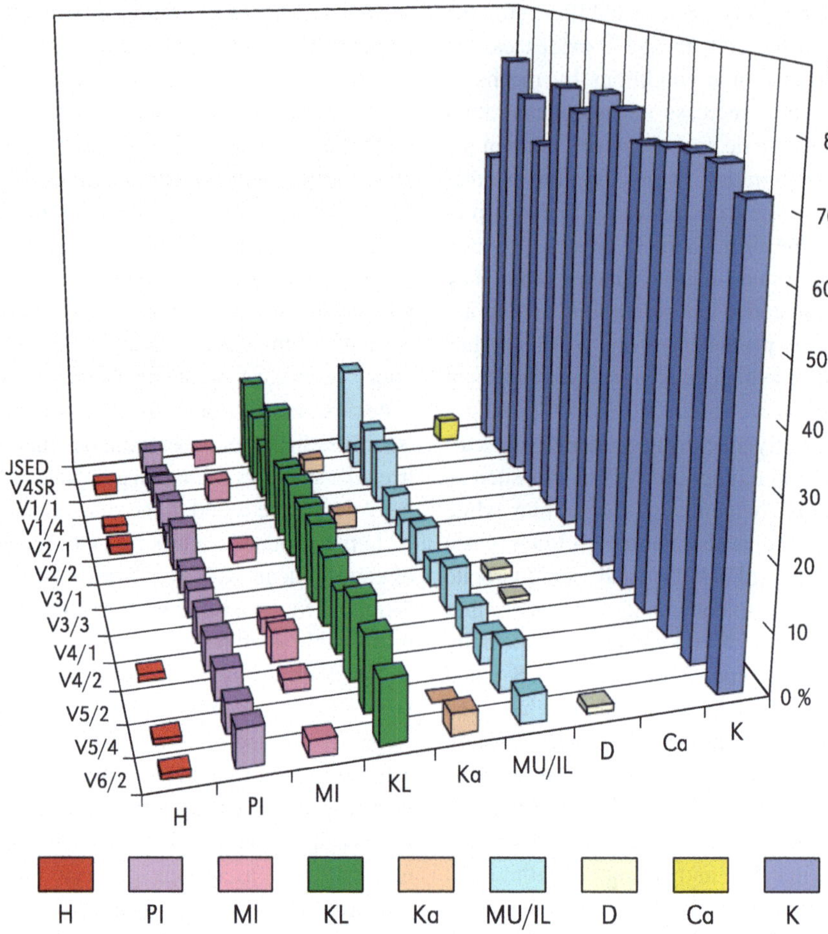

Fig. 10.10 Mineral composition of clastic sediments from the dolines and a cave on the motorway section Divača–Kozina. *Legend K* Quartz, *Ca* Calcite, *D* Dolomite, *Mu/IL* Muscovite/Illite, *Ka* Kaolinite, *KL* Chlorite, *Mi* Microcline, *PL* Plagioclase, *H* Hematite. The number next to the doline number designates the soil from which the sample was taken

The mineral composition of all samples examined by means of the powder x-ray diffraction method is presented in Fig. 10.10. The results of analyses conducted on selected yellow sediments from the dolines, i.e. their mineral composition, are presented in Fig. 10.11. It is evident that the samples mostly contain quartz (57–77 %), quite a lot of chlorite (9–15 %), muscovite/illite (4–16 %), plagioclase (4–7 %) and microcline (2–5 %). Individual samples also contained minor amounts of calcite, dolomite, kaolinite and hematite.

The mineral composition of red sediments is presented in Fig. 10.12. The diagram shows that the samples mostly contain quartz (64–77 %), quite a lot of chlorite (12–19 %), muscovite/illite (3–10 %), plagioclase (3–5 %), and that all samples, except V3/3, also contained hematite. Two of the samples also revealed to contain kaolinite, while dolomite was found in one of the samples.

The yellow loams contained more muscovite/illite as well as a higher percentage of plagioclase, with microcline also present. Calcite was found only in a sample of cave sediment from weathered flowstone. The red-coloured sediments contained more chlorite and less muscovite/illite and plagioclase, while microcline was not present in these samples.

In dolines, red and yellow sediments prevailed, occurring in several shades and transitioning from red to yellow and then to brown. The same layer of clastic sediment could transition upwards to darker, browner shades, which showed the sediment's contact with the surface. Such spots reflect both the soil formation processes and the impact of human intervention. If the

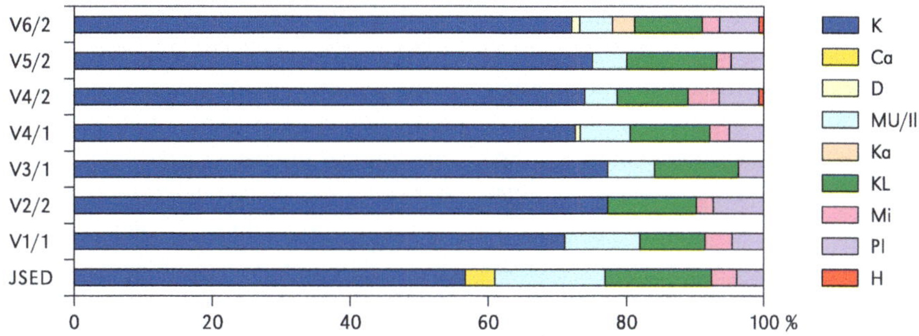

Fig. 10.11 Mineral composition of yellow, reddish-yellow and brown sediments. *Legend K* Quartz, *Ca* Calcite, *D* Dolomite, *Mu/IL* Muscovite/Illite, *Ka* Kaolinite, *KL* Chlorite, *Mi* Microcline, *PL* Plagioclase, *H* Hematite. The number next to the doline number designates the soil from which the sample was taken

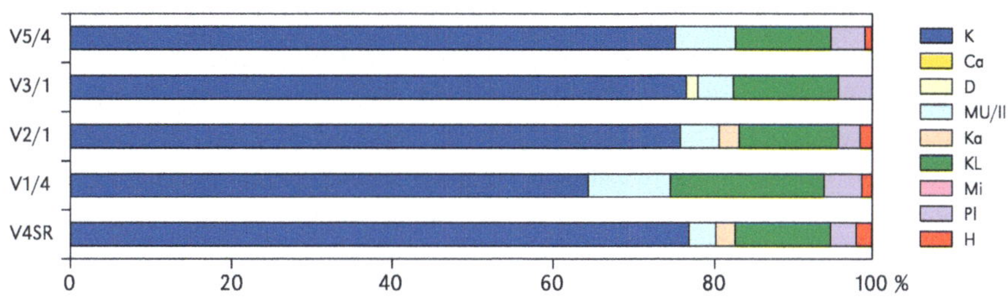

Fig. 10.12 Mineral composition of red sediments. *Legend K* Quartz, *Ca* Calcite, *D* Dolomite, *Mu/IL* Muscovite/Illite, *Ka* Kaolinite, *KL* Chlorite, *Mi* Microcline, *PL* Plagioclase, *H* Hematite. The number next to the doline number designates the soil from which the sample was taken

upper layer sediments of the doline were in lengthier direct contact with the surface, they were exposed to the pedogenetic processes. The upper layers began forming soil which people probably cultivated, as evidenced by the small fragments of prehistoric ceramics discovered in the course of archaeological excavations. Later on, they were washed off the slope and covered with new layers; alternatively, they were covered by people who levelled the bottoms of dolines. Whenever the new layers of sediments were in contact with the surface for a longer period of time, the pedogenetic processes took place again, causing another colour change of the upper layer. That this is soil formation is evidenced by the undefined transitions from the fresh, lighter-coloured sediment to darker shades, which also points to the increased presence of organic substances in the sediment. The yellow sediments featuring shades of yellow and brown get their colour from goethite, while hematite colours the red sediments. The black and dark brown colour is given to the sediments by organic substances, which also indicate the formation of humus in the soil.

All samples were similar in terms of their mineral composition, but the amount of individual minerals differed from one sample to another. The results obtained could provide the grounds to infer the same origin of sediments and thus of the minerals in them, i.e. that they all originate from the weathered remains of flysch rocks, namely because the samples contain microcline, plagioclase and muscovite, all of which are typical of flysch rocks. Chlorite can also originate from flysch rocks. The analysed sediments contain no amphiboles that are typical of the loess sediments in Istria (Durn and Aljinović 1995), which prompts us to eliminate their Eolian origin. Microcline and plagioclase, which can be found in these sediments, indicate that they cannot represent insoluble remains of limestone because the latter don't contain these two minerals.

Based on the examination of the sediment samples, we assume that both yellow cave sediments and red-coloured surface sediments originate from flysch rocks and their weathered remains. Their colour depends on the presence of iron minerals in the sediment. The colour of the weathered flysch is yellow to yellow-brown, while the surface and fissures, where the deposited sediments came into lengthier contact with atmospheric conditions, feature a typical transformation of yellow to red. This indicates the transition of iron hydroxides to oxides, e.g. hematite, which does not necessarily require tropical climate conditions.

The bottoms of depressions and caves, where the deposited sediments were protected from atmospheric influences, retained their yellow colour. Yellow therefore indicates cave sediments which are now located on the surface due to the denudation of cave ceilings. We lack sufficient evidence to state that the yellow sediments at the bottom of these dolines are also cave sediments. Since no flowstone or other typical cave formations could be found, it was impossible to deduce from the bottom of dolines whether a particular doline would reach the cave space. The sediments' composition points to their origin in the weathered remains of the Eocene flysch. The latter could have been deposited from flysch areas to the Karst by smaller streams in Pleistocene as flysch and limestone come into contact near the examined territory.

The red soil can originate not only from flysch rocks, but also from the weathered remains of limestone abundant in chert. Eolian sediments can also be found in them. A greater thickness of red soil can be attributed to its origin in the tropical climate where the rock weathering and soil formation processes are much faster. In the Slovenian Karst, some of the red-coloured sediments and red soil have passed the tropical climate period, which is reflected by the bauxite minerals contained by some of them. The red soil differs in its mineral composition and thus also in its origin and time of formation.

The widespread extent of sediments that originate from flysch rocks indicates the occasional distinct weathering of flysch rocks and the heavy transport of their weathered remains in the then existing caves and into the existing depressions across the karst surface (especially close to the contact between the karst and the flysch territory). Naturally, this process of flysch transport continues today, but not to such an intense extent. The increased erosion of flysch rocks could have occurred due to the colder climate, increased amount of precipitation or due to the intensive tectonic uplifting of the area. The sediments deposited in the Slovenian Karst at that time were definitely yellow. Those located on the surface and mixed with non-soluble remains of limestone, loess or water percolating through fissures, subsequently turned red when coming into contact with the surface. The red soil formed on the surface was possibly transferred to the caves later on.

11 Cave Sediments from the Infilled Cave Near Divača

The Divača karst is part of the original Karst, namely its southeastern edge around Divača and extending to the Škocjanske jame Caves. The Divača karst typically features numerous dolines, deep collapse dolines and large caves. The surface is formed at the altitude of 420–450 m, while the ponor of the Reka River leading to the Škocjanske jame Caves is located at 317 m. The Škocjanske jame Caves and the Kačna jama Cave are the largest caves through which the underground Reka River flows. In total, the passages are 12.5 km long. The Divaška jama Cave, whose entrance is situated at the altitude of 427 m, is also large, but its passages descend to the altitude of 350 m.

South of Divača, between the Divača exit and the Škocjan–Lokev road, a cave previously completely filled with yellow fluvial clastic sediments was exposed on the eastern edge of the motorway in 1997. The cross-section of the completely filled cave was app. 6 m high, with a few metres of ceiling above it. In view of the knowledge available back then, we assumed that applying the palaeomagnetic method is the only way to undertake the dating of these sediments. The flowstone analyses conducted with the U/Th method revealed that a lot of sediments present in the Slovenian Karst are older (Zupan 1991; Zupan Hajna 1992; Mihevc 1996) than the method is able to date.

During the sampling of sediments for palaeomagnetic analyses, we also took samples for mineralogical and palynological analyses. The palaeomagnetic analyses were conducted at the Institute of Geology, Academy of Sciences of the Czech Republic. The mineralogical analyses were carried out at the ZRC SAZU Karst Research Institute and at the Department of Geology at the Faculty of Natural Sciences and Engineering, University of Ljubljana, while the palynological analyses were conducted at the ZRC SAZU Jovan Hadži Institute of Biology in Ljubljana.

The palaeomagnetic sample analyses were carried out by D. Venhodová, J. Slepičková and J. Drahotová. The drawings are the work of V. Havliková and J. Forman, while the tables were put together by J. Čadková (all from the Institute of Geology, AS CR in Prague). Palynological analyses were conducted by M. Culiberg, PhD (ZRC SAZU Institute of Biology, Ljubljana), and the x-ray samples were taken by M. Bole, PhD (Department of Geology, Faculty of Natural Sciences and Engineering, Ljubljana).

11.1 Methods

The mineral composition of the sediment samples was determined with the powder x-ray diffraction method. The samples were analysed in line with the procedures described in Chap. 9. The amount of minerals in the samples is expressed in percentages which do not represent the absolute quantity of an individual mineral in the sample; rather, they express a calculated share of the mineral in the sample with regard to the height of its main reflection. The percentages are therefore only intended to provide certain information and enable comparisons. This method does not detect minerals whose content in the sample amounts to less than 3 %.

The palynological analyses were conducted on two larger samples of fine-grained fluvial sediments. The samples were processed with the usual method of palynological maceration (processing with HCl, KOH,

M. Knez and T. Slabe (eds.), *Cave Exploration in Slovenia*,
Cave and Karst Systems of the World, DOI 10.1007/978-3-319-21203-6_11

$ZnCl_2$ and $C_4H_6O_3$+H_2SO_4), but four sections of samples contained no spores or pollen.

For the purposes of palaeomagnetic analyses, 15 oriented samples of fine sediment and cemented sandstone were taken and divided into 29 laboratory samples. The laboratory analyses enabled the determination of magnetic remanence in different temperature intervals during gradual thermal demagnetization (TD) and alternating field demagnetization (AFD), the determination of modules and the directions of remanent magnetization, as well as the determination of minerals that carry remanent components in terms of individual mineral phases and their changes. In the course of conducting thermal laboratory testing, the phase and mineralogical changes of magnetically active minerals (mostly iron oxides) occur frequently, especially when the temperature intervals are small. These changes can be derived from the graphs of normalized values $k_t/k_n = f(t)$; k_n denotes the scope of magnetic susceptibility in specimens in their natural state, while k_t denotes the susceptibility of specimens demagnetized at t °C. The k_t and k_n values were measured using the KLY-2 KappaBridge instrument (Jelínek 1973).

In the course of laboratory analyses, the samples were subjected to gradual thermal demagnetization on the MAVACS (Magnetic Vacuum Control System) apparatus which creates a powerful magnetic vacuum in demagnetized samples (Přihoda et al. 1989). All samples were demagnetized through the alternating field procedures (Schonstedt GSD-1) up to the field of 1000 Oe (14 steps).

The remanent magnetization of samples in their natural state (NRM) is marked with J_n, while the corresponding remanent moment is marked M. The graphs of normalized values $M/M_0 = F(t)$ were made for each analysed sample. The direction of J_n and the remanent magnetization of samples demagnetized through the thermal or AF procedure during gradual demagnetization are shown in stereographic projection. The remanent magnetization components were separated by the multi-component Kirschvink analysis (Kirschvink 1980). The mean directions of remanent components

Fig. 11.1 A filled cave south of Divača in a doline on the eastern edge of the construction site of the motorway section Divača–Kozina

derived from the multi-component analysis were calculated by means of the Fisher statistics (1953).

11.2 Infilled Cave Located South of Divača

The infilled cave passage (Fig. 11.1) was located in the southern slope of a larger doline (x = 5058400, y = 5420100, z = 453 m). The cave was developed in the Lower Palaeocene limestone, with layers dipping by about 10° to the south.

The fluvial cave sediments in the profile were divided into four sequences. The lower sequence, No. 1, is separated from the sediments lying above it by means of a thicker limonite crust (Fig. 11.2) that consisted of various layers of clay and silty clay, with an admixture of fine sand in some places. The middle sequence, No. 2, consisted of multi-coloured clayey silts and clays, also with an admixture of sand. This layer concluded with a thin limonite crust which represents an even more noticeable erosion base of the third sequence (No. 3). The third (upper) sequence is marked by typical fluvial cycles with 4–40 cm thick white or greyish-brown, and greyish-brown or ochre sands with light-coloured clays and a silty finish to each cycle. The sands are mostly fine-grained. In them, it is possible to observe cross-stratification, sometimes also cross-lamination (Figs. 11.3, 11.4). The bases of individual cycles are sharp, usually featuring noticeable erosion. The profile concludes with an app. 30 cm thick layer of resedimented red soil (the terra rossa type) which is marked as the fourth sequence (No. 4). The red loam (Fig. 11.3) was partially covered with blocks of grey Palaeocene limestone from the disintegrating cave ceiling.

Fig. 11.2 The figure shows the sequence No. 1 in the lower section of the profile which ends with a limonite crust (along the scraper's handle), and the sedimentation of the sequence No. 2 above the crust

The sediment profile is shown in Fig. 11.5. The individual sediment areas were partially connected, i.e. cemented into irregularly shaped bodies, with a sharp contact with unconsolidated sediment. Cementation occurs especially in white and greyish-brown sands which indicate the contact type of carbonate cementation (Fig. 11.4). Some parts are heavily cemented with calcite crystals measuring several centimetres in length.

All sediments in the profile are characterized by secondary ferritization, especially in the form of iron crusts appearing in drained fissures and layers where erosion-related events are evident in spots where sedimentation was interrupted.

The sediments in the profile were heavily fractured as a result of individual sediment parts subsiding due to drying or tectonic events. The profile contained a prominent network of parallel fissures with spaces between them ranging from a few millimetres to several centimetres. Collapsed zones were also apparent in some parts of sediments, subsequently filled with reddish clays and calcite cement by infiltrated solutions oversaturated with calcium carbonate.

11.2.1 Results of Mineralogical Analyses

From the profile of clastic cave sediments, 21 samples were taken, of which 6 were analysed with the x-ray diffraction method. All samples contained quartz

(62–93 %), muscovite/illite (4–11 %) and chlorite (4–15 %). A relatively large amount of microcline (8–10 %) was detected only in one sample, while two samples revealed microcline content in traces. Traces of plagioclase were found only in one sample. Mixed-composition minerals were detected in one of the samples (2 %), while another sample contained traces of them. Goethite was present in two samples (4 and 5 %), and a rather large amount of calcite (24 %) was found in one sample.

Mineralogical analyses conducted on cave sediments show a relatively uniform composition of the light fraction (quartz, muscovite, microcline and plagioclase, also illite and chlorite), indicating their origin in the weathered remains of the Eocene flysch. Goethite appears in both limonite crusts found in two layers at the bottom of the profile, which formed when sedimentation was interrupted. The red loam on the top of the cave sediment profile originates from red soil which subsequently eroded and was transferred to the cave. Calcite represents cement which, in certain spots, bound clastic sediments (especially quartz sand) into rather porous sandstone.

11.2.2 Results of Palaeomagnetic Analyses

All selected samples underwent complete alternating field demagnetization (partially thermal). Basic magnetic parameters were determined for the profile measuring 6 m in height.

In general, the analysed samples revealed low magnetism with mean values of $J_n = 0.844 \pm 0.735$ (nT), $k_n = 103 \pm 26 \times 10^{-6}$ (SI), $n = 29$. The n symbol represents the number of samples used for calculations in the statistical processing of results. The J_n values

Fig. 11.3 The figure shows the sequence No. 3 in the bottom segment of the figure and above it the *red loam* of the sequence No. 4. The cave was filled up to the roof. The limestone above the cave sediments crumbled into single blocks

Fig. 11.4 Harder, cemented areas of the sediment protruding from the profile were located in the sequence No. 3

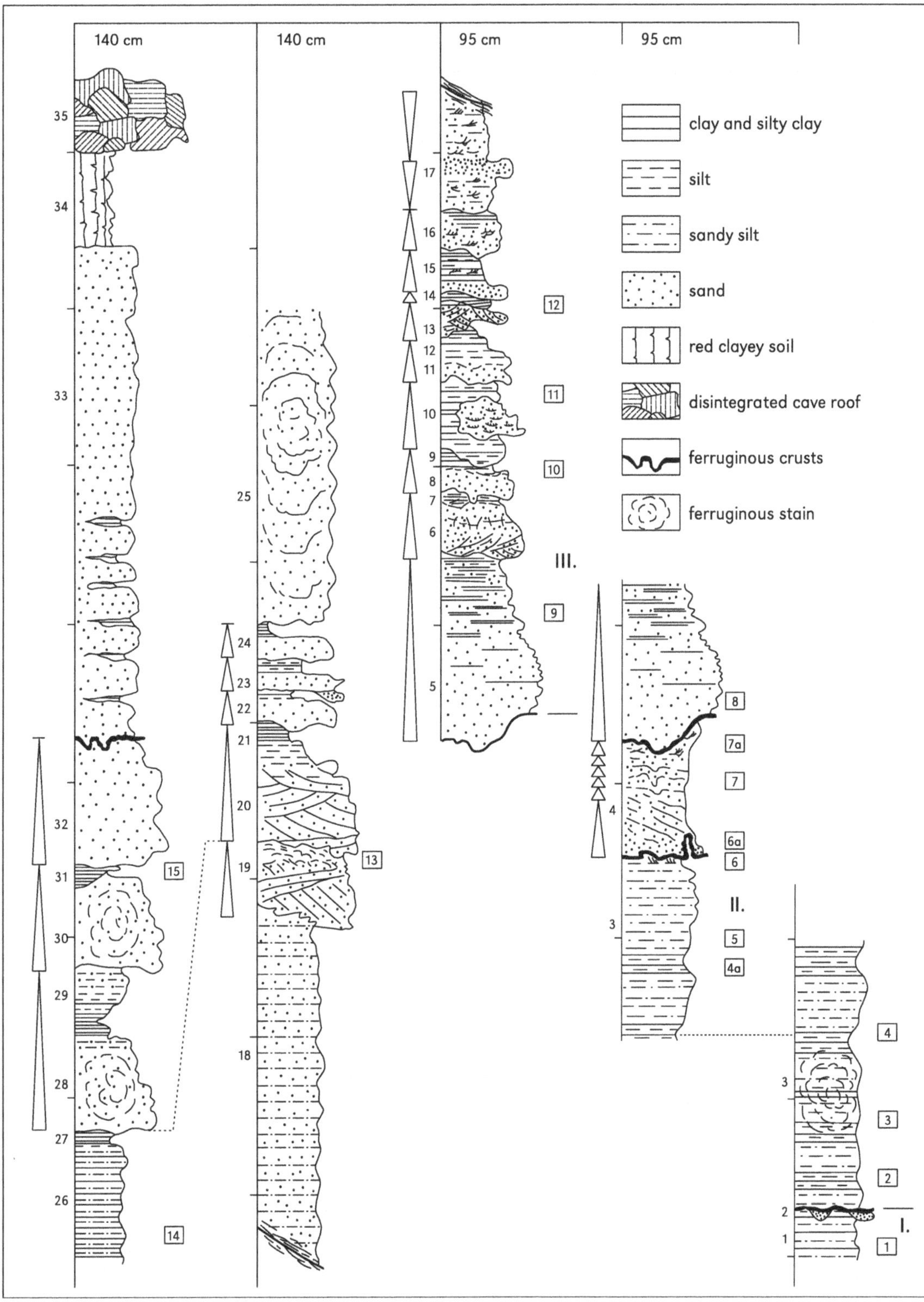

Fig. 11.5 The lithological column of the profile of cave sediments from the filled cave south of Divača. The numbers in squares designate the palaeomagnetic samples, whereas the small numbers designate layer numbers

were measured in the natural state with an adequate degree of confidence, and were very low for the processed samples, mostly depending on the source of magnetism. The values of the volume magnetic susceptibility are low as well, but point to a lower dispersion than the J_n values. The directions of remanent magnetization ascertained through the above-referenced procedures were tested with a multi-component analysis (Kirschvink 1980). Three components of remanence were determined, namely A, B and C. The A components are mostly of viscous or chemoremanent origin (weathering) and can be removed by means of an alternating field with the power of 10–30 Oe. The normal and reverse directions of the C component stem from two separate groups of samples in terms of the Fisher distribution (95 % probability) and are shown in Fig. 11.6. The basic magnetic sample parameters from the infilled cave near Divača are shown in Fig. 11.7. In the profile that represents 6 m of layers of fluvial cave sediments, we observed two zones with normal polarity in the bottom part of the profile, while all other samples from the profile exhibited reverse polarity. The remanent components were not determined for two samples in the profile, and two samples disintegrated too much.

The palaeomagnetic and magnetostratigraphic research provided data on the basic magnetic characteristics and enabled the determination of palaeomagnetic directions:

- the magnetostratigraphic research conducted on the profile near Divača yielded data on the presence of two normal and one reverse magnetozone,
- the magnetostratigraphic results of samples obtained from the profile near Divača and Kozina (Bosák et al. 2000a, b) exhibit an adequate correlation between both profiles [in both profiles, two normal zones were discovered in the reverse magnetozone; an adequate correlation of the values of remanent magnetization modules (J_n)].

The magnetostratigraphic results from samples of clastic fluvial sediments obtained from the infilled cave south of Divača thus indicate the presence of two narrow normal magnetozones in a long reverse magnetozone, which probably coincide with the Olduvai or Reunion (1.67–1.87 million years) normal magnetozones dating to the period of reverse Matuyama epoch, or with the normal magnetozones (app. 3.8–5.0 million years) of the reverse Gilbert epoch.

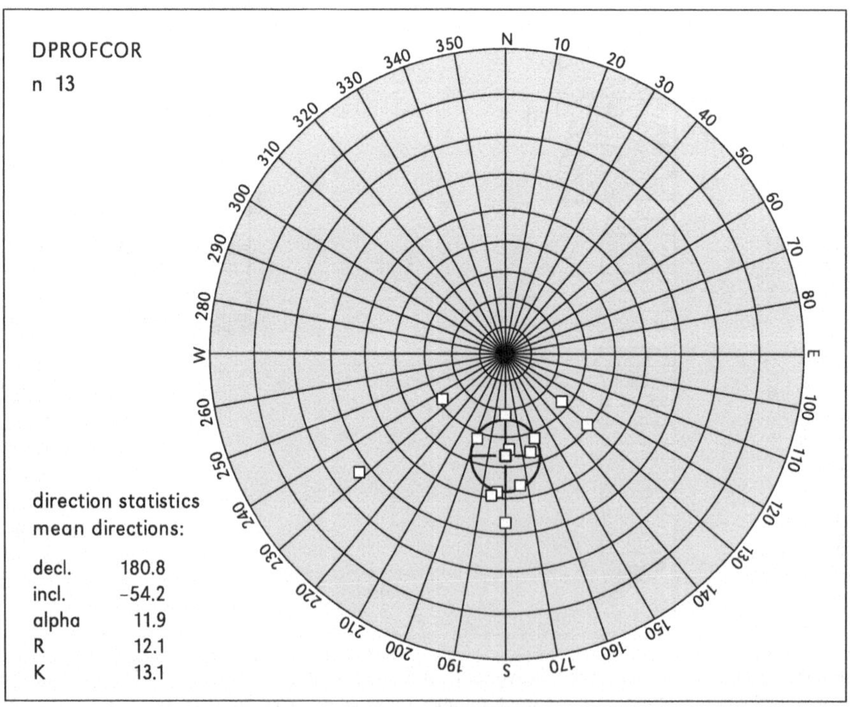

Fig. 11.6 Figure showing the distribution of the palaeomagnetic direction of C-elements of the remanent magnetization in samples

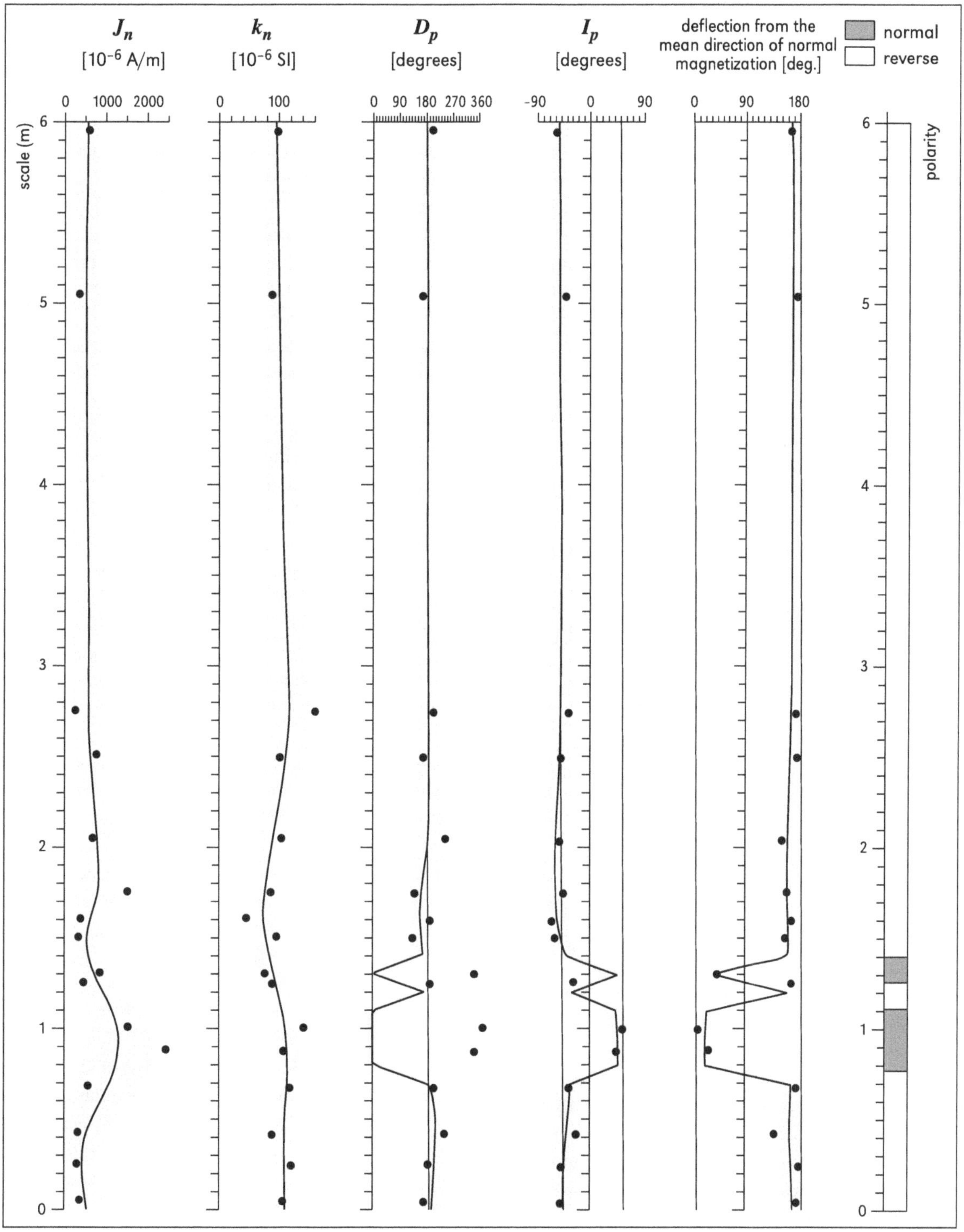

Fig. 11.7 Basic magnetic parameters of samples from the filling of the cave at Divača

The results of palaeomagnetic analyses obtained in relation with the infilled cave near Divača are thus comparable with the results relating to the unroofed cave near Kozina (Bosák et al. 2000a, b). The lithological composition of the lower sequence in the profile is comparable with the Kozina profile. The magnetostratigraphic results are also comparable as both cases reveal two shorter normal magnetozones within the reverse magnetozone. A few differences occur only in the distribution of normal magnetozones in terms of the position in the course of sedimentation, which can probably be attributed to the different velocities of depositing sediments in both caves.

11.3 Conclusion

The construction of the Divača–Kozina motorway has uncovered numerous infilled and unroofed caves, one of which was discovered south of Divača. The infilled cave passage was found on the southern slope of a larger doline and measured 6 m in height, while the motorway construction works did not reach the rocky bottom of the cave. The passage exhibited yellow layers of fluvial cave sediments ranging from sand and silt to loam, while a layer of red loam was present on top of the profile. Several different textures were evident in the profile which intersected the entire cave, with cross-stratification prevailing, while other erosion surfaces were also visible and divided the profile in individual sequences.

All sediments in the cave underwent heavy ferritization, especially along the fissures which formed in the sediment due to their drying or shifting on account of subsidence or tectonics. Certain parts of the profile were also heavily cemented with calcite, namely in the form of randomly positioned irregular bodies. Two iron crusts were present in the lower part of the profile, separating the first and the second, and the second and third sequence respectively. The iron crusts undoubtedly represent interruption in the sedimentation process. The first and the second sequence (at the bottom of the profile) allowed us to observe the layers which indicated sedimentation in a peaceful environment as evidenced by their relative thickness and even graininess. It is possible that the sediment layers from these two lower sequences were deposited under phreatic conditions, and the gradient changed significantly later on. We may deduce the velocity of sediment depositing in the lower part of the profile on the basis of palaeomagnetic results which indicate the slowness of depositing, namely 1 mm/100 years and 2.65 mm/100 years. In the third profile sequence, typical fluvial cycles developed, featuring different layers, which points to the changes in the hydraulic gradient and/or changes in the climate conditions during the settling phase.

From a total of 29 samples taken from the profile, several were cemented. The samples were demagnetized by means of an alternating field (AF) at 10–1000 Oe. Individual components of remanent magnetization following demagnetization with alternating/thermal field were ascertained through a multi-component analysis in line with the Kirschvink method. The profile contains two normal magnetozones in a wider reverse magnetozone. The sediments in the infilled cave are therefore definitely older than the Brunhes/Matuyama boundary (0.78 million years), and the distribution of individual magnetozones enables us to state that the sediments are older than the end of the Olduvai period (1.77 million years) or even older than 3.8 million years. The profile is comparable with the profile near Kozina (Bosák et al. 2000a, b) both from the palaeomagnetic and lithological aspect.

In terms of their mineral composition, the sediments found in the infilled cave south of Divača are weathered remains of the Eocene flysch as attested by the high content of quartz and the presence of feldspars.

The cave was formed in the Palaeocene limestone and is of unknown age. In view of the results obtained, its formation can be linked with the period when the sea level declined drastically, and with the development of deep karst (Perna 1996) in the Mediterranean and its vicinity upon the Messinian Event (Hsü et al. 1977). The infilling of the cave can be related to the rapid rise of the sea level and the resulting lowering of the gradient as the Mediterranean Basin filled up again. According to the results obtained, this supposedly happened 5.2 million years ago.

12 Palaeomagnetic Research of an Unroofed Cave Near Kozina

We are conducting palaeomagnetic research of the cave alluvium in the Karst in the framework of scientific collaboration established between the Institute of Geology, Academy of Sciences of the Czech Republic, and the ZRC SAZU Karst Research Institute. The research covered some very interesting locations in the Karst and southeastern Slovenia, and yielded important findings. We studied cave sediments found near Divača, in the Divaška jama Cave and the Trhlovca Cave, in Črnotiče near Črni Kal and the unroofed cave near Kozina. The palaeomagnetic analyses were carried out at the Czech Institute.

12.1 Morphological and Geological Circumstances

The area is cavernous, featuring old caves intersected in places by younger shafts. The formation of shafts is linked with the decline of the piezometric level which is now 200 m below the surface. The shafts are empty or filled with younger (Pleistocene) alluvium (Rakovec 1958; Brodar 1958). Formations on the Karst surface that are similar to large valleys supposedly represent original river valleys as they contain the remnants of fluvial alluvium. Nevertheless, recent research indicates that, rather than being the remains of surface river systems, fluvial alluvium probably originates from fossil caves. The cave fills appear on the surface due to their lowering. These caves are called unroofed caves and were first described during the construction of the motorway across the Karst (Knez and Slabe 1999a, b, 2000, 2001c, 2002a, 2004a, b, 2005, 2006a, 2008, 2009b, d, 2010b, c, d, f, 2011, 2012b; Knez and Šebela 1994; Knez et al. 2012; Kogovšek et al. 1997; Mihevc 1996; Mihevc and Zupan Hajna 1996; Mihevc et al. 1998; Slabe 1996, 1997a, b, 1998a; Šebela and Mihevc 1995; Šebela et al. 1999).

Some of them are parts of the same cave system or several such systems. It seems such caves can be located all over the Karst (e.g. Šušteršič 1998; Mihevc 1998; Stepišnik and Šušteršič 1999; Geršl et al. 1999).

Mihevc with coworkers (1998) as well as Knez and Slabe (1999a, b) attempted to explain the typical formations of unroofed caves. They have been reshaped through surface processes and are an important feature of the epikarst zone (Knez and Slabe 1999b). The formation of unroofed caves results from the formation of the present surface, the formation of old caves, distinctiveness of more recent karstification (speleogenesis) and the level of the recent denudation of the cave fill. On the terrain, they are manifested as shallow, oblong notches and doline-like formations. The unroofed caves represent a typical example of palaeokarst, i.e. excavated or rejuvenated karst (Bosák et al. 1989, 32) that is partially integrated in the present karst landscape and the hydrological system.

Dating the sediment by means of the palaeomagnetic method in some unroofed caves (Bosák et al. 1998a, b, c, 1999a, b, 2000a, b; Pruner and Bosák 1999) indicated a significant age of the cave fills, certainly over 1.77 million years.

12.2 Site Location and Characteristics

The site is located to the northeast of Kozina, near the existing Ljubljana–Koper main road, in the cutting which was formed during the construction of the Divača–Klanec motorway (Fig. 12.1).

M. Knez and T. Slabe (eds.), *Cave Exploration in Slovenia*,
Cave and Karst Systems of the World, DOI 10.1007/978-3-319-21203-6_12

Knez and Slabe (1999b) provided a detailed description of the typical features of unroofed caves near Kozina.

The caves were formed in the Turonian to Thanetian shallow-marine limestone. The tectonic contact with the Eocene flysch is located near the construction site. The largest cave system (measuring 400 m in length) lies on the right side of the construction site near Kozina. The system appeared on the surface as more or less a distinctive oblong depression (Fig. 12.2) which formed a connected string of doline-like depressions. Morphologically, the unroofed caves are more obvious when they are located near dolines where the erosion of the cave fill into the doline is more intense. For the most part, the depressions were small and shallow. At the bottom, they were filled with red and brown soil in layers that were up to a few metres thick. We noticed traces of water inflows at the contact of soil and limestone. At the bottom, entrances into narrow and inaccessible shafts were found.

The cave passages were both hollow and filled, with very thin ceilings that were removed in some places. The caves were mostly filled with fine-grained sediments of an underground river and, in some places, with layers of rubble from the flysch sediment. In the southwest, the layers of flowstone and stalagmites intertwined with river cave fills. Some parts of the sediment were covered with angular blocks, rocks and rubble originating from the collapsed cave ceiling made of limestone. The rubble is said to be the result of weathering and deterioration that occurred in the cold Pleistocene climate (Knez and Slabe 1999b).

12.3 Description of the Profile

The sampled profile is shown on Figs. 12.3, 12.4 and 12.5. The entire profile consists of more than 5 m of alluvium. The bottom was uncovered. The fill comprised two main sequences. The lower sequence comprised

Fig. 12.1 Site location

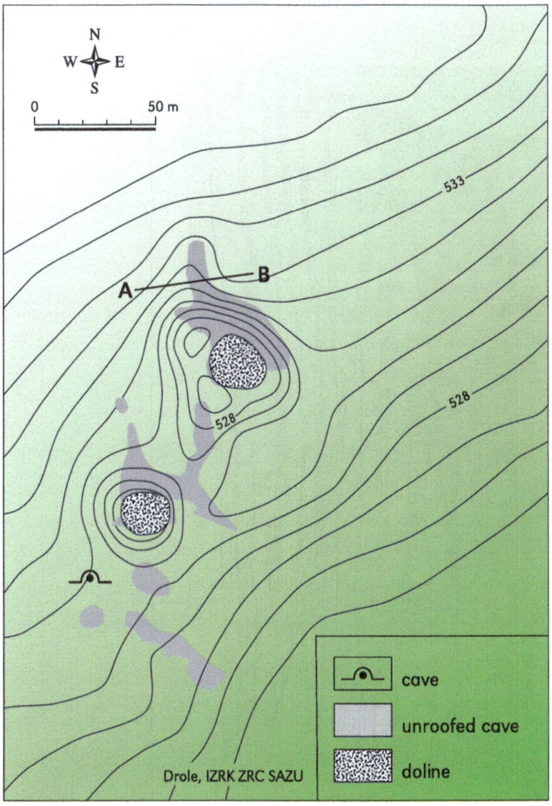

Fig. 12.2 Cave system at Kozina with caves, unroofed caves, and dolines

ochre-coloured sandy to clayey sediments that were app. 3 m thick. These sediments were sampled to undergo the palaeomagnetic method. The lower sequence was covered with collapsed breccia and limestone blocks (ranging from 1 cm to 1 m in size) and brown loam. In the upper part of the collapsed breccia, the matrix was ochre, with smaller stone fragments. Due to the possibility of collapses, including the possibility of shifts following the deposit as well as slumping and rotation of the sediment, no samples were taken from the upper sequence for the purposes of palaeomagnetic analysis.

Close to the contact of the sediment and limestone, a narrow and inclined cavern formed in the bottom part of the sediments. The cave walls were covered in flowstone which cemented the surrounding sediment. A minor water flow ran through the cavern, forming a shallow notch on the surface.

The lower sequence was around 3 m thick. Palaeomagnetic samples were marked in 1 cm intervals and were measured from the bottom of the profile (2–295). The following lithological units were determined:

- sand, yellow, black-purple strips, very fine-grained, silty, poorly visible lamination with a high content of clayey admixture, angular stone fragments;
- clay, silt, diverse (ochre-coloured, light brown with dark grey and purple strips and lamination), slightly finely sandy, more at the bottom, laminated (dynamic lamination), erosion base with secondary colouration with iron oxides (2–28);
- clay, silty, ochre to light brown, yellow and white-yellow, laminated, with laminas of fine-grained sand and fine sandy silt, thin laminas coloured with iron oxides, erosion base (29–93);
- clay, light brown, purple-brown, with a thin strip of white sand on the top, erosion base (98–109);
- clay, silty, ochre to light brown, yellow and white-yellow laminated, with laminas of fine-grained sand and fine sandy silt, the upper part featuring clasts of brown clays, calcified in places, erosion base is disproportionately on layer No. 4 (incline lower by about 10°) (116–212);
- clay, silty, light brown, erosion base (219–231);
- clay, silty, light brown, slightly finely sandy in indistinct laminas, with larger chunks of mica, irregularly-shaped clasts, erosion base;
- breccia with light ochre clay matrix, erosion base;
- clay, silty, brown, with manganese content, erosion base (290);
- sand, yellow, fine-grained with cross-stratification, erosion base (295).

Five samples were taken from the profile for the purposes of palynological analyses (Fig. 12.5), namely at 30, 30–45, 70–80, 130–150 and 180–200 cm above the profile bottom.

The sample taken at 70–80 cm above the profile bottom contained two very corroded grains of pollen originating from the herb vegetation (the *Dipsacaceae* and *Apiceae* families). The sample taken at 130–150 cm above the profile bottom contained one (fern) spore.

Tree pollen was not found. The pollen present belonged to herb vegetation typical of dry areas similar to steppes.

12.4 Palaeomagnetic Analyses

In total, 38 oriented specimens of cave sediments were examined for their palaeomagnetic properties.

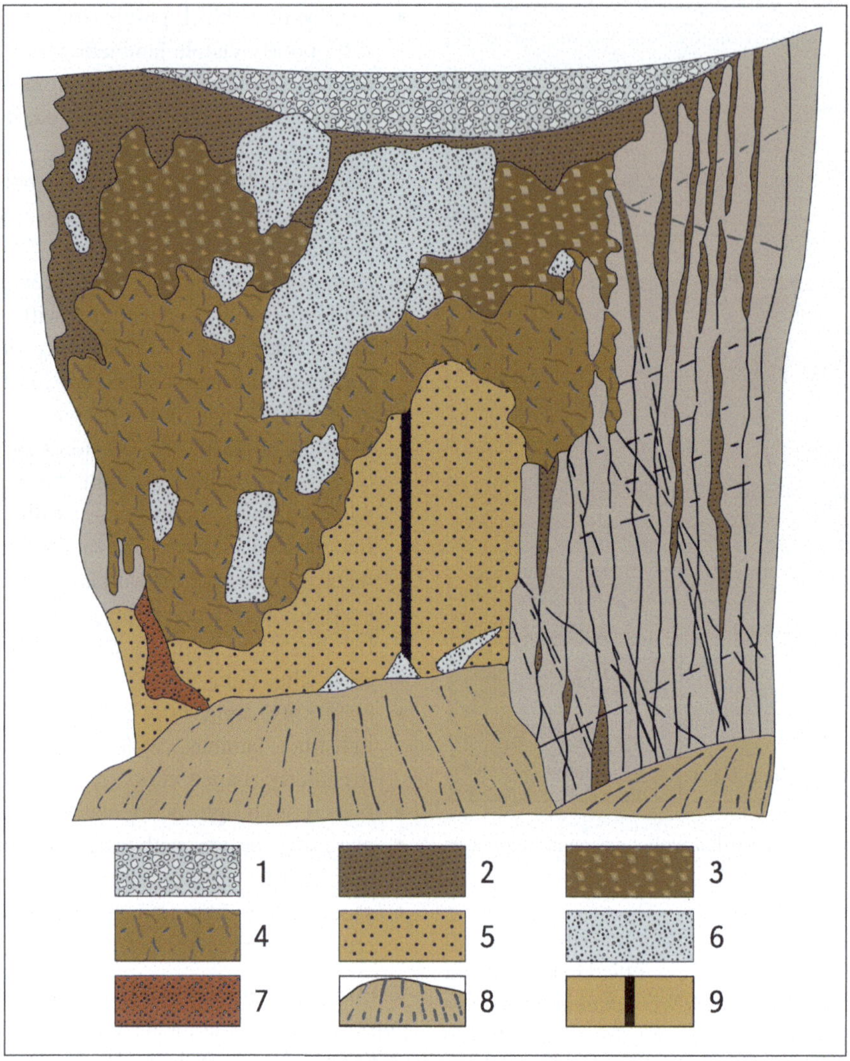

Fig. 12.3 Sketch of the profile and roadcuts at Kozina. *1* rubble due to mining; *2* sediments similar to red ground; *3* collapsed material with matrix coming from sediments similar to the lower sequence of the profile; *4* collapsed material with brown matrix; *5* lower sequence; *6* blocks; *7* cave; *8* heap; *9* sample profile (see Fig. 12.5)

12.4.1 Laboratory Procedures

The laboratory procedures enabled the determination of the structure of the appurtenant magnetic remanence in various temperature intervals in the course of gradual thermal demagnetization (TD) and alternating field demagnetization (AFD), as well as the determination of modules and the directions of remanent magnetization. The procedures are described in the Chap. 9.

The oriented samples were taken in the field from various layers. The laboratory samples in the shape of cubes measuring 20 × 20 × 20 mm in size were prepared either on-site or from samples. They were measured on the spinner magnetometer (JR-4 and JR-5; Jelínek 1966).

12.4.2 Palaeomagnetic Research Results

All samples taken (38 of them in total) underwent AF demagnetization, while one sample was subjected to thermal demagnetization. The J_n module values of rocks in their natural state indicate that they are

Fig. 12.4 The entire profile in the roadcut near Kozina

widespread. The mean values of remanent magnetization J_n modules and k_n magnetic susceptibility in natural state from 38 samples amount to $J_n = 7.005 \pm 8.391$ (nT), $k_n = 267 \pm 216 \times 10^{-6}$ (SI). The rocks exhibit low or medium level of magnetization.

The directions of remanent magnetization ascertained through the above-referenced procedures were tested with a multi-component analysis (Kirschvink 1980). In general, the samples showed three remanence components, namely A, B and C.

The A components are mostly of viscous or chemoremanent origin (weathering) and can be removed by means of an alternating field with the power ranging from 10 to 30 Oe. The detected remanent magnetization in natural state ranges between 95 and 36.470 pT, while the values of the volume magnetic susceptibility moved between 55 and 998×10^{-6} SI. Some samples contained a distinct viscous component (up to 90 %), making it impossible to state the original magnetization component and polarity.

The normal and inverse directions of the C component in samples (Fig. 12.6) form two separate groups of samples by means of the Fisher distribution. The mean direction of remanent magnetization of the profile near Kozina is stated in the Table 12.1.

The upper and lower parts of the profile indicate an inverse magnetozone. The middle part of the profile contains two normal zones.

12.4.3 Magnetostratigraphic Research Results

The palaeomagnetic and magnetostratigraphic research provided data on the basic magnetic characteristics and enabled the determination of palaeomagnetic directions:

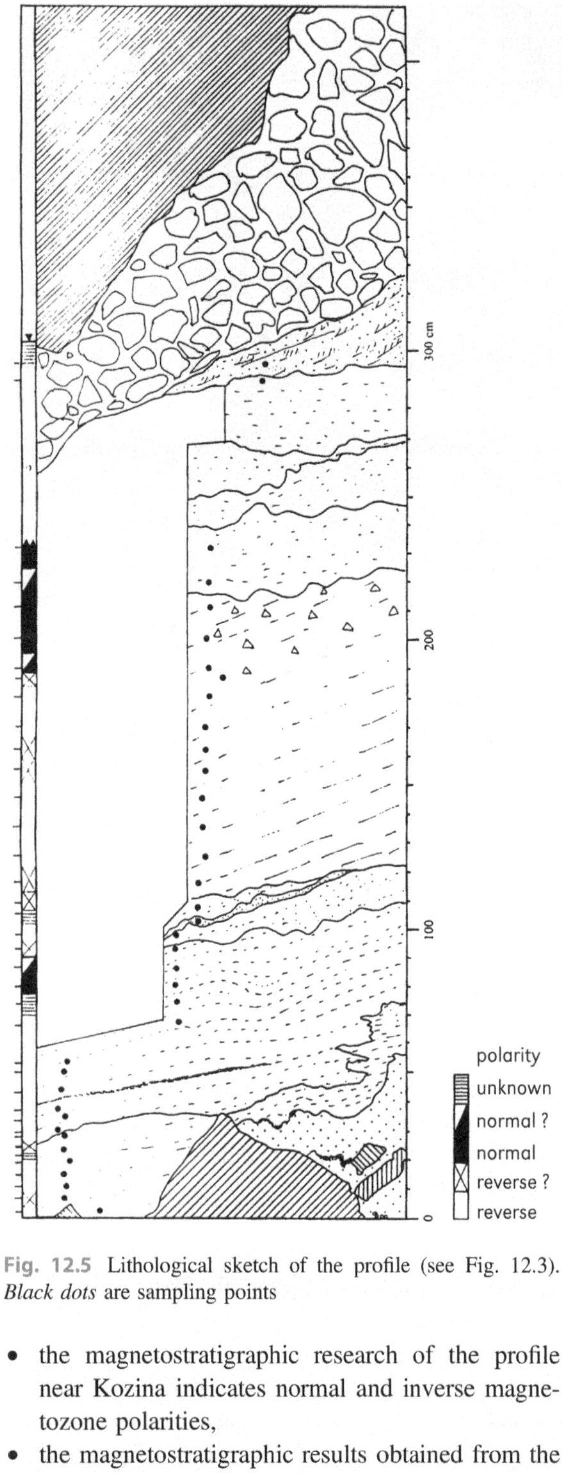

Fig. 12.5 Lithological sketch of the profile (see Fig. 12.3). *Black dots* are sampling points

- the magnetostratigraphic research of the profile near Kozina indicates normal and inverse magnetozone polarities,
- the magnetostratigraphic results obtained from the samples taken from the profile near Kozina and Divača show an adequate correlation between both profiles; two normal sub-zones within the inverse magnetozone were ascertained for both profiles and a good correlation in the values of remanent magnetization modules (J_n).

12.5 Findings

The profile's lithology clearly manifests a two-phase depositing in the past. The lower sequence underwent erosion after having been deposited. The erosion channel was more developed on the left side of the passage. Subsequently, during the collapse, the empty space in the cave filled with rubble ranging in size from rocks to blocks mixed with brown karst soil. The ochre-coloured intercalations in the upper part of the upper sequence may indicate the presence of eroded sediment comparable with the lower sequence. The thinning of cave ceilings through erosion and karst denudation triggered collapsing.

The lithological composition of the lower sequence is comparable with the Divača profile, namely that from the fossil cave near Divača (Bosák et al. 1998a, b, c), especially in relation to sequences Nos. 1 and 2. Layer No. 10 of the Kozina profile can be correlated with the base of sequence No. 3 of the Divača profile. It seems that the sediment originates from a similar rock source, most probably from the weathered Eocene flysch.

The relevant erosion boundaries of the main lithological units within the lower sequence are located between sample Nos. 28 and 29, 93 and 98, 109 and 116, 212 and 219, and 290 and 295. Unlike other examined profiles (Bosák et al. 2000a, b), the erosion boundaries do not correspond with the boundaries of normal and inverse polarity zones, as they are located within them. This could mean that the pauses between deposits were not long.

The magnetostratigraphic image obtained from the Kozina profile is entirely comparable with the magnetozones observed in the Divača profile (Bosák et al. 1998a, b, c, 2000b), both in the occurrence of normal and inverse magnetozone polarity (Fig. 12.7) and in the character of remanent magnetization modules (J_n; Fig. 12.8). Inverse magnetozones are the prevailing part of both profiles. Two relatively narrow zones of normal polarity are present. Unfortunately, there is a gap between samples 231 and 290 from the Kozina

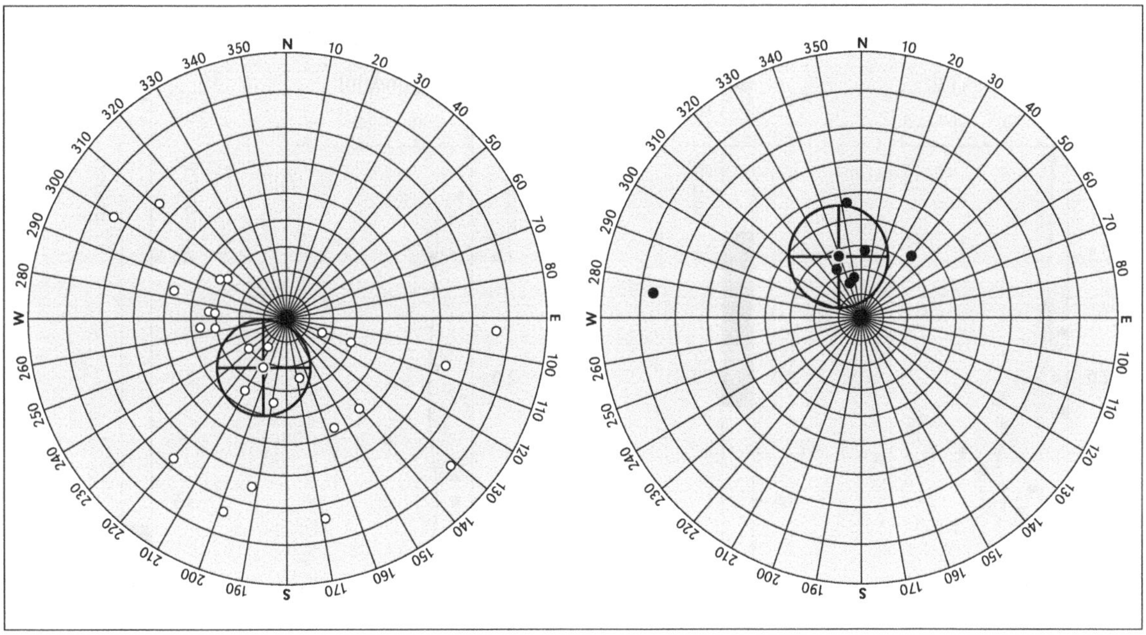

Fig. 12.6 Samples with normal palaeomagnetic polarity (*left*) and samples with inverted palaeomagnetic polarity (*right*). Mean directions were calculated using the Fisher's method (1953)

Table 12.1 Mean palaeomagnetic directions in the area in question

Kozina	Polarity	Mean palaeomagnetic directions		α_{95} (°)	k	n
		D (°)	I (°)			
	N	338.2	62.3	20.7	8.1	8
	R	206.0	−67.6	20.1	3.1	25

profile due to the rock petrography that is unsuitable for sampling. Some of the differences in the distribution of normal polarity magnetozones in both profiles could stem from the differing depositing velocities during the fossilization of both channels.

The Kozina profile is older than the Brunhes/Matuyama boundary (0.78 million years). The distribution of individual magnetozones supports the claim that the sediment is older than the end of the Olduvai epoch (1.77 million years) as the magnetostratigraphic profile near Kozina concludes with the inverse polarity magnetozone and contains two normal polarity zones (Table 12.1). The highly comparable natures of the remanent magnetization modules values (J_n; Fig. 12.7) strongly support the age correlation of profiles obtained from the fossil caves near Divača and Kozina.

We believe that the cave, much like the Divača profile, is the result of the Messinian speleogenesis (especially if the normal polarity magnetozones can be correlated with those from the inverse Gilbert period, Fig. 12.8).

At that time, the level of the Mediterranean Sea declined quickly (Hsü 1973; Hsü et al. 1973, 1977), which was related to the deep entrenchment of valleys in the areas surrounding the Mediterranean Basin (valleys of the Ebro, Durance, Var, Po or the Orontes rivers; the Rhône river valley in Southern France with an exceptional thickness of the Pliocene-Quarternary fill—Clauzon 1973, 1979; Clauzon et al. 1997; the Nile river valley in Egypt—Khumakov 1967, 1971; valleys on the carbonate plateau of Cyrenaica in Libya or in the carbonate-flysch region of Istria, Croatia). The deep karst featuring thickness of 1–3 km

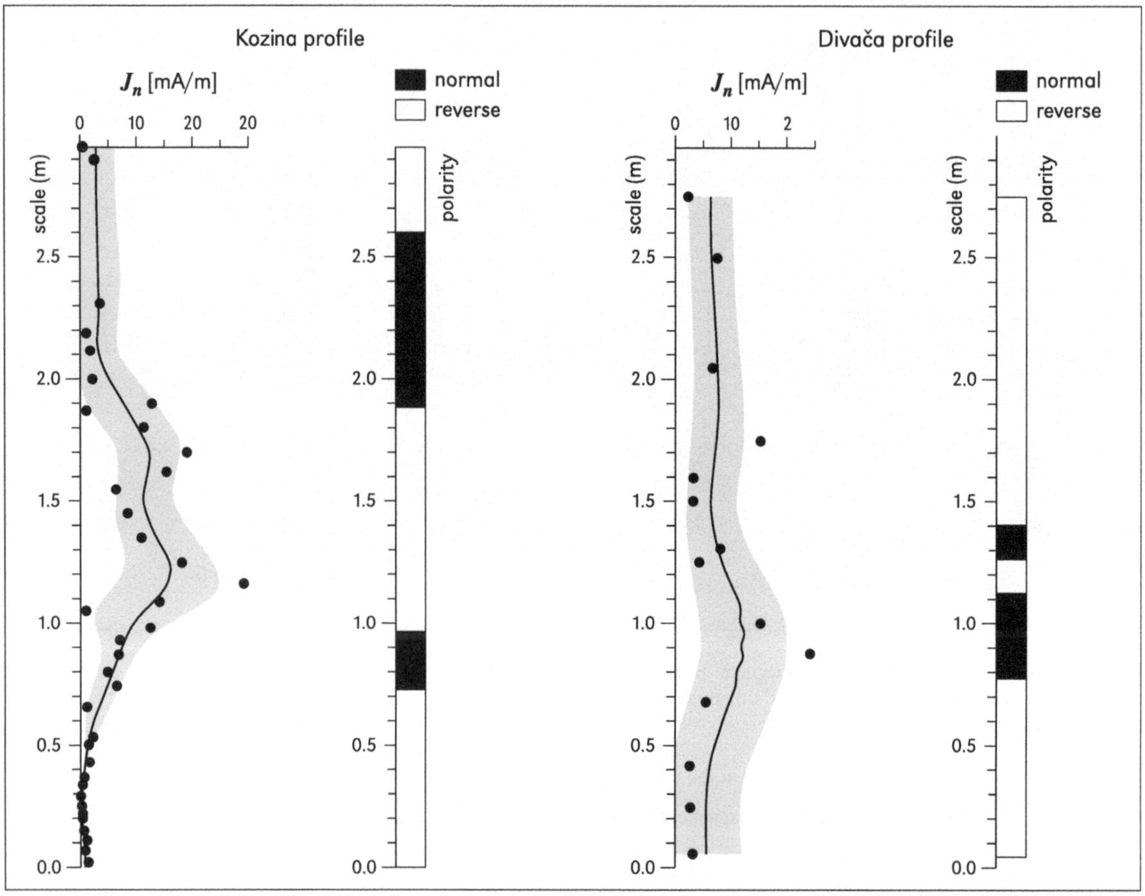

Fig. 12.7 Comparison of basic and magnetostratigraphic parameters at Divača (Bosák et al. 1998c) and of the profile at Kozina

developed in the entire Mediterranean area (Perna 1996) as a result of underground outflow from the hinterland into the Mediterranean Basin (Głazek 1993). Perna (1996, 12) denoted the resulting karst formations as the Messinian karst cycle. The now submerged parts of these systems are frequently manifested as large submarine springs (the *vrulja* submarine karst spring on the Adriatic coast, in Cyrenaica, in Apulia, Southern France etc.) or other phenomena (Sardinia). It's highly likely that the fossil caves near Kozina and Divača are the result of this phase.

Fossilization of the cave systems was related to the quick uplift of the base plain after, around 5.2–5.3 million years ago, the Strait of Gibraltar opened up and the Mediterranean Basin filled again (Hsü 1973; Hsü et al. 1973). Further fossilization resulted from the changes at the regional base plain and hydrological situation which came about with gradual development of the surface and neotectonics of this part of the Karst, as well as due to the changes in the sea water level in the Mediterranean Basin.

12.6 Conclusion

The construction of the Divača–Klanec motorway has uncovered numerous fossil and unroofed caves, one of which was discovered near Kozina. The cave passage had no ceiling, with remains of the collapsed ceiling only present in the upper part. A smaller notch was observed on the terrain. The profile of the cave sediment measured around 5 m in height. It mainly comprised sandy sediment ranging in colour from light brown to ochre, with clayey and silty intercalations. The sediment contained dynamic structures and textures (lamination, cross-lamination etc.). The erosion surfaces divided the profile into individual sequences.

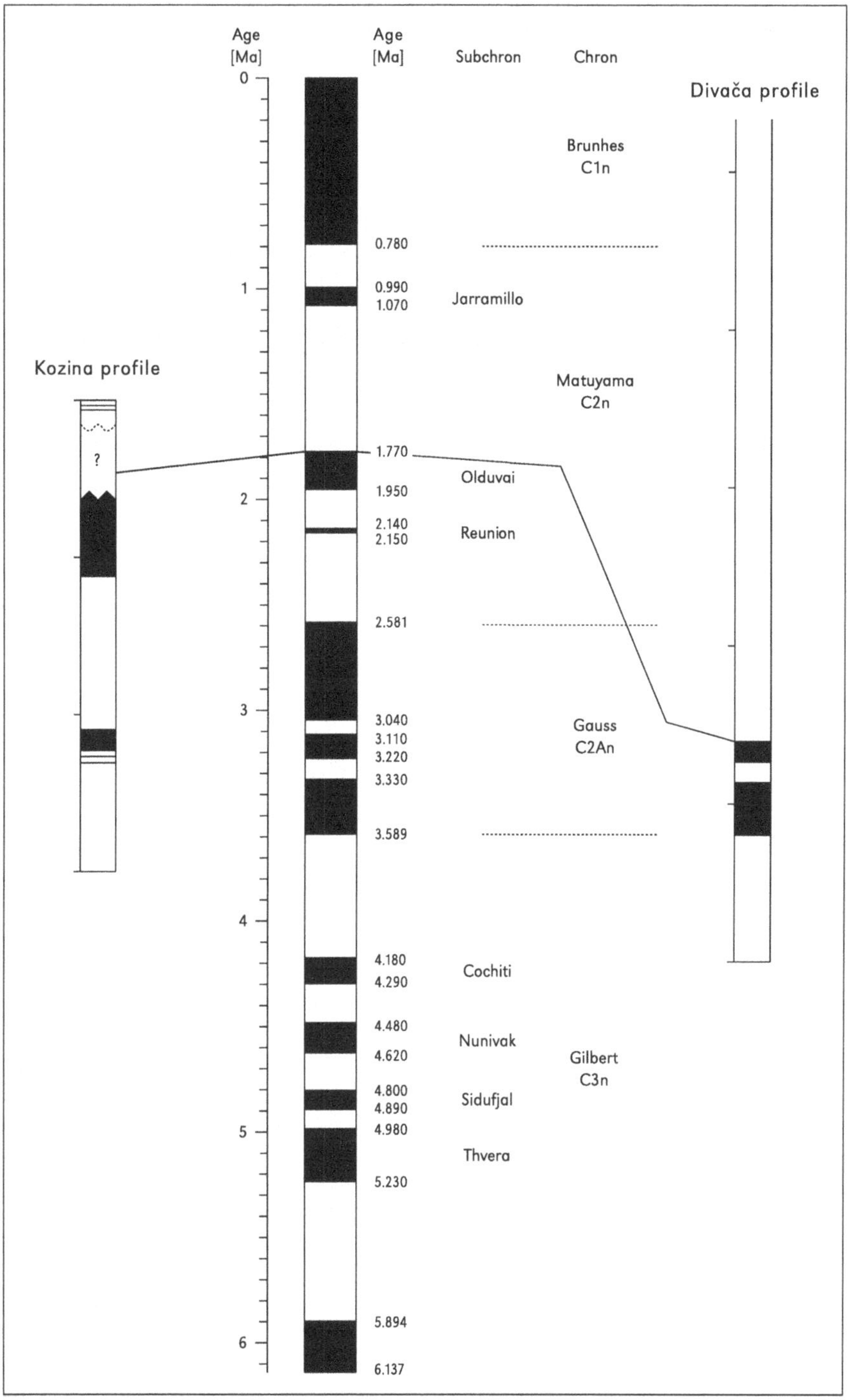

Fig. 12.8 Correlation of acquired magnetostratigraphic results with standard paleomagnetic scales (Cande and Kent 1995)

From a total of 38 samples taken from the profile, only one was cemented. The samples were demagnetized by means of an alternating field (AF) at 10–1000 Oe. The cemented sample was demagnetized through a gradual thermal process at 80–560 °C in the MAVACS apparatus. Individual components of remanent magnetization following demagnetization with alternating/thermal field were ascertained through a multi-component analysis in line with the Kirschvink method. The detected remanent magnetization in natural state ranged between 95 and 36.470 pT, while the values of the volume magnetic susceptibility moved between 55 and 998 × 10^{-6} SI. The rocks exhibited low to medium magnetization. The normal and inverse polarities were ascertained after demagnetization. Some samples displayed a distinct viscous component (up to 90 %). It was impossible to state the primary magnetization component and the resulting polarity.

The profile contains magnetozones of inverse and normal polarity. The typical distribution of magnetozones is similar to the location of the profile near Divača (fossil cave in the road cutting at Divača). The Kozina profile is older than the Brunhes/Matuyama boundary (0.78 million years). The distribution of individual magnetozones supports the claim that the sediment is older than the end of the Olduvai period (1.77 million years) as the magnetostratigraphic profile near Kozina concludes with the inverse polarity magnetozone and contains two normal polarity zones.

The profile correlates with the Divača profile both from the palaeomagnetic and lithological aspect. We believe that the cave, much like the Divača profile, is the result of the Messinian speleogenesis, its fossilization related to the quick uplift of the base plain after the Mediterranean Basin filled again. If our conclusions are correct, the cave began filling with alluvium 5.2 million years ago.

Sediments in the S-647 Cave in the Kastelec Tunnel 13

The process of constructing the left tube of the tunnel uncovered a cave that was unknown until then, over 550 m long and 50 m deep. The eastern part of the cave descends evenly and reaches its lowest accessible point at the altitude of 315 m, i.e. 47 m under the bottom of the tunnel (Zupan Hajna and Drole 2003). The western part ascends gently so that the final point lies at the altitude of 356 m, 6 m under the tunnel's bottom. There is no concern as regards to any subsidence occurring in this section of the tunnel as most of the cave runs between both tunnel tubes. Since the cave is located under the tunnel, the construction operator additionally reinforced the ceiling above it by means of a reinforced concrete plate. At the 590 chainage, an entrance opened up in the left tunnel tube, leading into a slanting passage measuring 85 m in length. The end point of the passage is located 25 m above the tunnel. The passage runs parallel with the tunnel towards the west, and the entrance to that passage is closed off by a tunnel wall. Despite the close proximity of the blasting work, the flowstone was only slightly damaged.

The area of the Kastelec tunnel belongs to the morphostructural unit of the Podgorje Karst (Mihevc 1991; Zupan Hajna 1997) comprised of the Palaeocene and Eocene limestone and the Eocene flysch (Pleničar et al. 1969, 1973). This unit is characterized by the strips of limestone and flysch rocks following one another in several overthrusting sections as part of the imbricate structure of the Čičarija plateau (Placer 1981).

The Podgorje Karst comprises the plateau area beneath the Škrklovica Hill (461 m) at the altitude above 400 m, and the areas of the Petrinje Karst, Upper Karst and Lower Karst, the latter being an area between the walls over the Osp Valley and the Kastelec tunnel. This karst area is special because the walls, also known as the Karst Edge, are located at the point where the karst world on limestone comes into contact with the impermeable flysch.

Several larger horizontal caves are known in the area of the Podgorje Karst (Podgorski Kras karst area), with entrance shafts leading to their interior. The caves are 15–150 m deep, and 30–300 m long. Now dry, they are only reshaped by the percolating water. The heavily karstified caves that are frequently filled with alluvium also contain preserved formations which indicate that the cave passages were created in the phreatic zone, i.e. the zone of saturation. The swift decline in the water level upon the lowering of the flysch barrier in the west, or possibly the uplift of this part of the karst has caused the traces of the original cave formation to remain preserved in some places (Bosák et al. 1999a, b). The closest sink caves that were originally formed already in the phreatic zone and are now being reshaped by the sinking streams are the caves belonging to the Beka-Ocizla cave system. This system of sink caves is developed along the northeastern edge of the Podgorje Karst area where several smaller streams sink into it from the flysch surface (Zupan Hajna 2004). The Podgorje Karst also features several rock

M. Knez and T. Slabe (eds.), *Cave Exploration in Slovenia*,
Cave and Karst Systems of the World, DOI 10.1007/978-3-319-21203-6_13

shelters, some of which carry great archaeological significance (Turk et al. 1992).

According to the Cave Registry kept by the ZRC SAZU Karst Research Institute and the Speleological Association of Slovenia, the following caves are known in the vicinity of the Kastelec tunnel: The Udor na Škrklovici Shaft—the Jama pri Črnem vrhu Cave (Reg. No. 1393, length of passages: 35 m, depth: 10 m) and the Kraljičevka Cave (Reg. No. 4531, length of passages: 46 m, depth: 7 m). The closest to the tunnel is the Brezno na Škrklovici Shaft (Reg. No. 1391), whose passages are 250 m long and 85 m deep. The cave was measured and outlined anew (Archives of the ZRC SAZU Karst Research Institute) for the purposes of the project when the tunnel was constructed. The cave passage runs parallel with the tunnel, its closest point app. 25 m away from the tunnel, and its final part almost 15 m away from the S-647 cave. The lowest point lies at the altitude of 355 m. The main passage is inclined and measures up to 17 m in height, with visible traces of a slow water flow (large scallops) on its walls. The cave is accessible because the surface above it has lowered to such an extent that the top part of the cave is open towards the surface. Beneath the 13-m entrance shaft, the passage descends steeply so that the heap of stones thrown into the shaft while cleaning the surface is very unstable. After a smaller passage at the end of the inclined passage, the cave becomes very similar to the S-647 cave in the left tube of the tunnel, to which it approaches very closely. The inclined passage then widens into a few smaller halls; it is intersected at transverse faults by chimneys down which flows very corrosive water. The chimney walls are eroded, with the foraminifera limestone (alveolinid-nummulitid limestone) so heavily corroded that the white shells of foraminifera (hole bearers) jut out from the dark-grey limestone. The main passage that descends at an incline is rather karstified, with flowstone covering the walls and the floor. The Brezno na Škrklovici Shaft features several stalagmites of various shapes, from cypress-shaped ones to long bats, as well as stalactites with their points mostly broken off. The older flowstone is reddish-brown, while the younger specimens are grey and white. This cave contains no helictites. In the lower parts of the cave, across the passage floor, an increasing amount of loamy sediments can be found (alternation of sandy and clayey layers of flysch origin), particularly at the bottom of the cave. The loamy alluvium is mixed with white quartz pebbles, iron pisolites as well as pebbles of flysch sandstone measuring even over 10 cm in size. In some places, the loam is covered by a black crust. The sediments date back from the time when the cave served as part of a sink system, but naturally not through the present-day entrance to the cave. The sediments are now washed through several smaller subsidences just before the bottom and at the very bottom of the cave. The walls contain the remains of older sediments which were located at least 2 m higher than the alluvium at the present. The bottom of the cave also contains a larger accumulation of fine rubble, which was probably formed near the surface (impact of frost) and was poured into the final part of the cave through a chimney measuring 40 m in height. Rubble is also sliding towards the subsidence at the bottom of the cave.

13.1 Speleomorphological Description of the Cave

The cave strikes south of the tunnel and is shaped like the letter V (Fig. 13.1). One section strikes to the southwest and is rising so that the highest point measured is located at the altitude of 377 m and around 7 m above the tunnel. This section can be found 10 m to the south of the left tunnel tube. This part of the cave is up to 5 m wide and up to 8 m high. The southeastern section is longer and gradually descends for the first 35 m, after which the 7 m step prevents uncomplicated visits to the cave.

After this step, the height of the passage lowers considerably, from 4 m to a little under 0.5 m. The low-ceilinged passage continues for about 20 m, after which a heavily corroded hall is reached via a 5 m deep step, which continues into a chimney. The bottom of this hall lies at the altitude of 334 m. Taking into account the height of the chimney, the last point measured in the southeastern section lies at the altitude of 354 m, which is the same height as that of the lowest-lying passages of the Brezno na Škrklovici Shaft. The cave is located 60–100 m below the surface.

The upper part of the cave follows the dip of layers to the northeast, while the lower part follows the fault zones. The passage walls feature many remains of sandy loam, indicating that, at some point,

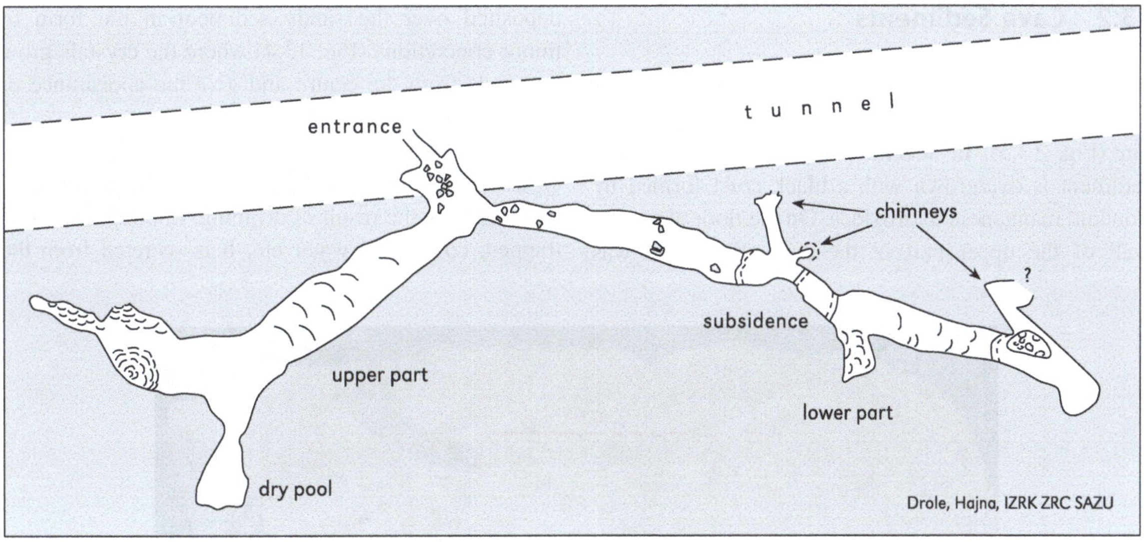

Fig. 13.1 Location of the cave S-647 in the Kastelec tunnel

the cave was completely filled with deposited sediments of flysch origin. Now, the alluvium is disappearing in a smaller subsidence which opened in the middle of the lower part of the cave. The subsidence is around 6 m deep, its bottom and walls covered in sandy sediments.

Next to two distinct fissures which cross the cave in the lower part, two larger chimneys are formed, down which rainwater flows into the cave. The percolating water is highly aggressive as evidenced by the limestone walls of these two chimneys, which are heavily corroded. The flowstone, which was previously deposited in these parts of the cave (Fig. 13.2), is also considerably eroded. In the middle part of the lower passage, next to the larger chimney, another smaller, side chimney is open along the fault plane that runs parallel to the direction of the main passage. The foraminifera limestone is corroded to such an extent that the shells of foraminifera and of up to 2 cm large sea urchins are jutting out from it. At the same time, rain flutes are forming along the fault plane that runs parallel with the main passage.

The cave was most probably formed already in the phreatic zone and was later filled with flysch alluvium, with a lot of flowstone deposited as well. The water now flows through the deeper parts of the karst (Beguš et al. 2003).

Fig. 13.2 Corroded flowstone (a curtain and a stalagmite) in the midsection of the cave

13.2 Cave Sediments

In addition to flowstone, the cave also contains larger quantities of sandy and loamy alluvium of flysch origin (Fig. 13.3). In several places, the sandy-loamy sediment is overgrown with a black crust formed by iron and manganese hydroxides. On the floor along the wall of the upper part of the cave, flowstone was deposited over the sandy sediment in the form of minor concretions (Fig. 13.4) where the crystals grow outwards from the centre and give the appearance of karstified pebbles.

Flowstone in the form of a fine-crystal aggregate is typical only in the vadose karst zone. Various forms of flowstone are the result of dripping, running, trickling, trapped, condensed water etc. It is secreted from the

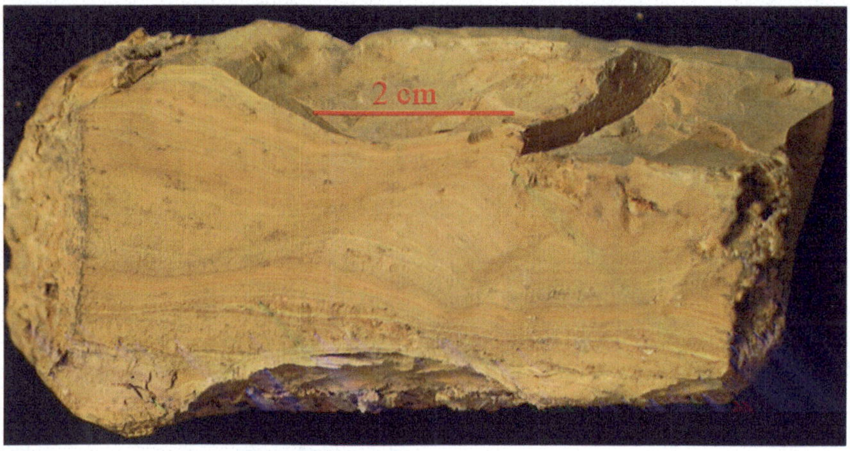

Fig. 13.3 Laminated yellow sand from the cave in the tunnel

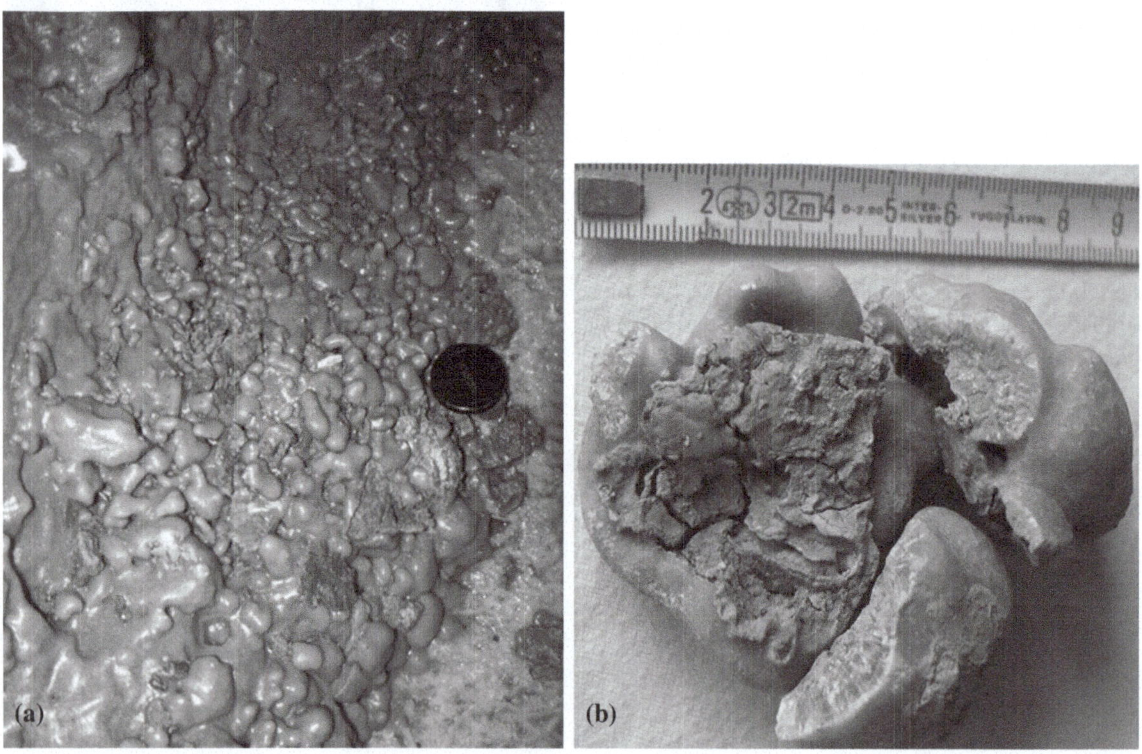

Fig. 13.4 Flowstone-covered alluvium (**a**) cross-section of the flowstone's crust over the alluvium (**b**)

water flowing down the walls or the floor in layers, creating coatings, waterfalls and canopies. The cave features quite a lot of flowstone in various forms, ranging from pools, curtains, stalactites and stalagmites. Different forms indicate different ways in which water flows and in which the flowstone is secreted. The flowstone is mainly yellow and brown, with the exception of translucent helictites and a few translucent stalactites which are in the process of growing.

In the upper part of the cave, translucent multi-shaped and multi-oriented helictites are growing from the ceiling, walls and stalactites (Fig. 13.5). Helictites are a less frequent form of flowstone and grow from capillary water that trickles through thin channels, usually through transverse fissures in stalactites. The growing helictites follow the principles of crystallization rather than gravity. Colourless, white or yellow, they come in various shapes, with two being predominant, namely thin, long straws that turn in different directions while growing, and short, thick-set monocrystals similar to smaller horns.

The floor of the upper part of the cave is covered in flowstone crust and a large number of pools in various sizes. Water is still flowing into most of the pools, while a large pool at the back of the cave has probably been dry for a while as indicated by the weathered flowstone at the edges. The dry pools also contain stalagmites with triangular cross-sections (Fig. 13.6), which are usually typical of the more peaceful parts of caves. Stalagmites with triangular cross-sections are those which do not display a concentric structure, i.e. individual growth layers, as the formation is building a single large crystal (Fig. 13.7a). Such stalagmites can already grow as monocrystals (Hill and Forti 1997) or are formed through re-crystallization of ordinary stalagmites as seen in our caves. The most frequent formations with triangular cross-sections are helictites or stalactites, while stalagmites are very rare (they were described in Romania, Texas, New Mexico, South Africa and Brasil). As monocrystals, they grow from a low-saturated solution at high moisture levels, partial pressure of carbon dioxide (CO_2) and minimum air flow, which is typical of closed caves (Hill and Forti 1997). Although their formation is yet to be fully explained, it is certain that they are very rare. In Slovenian caves, stalagmites with triangular and rectangular cross-sections are fairly frequent, especially in the more peaceful parts of caves, but, according to the first analyses, their formation cannot be attributed to the growth of calcite monocrystals; rather, they are the result of regular stalagmites undergoing recrystallization.

Fig. 13.5 Helictites growing out of stalactites in all directions

In most cases, recrystallization begins in the middle of a stalagmite, with the newly growing crystals being a lot larger, translucent and colourless (Fig. 13.7b).

In time, recrystallization continues towards the stalagmite edges so that the newly growing crystals reach all the way to its outer edges, thus affecting the morphology of the entire stalagmite. The external layer of the stalagmite remains brown, as the impurities from the initially coloured calcite crystals are pushed to the edge of the stalagmite during the recrystallization process. In the stalagmites that have not been fully recrystallized yet, one can still see the initial concentric structure; sometimes, only the top of stalagmites has undergone full recrystallization.

The upper part of the cave features a large pool, in and around which flowstone was secreted in various forms and in the course of several generations (Fig. 13.8). It was secreted into the pool and above it in the following order of sequence: (1) growth of stalactites, (2) growth of pools, (3) growth and thickening of stalactite points under the water level,

Fig. 13.6 Dry pools in which stalagmites with a triangular cross-section grow

Fig. 13.7 Cross-section of a monocrystalline stalagmite (**a**) and common stalagmite (**b**); the latter shows an onset of recrystallization

(4) formation of floating calcite, (5) settling of floating calcite and formation of cave milk, (6) thickening of attached floating calcite plates, (7) growth of helictites, (8) flooding and simultaneous secretion of grey flowstone and (9) growth of translucent straws and stalagmites.

Stalactites grow into this pool; they feature a special lower part and their onion-like points have grown under the water level (Fig. 13.8). Fibrous calcite was deposited on them, looking like fine crystals of cave milk. Cave milk is secreted in the form of a micro-crystallized white and soft matter on walls and flowstone, containing up to 70 % of water between the mineral crystals; it is rather crumbly when dry. The mineral composition of cave milk can vary but mostly comprises calcite or aragonite; alternatively, it can also contain non-carbonate minerals. It is formed through chemical precipitation from over-saturated solutions and with the help of microorganisms. Upon the lowering of the water level, individual floating calcite crusts (Fig. 13.9) adhered to these onion-like points of stalactites covered in cave milk. The floating calcite crusts grow from the water trapped in puddles and pools, and are secreted on the surface due to the change in the partial pressure of carbon dioxide (CO_2). Larger crusts sink because they become too heavy, especially because calcite crystals grow on their bottom, weighing them down even more. They also sink to the bottom if the water in the pools evaporates.

Stalactites in the upper part of the cave are mostly broken, primarily missing their lower parts. In view of the fact that the cave was not open, the possibility of anthropogenic influence is out of the question. New straws and helictites are already growing at breakage points, indicating that not all breakages are fresh. In view of their position, these breakages are in no way related with sliding of alluvium at the bottom of the passage, due to which most of stalagmites and flowstone crusts in the central part of the cave were broken off or fissured. The fissures are quite distinct and mostly already karstified with white flowstone. The fissures in flowstone probably point to the period during which a subsidence opened in the centre of the lower part of the cave, causing the sediments to wash away more intensely. Breakages of lower parts of stalactites have yet to be explained. Possible causes of their fracture include a larger flood surge, perhaps involving water that was full of sediments and thus

Fig. 13.8 Various types and colours of flowstone above the now dry pool in the upper section of the cave and inside it

Fig. 13.9 Lower section of the stalactites which were submerged into the water in the pool, have been overgrown with fibrous calcite onto which platelets of floating calcite have settled

powerful enough to break off the stalactite points which jutted out. The stalactite breakage edges don't seem to point to weathering or corrosion. The lower part of the cave features typical larger columns and stalagmites which are heavily fissured, some even broken off. Stalagmites and stalactites of various sizes are lying around on the floor. Certain breakages already have individual straws and helictites growing out of them.

13.3 Conclusion

The cave is interesting because it is located deep under the surface and has been inaccessible until the tunnel began being constructed. What is more, this is not a sink cave in terms of origin, but it contains a lot of alluvium of flysch origin as well as vast quantities of collapsed stalagmites and stalactites. The breakages and fissures can partially be connected with the sliding and subsiding of alluvium, while the broken points of stalactites have yet to be explained. The cave's special feature is its large quantity of helictites and the presence of stalagmites with a triangular cross-section.

The special character of the cave, its position and speleomorphological features prompted us to propose that access to the cave remain open for research purposes, which the tunnel constructors complied with, enabling an entrance through a concrete tube.

The History of Karstification on the Upper Cretaceous and Lower Paleogene Limestones in the Wider Kozina Area 14

While constructing motorways in the Karst region, the excavation of roadcuts revealed the upper part of the karst aquifer's vadose zone, the epikarst, in several places (*sensu* Klimchouk 2000). Lengthy denudation combined with simultaneous spatially and temporally uneven tectonic uplifting of the area brought the traces of various karst phases (*sensu* Bosák et al. 1989) of the actual karst aquifer and of the paleokarst (*sensu* Osborne 2000) phenomena that emerged in some previous karst period (*sensu* Bosák et al. 1989) closer to the present-day karst surface.

The research covered the relatively shallow motorway cuts in the vicinity of Kozina, where it is possible to see the intertwining of karst features that developed through the geologic history in greatly changing climate, geotectonic, geochemical and hydrologic conditions, as well as in lithologically different rocks which, by the time they became subjected to the karstification, passed over different stages of diagenesis and structural deformations.

The aim of this chapter is to show what the development of a certain karst system is influenced by and in what way, what forms emerge in this process and how the karst features of previous periods and phases can impact the geomorphological appearance of the later karst. Our primary focus will be on the consequences of development of two different, relatively extensive karst systems that underwent karstification for several million of years. The older of the two systems developed in a warm tropical climate of the Upper Cretaceous, in diagenetically immaturate carbonates located slightly above the sea level, while the other system developed in the tectonically active period extending from Miocene to present-day, in diagenetically mature and structurally deformed carbonate rocks in a constantly changing climate.

14.1 Geology of the Area

The present geological and landscape appearance of the researched area is largely the result of geodynamics occurring in the peripheral parts of the Dinaric Orogene (External Dinarides) or Istria underthrusting below the Dinarides, especially after the Middle Miocene (Placer 2008; Placer et al. 2010). In this area, such processes mainly involved the Mesozoic sedimentary successions of the passive margin of the Adriatic/Apulian microplate (*sensu* Stampfli and Mosar 1999) and the Upper Cretaceous and Paleogene synorogenic depositional areas. Istria's underthrusting resulted from the counter-clockwise rotation of the Adria microplate (*sensu* Stampfli et al. 1998) and the related pressures (Marton et al. 1995; Bressan et al. 1998; Placer et al. 2010), mostly directed from north to south, which led to the reactivation and horizontal shifts along the Dinaric-oriented (NW–SE) faults (i.e. dextral strike-slip faults) (Jurkovšek et al. 1996; Bressan et al. 1998). The underthrusting of Istria below Čičarija (Placer 2002) is reflected in the uneven uplifting of its hinterland

M. Knez and T. Slabe (eds.), *Cave Exploration in Slovenia*,
Cave and Karst Systems of the World, DOI 10.1007/978-3-319-21203-6_14

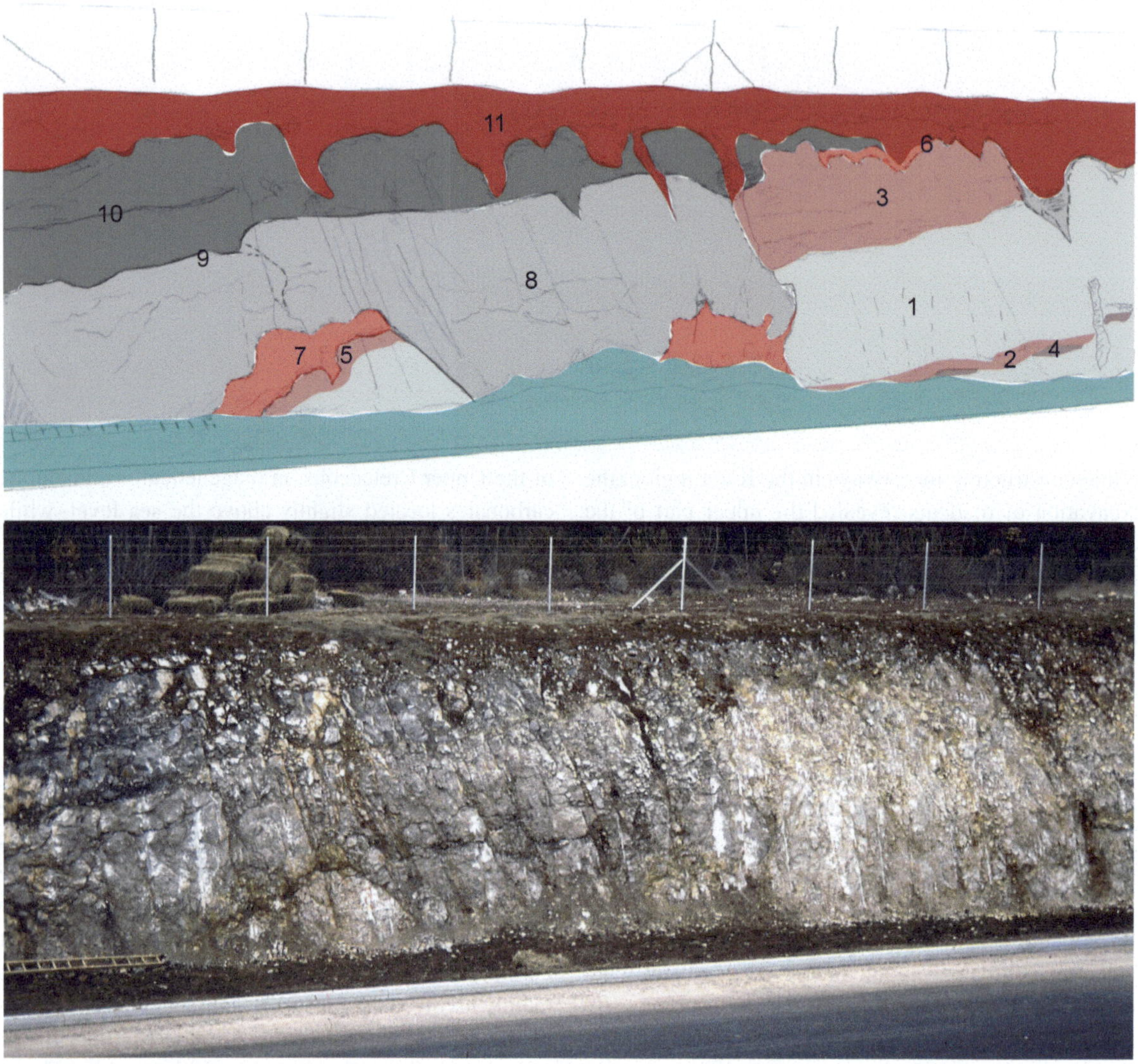

Fig. 14.1 The profile along the Divača-Kozina motorway section. *Legend 1* limestones of the Lipica Formation (Upper Santonian); *2* bottom red horizon; *3* upper red horizon, direct base of the paleokarst surface; *4* series of horizontal centimetre- to decimetre-scale cavities, filled with brown silt; *5* paleokarst surface, contact area between the altered limestones of the Lipica Formation and the weathered zone; *6* amalgamated paleokarst surface; *7* weathered zone: bauxite, calcite-bauxite-clayey sediments and breccias; *8* limestone of the Liburnia Formation, lithofacies of brackish closed bays/lagoon; *9* micro-paleokarst surface in the Liburnia Formation; *10* pedogenically modified palustrine limestone (pseudo-microkarst features); *11* present karst surface and soil

territory (Mihevc 2007) and in the recent seismic activity of the area.

Paleogeographically, the area belonged to the northwestern part of the Cretaceous Adriatic Carbonate Platform (AdCP) and the synorogenic Upper Cretaceous/Paleogene Carbonate Platform, whose sedimentary successions are separated by a stratigraphic gap of several millions of years. In this area, the last megasequence of the AdCP below the Upper Cretaceous paleokarst surface is composed of around a

Fig. 14.2 The interwined system of small channels/vugs created through non-selective dissolution of diagenetically immature limestone is filled by geopetally deposited red micrite and coarse-grained sparite

1000-m thick Upper Albian to Upper Santonian/Lower Campanian(?) succession of shallow-marine carbonate rocks.

The Maastrichtian and Paleocene-Eocene shallow-marine carbonate rocks of the Adriatic-Dinaric area, which are located between the thick Mesozoic carbonate successions and Paleogene clastites, represent the youngest (terminal) carbonate megasequence in the area of the previous AdCP (Košir and Otoničar 1997). It consists of three lithostratigraphic units of the higher order, united to form the Karst Group (Košir 2003): the Liburnia Formation, the Trstelj Beds and the alveolinid-nummulitid limestone. The carbonates of this megasequence were deposited during intense tectonic activities in the Upper Cretaceous and the Paleogene, and can be defined as sedimentary successions of synorogenic carbonate platforms (Košir and Otoničar 2001, 2002). In SW Slovenia, and elsewhere on the former AdCP, sediments of various lithofacies, members and formations, varying in their age, were deposited on the paleokarst surface. This variance results from deposition over an uneven paleokarst surface and the specific uplift and subsequent subsidence of the platform related to the foreland basin evolution, whereby an important role also played the local/regional structural and tectonic conditions (Otoničar 2006, 2007, 2008).

In the wider Kozina area, the alveolinid-nummulitid limestones are overlain by the Eocene carbonate, mixed carbonate-clastic and siliciclastic rocks (flysch) deposited in the deeper parts of the foreland basin.

The area discussed is located on a tectonically heavily affected transition between the External Dinarides and its foreland, and belongs as such to the External Dinaric Imbricated Belt, part of which are the Istrian-Friuli Underthrust Zone or the North Istria Structural Wedge and the Kras-Notranjsko Folded Structure (Placer et al. 2010). In the narrower sense of geotectonic subdivision, the Karst and the Materija corrosional surface (Matarsko podolje) are part of the Čičarija Anticlinorium (Placer 2005; Placer et al. 2010) which, as part of the Komen Thrust Sheet (Placer 1981, 1998b), belongs to the Kras-Notranjska Folded Structure (Placer et al. 2010).

14.2 Upper Cretaceous Paleokarst

14.2.1 Description

Traces of subsurface paleokarst features occur just below the paleokarst surface, in the Upper Santonian shallow-marine subtidal limestones of the Lipica Formation; in some places, they are manifested as three sub-horizontal reddish belts measuring up to 1 m in height, with secondary dissolution vugs (Fig. 14.1). The millimetre- to centimetre-scale vugs were later filled with several generations of meteoric cements alternating with muddy to silty sediments (Figs. 14.2 and 14.3). Directly under the paleokarst surface and even on it, remains of larger caves filled with sediments or flowstone can be found. Flowstone (Fig. 14.4) indicates precipitation in the vadose zone, while the frequent presence of floating calcite (Fig. 14.5) and, in some places, of cave pearls points to the precipitation from the cave pools of the epiphreatic or vadose zone. The karst surface approaching to the phreatic caves is also seen in the intense filling of caves with soil products (ferruginous-bauxite material), selectively leached cements, fossils and flowstone as well as carbonate micrite and silt (Fig. 14.6). Cave sediments also occur frequently as clasts in paleokarst breccias (Fig. 14.7) deposited at the bottom of paleokarst depressions.

Fig. 14.3 Dissolution vug enveloped by meteoric sparite and filled with laminated light brown calcite silt as well as individual grains of sand-sized calcite crystals

The presence of boehmite can be observed among the bauxite minerals; the purest bauxite contain around 80 % of it, while the clay minerals contain the greatest weight percent of kaolinite (up to 10 %). Bauxite is usually the last sediment to have filled the remains or the newly formed pores/voids both in the host rock and cave sediments.

Fig. 14.4 Flowstone-filled paleokarst cave cut off by the recent karst surface

In two-dimensional motorway cuts, the paleokarst surface is manifested in the form of pocket-like and gently undulating depressions measuring a few decimetres to several metres. (potholed paleokarstic depressions and hummocky paleokarstic depressions; *sensu* Vanstone 1998) (Fig. 14.8), and larger, distinctly irregular depressions and shafts measuring a few metres (Fig. 14.9) to as much as several tens of metres in size.

Depressions are filled/covered with paralic carbonates, while ferruginous-bauxite-carbonate sediments and karst breccias (Fig. 14.1) often directly overlie the paleokarst surface. The highly unevenly thick, non-homogeneous clast- and matrix-supported breccias are up to a little more than 1 m thick. The matrix is mostly "ferruginous-bauxite-calcite", while grains of crystal calcite are also frequent (Fig. 14.6). Certain cements in breccias exhibit characteristics typical of biodiagenesis (Figs. 14.10 and 14.11) which is the result of the activities of various microorganisms in the vadose diagenetic zone (Jones and Khale 1985). Due to various pedogenetic, diagenetic and biodiagenetic processes as well as due to the similar colouration of the matrix, weathered clasts and the host rock, it is often difficult to draw a boundary line between the individual component parts of breccias.

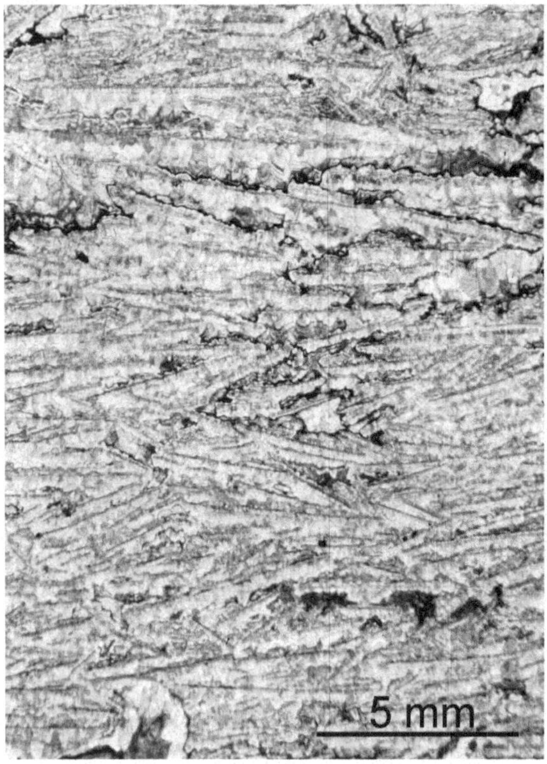

Fig. 14.5 Calcite rafts that underwent secondary cementation. The primary intergranular pores that, in places, underwent a secondary enlargement through dissolution are filled by sparite and a yellowish-brown silty sediment

Only rarely are the karst shafts/pockets filled with polymict breccia in which the remains of vertebrates as well as crushed bones and especially teeth of dinosaurs and crocodiles occur locally (Debeljak et al. 1999, 2002) (Fig. 14.9).

In some spots, directly under the paleokarst surface, it is possible to observe up to a few decimetres wide and up to several metres deep dissolution-enlarged subvertical vadose fissures filled with "ferruginous-bauxite-carbonate" sediments.

The facies and sedimentary successions of limestones of the Liburnia Formation, which deposited over the markedly uneven karst surface, change quickly in the lateral direction (Fig. 14.1). The oscillating transgression over the paleokarst surface is indicated by the sedimentary succession arranged in the shallowing upwards cyclical parasequences (Fig. 14.12). The limestone is frequently interwoven with pseudo-microkarst voids (*sensu* Freytet and Plaziat 1982) (Fig. 14.13) that show the breccia-like and "pseudo-breccia-like" texture (Fig. 14.14). In the bottom few metres of the succession, only one cycle with a "true" paleokarst surface (Fig. 14.1) covered with regolith (Figs. 14.14 and 14.15), which also includes karst pockets that are up to a few metres deep ("pit cave" or "dissolution pit"; *sensu* Mylroie and Carew 1995) (Fig. 14.16) occurs.

Somewhat higher up, but still in the lower parts of the Liburnia Formation, dark grey, bedded, locally also laminated micrite limestones prevail, containing ostracodes, gastropods and foraminifera, along with rudist in some places. In places, limestones are pedogenically modified, and thin inserts of coal also appear (Otoničar and Košir 1998; Ogorelec et al. 2001). Although in the Kras traces of pedogenic alteration and karstification still occur occasionally even higher in sedimentary successions of "synorogenic carbonate platform", they, in terms of extent and duration, do not reach the paleokarst located on the boundary between the AdCP and its successor on the periphery of the foreland basin.

14.2.2 Interpretation

In the area of the Julian Alps, the Slovenian Basin and the northern part of the AdCP, a collision in the area NE of the present Periadriatic Lineament caused the creation of a synorogenic depositional system with a foreland flexural basin and the related peripheral bulge as well as the intermediate transitional area (i.e. hinge line) already in the Upper Cretaceous (Otoničar and Košir 2002; Otoničar 2007). The primary, probably

Fig. 14.6 The fine-grained breccia that fills the smaller vadose channels under the paleokarst surface, is composed of partly rounded clasts of recrystallized host rock and weathered flowstone embedded in silty matrix

Upper Campanian, uplift of the peripheral bulge placed shallow-marine, diagenetically immature and unaltered carbonate sediments, which are deposited in the internal parts of the northern sector of the AdCP, into the area of the meteoric and mixing diagenetic environment (Otoničar 2006, 2007). Here, they were subjected to hardening (neomorphism, cementation) on the one hand, and selective dissolution (formation of vadose and phreatic channels/vugs) on the other (Otoničar 2006). During these processes, porosity decreased, while the karst system's permeability increased. A similar duality of the constructive and destructive diagenesis and karstification can today be observed, for example, in Tertiary carbonates of the Floridan Aquifer (Cander 1995).

The creation of the secondary dissolution porosity, the reddish rock (from oxidation of iron sulphides), and occasional precipitation of carbonates and pyrite could be the result of reduction/oxidation processes that take place in the mixing zone area (Stoessell 1992). On the other hand, the reddish staining directly under the paleokarst surface could be the result of the

Fig. 14.7 Weathered zone, pedogenic-karstic breccia: altered limestone clasts of the Lipica formation and of flowstone embedded in the calcite-bauxite-clay matrix

fine soil material infiltrating into the porous carbonate (Foos 1991; Rossinsky and Wanless 1992).

Diagenetic and speleogenetic features of the area in question indicate processes that took place in the young limestones that experienced only early diagenetic alteration prior to karstification, mainly in the meteoric and mixed meteoric/marine diagenetic environment (Fig. 14.17a). In view of the overall extensiveness of otherwise low and preferentially horizontally oriented (originally) in-filled caves or horizons of smaller, connected vugs, along with general geological characteristics (dip similar to layers below the paleokarst surface, characteristics of sediments and cements of fills etc.), we suggest that they were formed in the area where the meteoric and seawater mixed on the periphery of freshwater lenses (Fig. 14.17b) during the eugenic stage of diagenesis (*sensu* Choquette and Pray 1970). With continuous uplifting of the area, the caves were placed into the epiphreatic and/or vadose zone, where they were gradually filled with sediments and flowstone (Fig. 14.17c–e). Meanwhile, the upper parts of the carbonate aquifer were constantly exposed to karstification and the lowering of the karst surface, which brought the partly filled underground karst caves closer to the surface. Simultaneously with the lowering of the karst surface and the formation of karst features, pedogenic modification of the residual and eolian transported sediments took place along with the formation of the ferruginous-bauxite soil which was occasionally resedimented into unfilled parts of paleokarst cavities and surface depressions, karst pockets and widened fissures (Fig. 14.17d, e). On the one hand, the carbonate micrite and silt, which often fill large parts of cavities, represent undissolved or neomorphically altered original carbonate sediment which bears great geochemical similarities with the neomorphised host rock; on the other hand, they represent flowstone disintegrated by weathering, especially calcite rafts. As the karst surface was getting closer to the phreatic caves, which were either partly or entirely filled, the ceilings of these caves disintegrated completely; the blocks formed during this process "floated" in the cave sediment (Fig. 14.17d). In the course of surface processes of weathering and pedogenesis, the blocks became rounded, recrystallized and coloured. At the same time, the paleokarst surface under the soil was subjected to subsoil karstification. Geomorphologically, the disintegrated caves were

Fig. 14.8 The undulating paleokarst surface separates the Upper Santonian limestone of the Lipica Formation from the Maastrichtian limestones of the Liburnia Formation

Fig. 14.9 A paleokarst shaft filled with breccia, in which remains of fossil vertebrates were found, i.e. the teeth and crushed bones of dinosaurs and crocodiles (the Maastrichtian) (from Košir et al. 1999)

Fig. 14.10 The clast of pedogenic-karstic breccia indicates several generations of dissolution and filling. The *dark brown* edge is the result of biodiagenetic processes (see Fig. 14.11) (app. image width 6 cm)

probably expressed as somewhat oblong, more or less shallow depressions, i.e. denuded or unroofed caves (*sensu* Mihevc 2001; Knez and Slabe 2002a). Breccias, which cover parts of irregular paleokarst relief, also frequently resulted from the disintegration of caves and cave deposits on the paleokarst surface or directly beneath it, which is especially obvious in places where the clasts are also formed of flowstone.

The humid tropical or subtropical climate during karstification and the processes of pedogenesis that lasted at least a million years are indicated also by bauxites (Birkeland 1984). Such climate can also be inferred indirectly, as the warm-water carbonates were still deposited on the submerged parts of the platform.

The quick lateral and vertical alteration of the facies of the lower part of the Liburnia Formation (Figs. 14.12, 14.13, 14.14, 14.15 and 14.16) can be explained by the oscillating transgression over the markedly uneven karst surface (Figs. 14.17f and 14.18). Some spatially very restricted lithofacies (only a few m^3 to a few tens of m^3) represent sediments of karst depression fills during the "blue hole" transgression phase (*sensu* Durn et al. 2003) (Figs. 14.17 and 14.18). Considering the entire area of the northern part of the AdCP (SW Slovenia and Istria), the paleokarst surface is covered with carbonates of the Liburnia Formation of the Maastrichtian age (Drobne 1977; Jurkovšek et al. 1996; Otoničar 2006, 2007) only in the Karst and in the vicinity of Kozina. Only here was thus possible, in relation to the discussed "main" paleokarst period, for some paleokarst pockets and shafts to be filled, in the initial transgression phase, with breccia that contained among other vertebrate remains also crushed dinosaur bones and teeth (Fig. 14.9). We suggest that the remains of fossil vertebrates, together with sediments that overlie the paleokarst surface and clasts of the host rock, were resedimented in some paleokarst pockets and shafts from the nearby freshwater and/or brackish marshes that were located between the karst mainland and the sea.

Although lenses of freshwater undoubtedly existed under the fairly extensive karst land in the hinterland of sedimentary environments, where the carbonate sequences of the lower part of the Liburnia Formation were deposited, the sedimentological and paleontological characteristics (e.g. foraminifera) of the limestone directly above the paleokarst surface indicate, at least in the Kozina area, at least some marine influence. Thus we suggest that the sediments of the lower part of the Liburnia Formation were deposited in peripheral marine to brackish, possibly (in some places) also freshwater environments of closed lagoons that were partially marshy (i.e. paralic environments) (Fig. 14.18).

14.3 Current Karst System

14.3.1 Description

In the wider Kozina area, the subsurface karst phenomena are manifested in the form of numerous partially accessible epiphreatic cave channels, relict

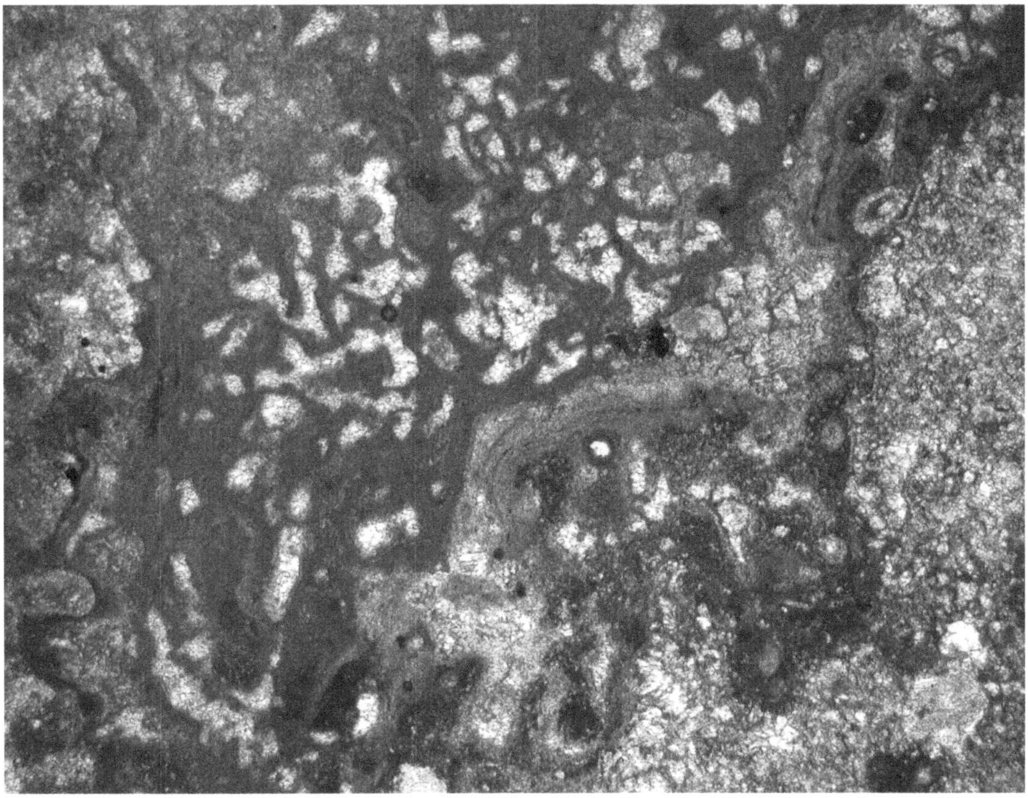

Fig. 14.11 Transition from the neomorphised host rock (microsparite) (left corner of the image), via the belt of brownish microsparite or clotted, partly laminated micrite, into a belt of fibrous micrite arranged into an alveolar-septal structure (app. image width 4 mm)

passages "perched" in the vadose zone and denuded caves. The average underground water level in this area is over 150 m deep under the surface, and the unknown phreatic karst channels drain water towards the springs in the valleys of the Rižana, Osp and Glinščica rivers and towards the Timavo River springs (Knez et al. 2015). A significant influence on the hydrology, evolution of underground cave features and cave sedimentation was and still is exerted by the vicinity of flysch and the related inflow of allochtonous waters and sediments into the underground. The contact area between the limestone and flysch or limestone and the "transitional beds" (marls) also influences the location and development of certain springs (e.g. the Osapska jama Cave) and ponor caves (e.g. the Beka-Ocizla cave system) (Knez et al. 2015). The "neotectonically" uneven uplifting of the area is most markedly reflected in the blind valleys of the Matarsko podolje (Mihevc 1994) which are cut to various depths and frequently bottoms of individual valleys are at different levels. The vadose cut of the canyon in the Škocjanske jame Caves under the initial phreatic channel is also very illustrative in this respect. The known karst shafts are up to a little more than

Fig. 14.12 A short geological profile of the direct cover of the paleokarst surface (*bottom left corner*) shows micrite limestones which, together with foraminifera in the *bottom part*, show at least a limited influence of seawater in a depositional environment. Higher up, the micrite limestones underwent several phases of pedogenic modification, and at least one rather prominent paleokarst surface interrupts them

100 m deep (see Chap. 1 Introduction, Fig. 1.8); like in this case of certain larger shafts that represent entrances to the more extensive cave systems (e.g. entrance shaft to the Kačna jama Cave), it is difficult to ascertain whether they are of vadose or, more likely, phreatic origin followed simply by subsequent alteration in the vadose zone (see Mihevc 2001).

The motorway cuts revealed upper parts of the vadose zone (Fig. 14.1) where, beside to entrances to shafts, mostly completely filled cave channels were discovered right below the surface. Profiles of the karst surface revealed that a large part of the yellowish-brownish sediments that are frequently located in the somewhat oblong dolines and notches (Fig. 14.17) share the same, flysch-based origin as those located in filled caves lying right below the surface (Zupan Hajna 1998). In places, flowstone occurs among the sediments. The upper parts of sediment profiles are often covered with limestone blocks and rubble formed during the disintegration of cave ceilings (Bosák et al. 2000a). The surface geomorphic features, in which yellowish-brownish sediments prevail, can be defined as denuded caves (Fig. 14.17) (*sensu* Mihevc 2001); dolines without any typical cave sediments and collapse dolines are present as well. In places, the bottoms of depressions continue into corrosion-widened fissures and narrow shafts.

Among the smaller surface karstic features, it is possible to observe karren which are especially prominent on Cretaceous limestones where they exhibit traces of fast modification and genesis both on the surface and below ground. The karren overlying Cretaceous limestones consist of stone teeth that measure 1–2 m in height (Fig. 14.19). The upper parts of teeth are dissected by rock features hollowed out by rainwater (flutes, solution pans) (Figs. 14.19 and 14.20), while on the lower parts subsurface rock features (subsoil channels, underground caverns etc.) (Fig. 14.21) were formed. Rock features hollowed out by rainwater gradually covered the subsurface rock relief, which points to the changing thickness of the soil due to the varying levels of overgrowth in the area. The process of change is fast, with results noticeable as early as throughout the historic periods rather than geological ones.

In the area of fissured zones, the rock is crisscrossed by a dense network of fissures where, under the karren and the bottom and sides of dolines, under the soil and

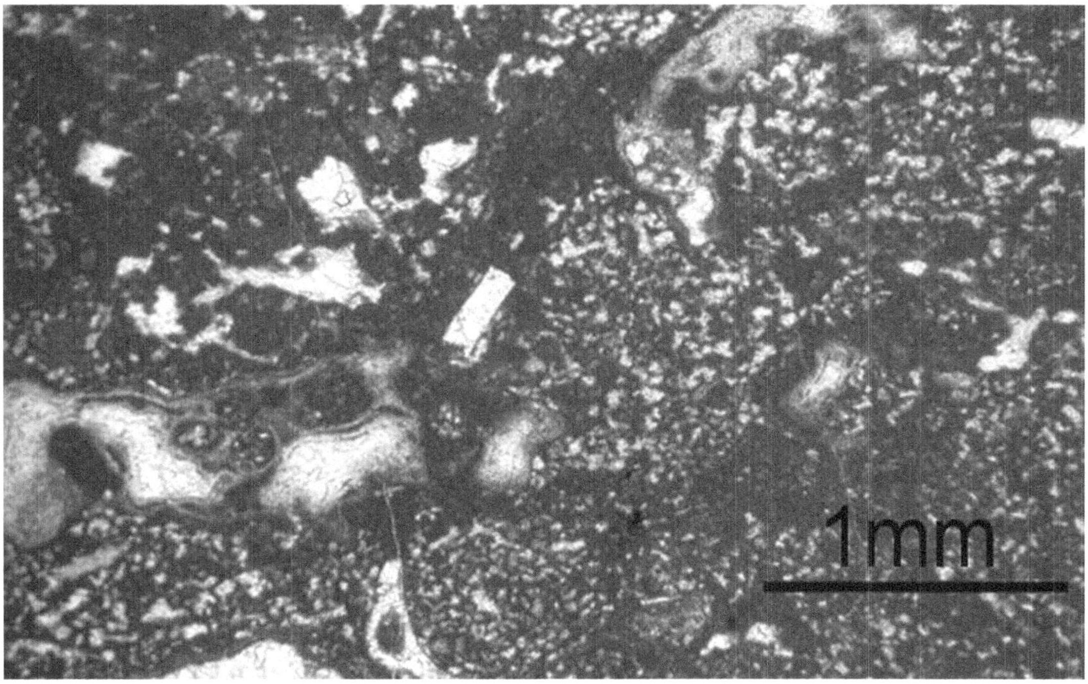

Fig. 14.13 Pedogenically modified micrite with typical risolites

sediments, individual and prominent crevices were formed, reaching as much as several tens of metres under the surface. The crevices are more frequent right below the surface (the epikarst zone), where especially the Paleogene foraminifera limestones disintegrate into rubble frequently, while the clasts undergo typical subsurface modification, i.e. they become rounded.

14.3.2 Interpretation

Karst features and the layout of hydrological zones of the current karst aquifer are the results of its continuous geological and hydrological development from the time when the Cretaceous-Tertiary limestones underwent the final uplift into the area influenced by rainwater, while the aquifer, being a dynamic system in time and space, was subjected to constant changes. Thus began the final period of karstification in the tectonically active, deformed orogenic area. Due to the unevenly uplifted thrusts and shifts along the prominent regional faults, karstification included carbonates that were very different stratigraphically, while the intensity and form of karst features as well as the karst profile characteristics also differ from one area to another, depending on the size of tectonic deformation and the relation between the carbonate and noncarbonate rocks.

In the subsequent karst phases, the oldest traces of karstification were probably completely blurred and cannot be discerned among the more recent karst

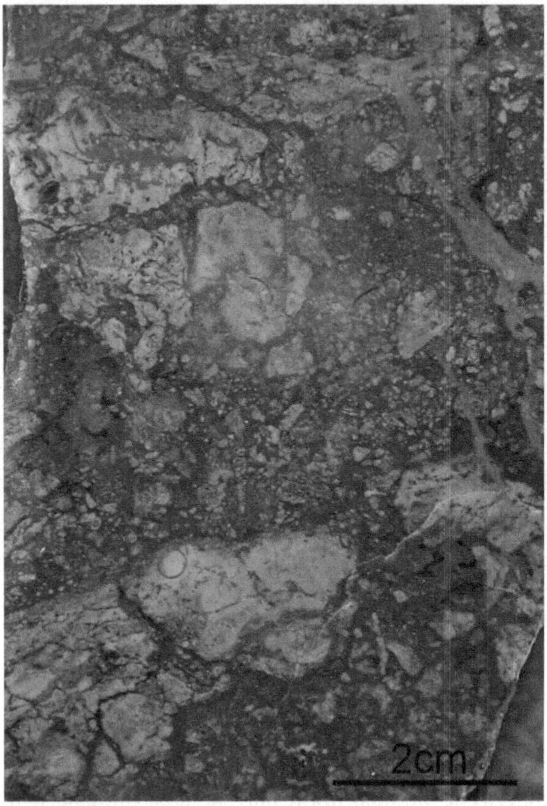

Fig. 14.14 Breccia-like or pseudo-breccia-like texture of pedogenically modified limestone with several centimetre-size nodules

features. Nevertheless, the preserved karst profile shows the history of karstification and modification that went on for several millions of years. The more or less continued lowering and, in certain tectonically calmer periods, levelling of the surface as well as the fluctuation or, generally speaking, relative lowering of the underground water level led to the exposure of the phreatic caves to the surface, whereby they were largely filled with sediments, mostly in the epiphreatic zone. On the karst surface, the processes of weathering and pedogenesis modified caves into surface geomorphic elements. While the denuded caves were still under the surface, in the vadose or the epiphreatic zone, in the phreatic zone, lower in the karst profile, new cave channels which are today placed in the vadose and epiphreatic zone were already being formed. The individual karst phases were thus not necessarily sharply separated in terms of time. The lack of information resulting from the limitations of absolute dating methods, the lack of fossils in cave sediments suitable for dating and the "intertwining" of karst features created in different karst phases make it very difficult to determine individual karst phases with greater precision. For the same reason, it is difficult to uniformly correlate certain speleogenic events (e.g. filling of cave channels, accelerated deposition of flowstone etc.) with "external" climate and tectonic changes/events. Nevertheless, the estimated rapidity of denudation of the karst surface and, above all, dating of allochtonous cave sediments of filled caves, including uncovered caves, of the highest parts of the karst profile in the area of the Kras, the Matarsko podolje lowland and the Podgorje karst enable us to estimate that the oldest known caves in this area began filling with sinking streams when they were located in the epiphreatic zone, even more than 4 Ma (Horáček et al. 2007) or 5 Ma ago (Bosák et al. 2004), while those located the highest (e.g. the Grofova jama Cave on the NW edge of the Kras) probably underwent this process much earlier (Zupan Hajna et al. 2008; Zupan Hajna and Mihevc 2008). Although it is accepted that the Tethys Ocean was finally closed in the area of subsequent Dinarides at the same time as the Paratethys Sea started to form, namely close to the Eocene/Oligocene boundary (Steininger and Wessely 2000), the lack of sediment records, also owing to the later erosion, makes it especially difficult to determine

Fig. 14.15 Regolith, which matrix, in addition to the calcite grains, is abundant in kaolinite (>40 wt.%), but does not contain bauxite minerals, covers the paleokarst surface in the *bottom part* of the Liburnia Formation, only a few metres above the main paleokarst surface

the precise timeframe of the primary Dinaride uplift above the sea level, and thus of the exposure of karst rocks to the rainwater. Most notably, it is difficult to precisely define the extent of originally uplifted areas as the Lower Miocene lake sediments of the Paratethys Sea, for example, are also located on the island of Pag, the present-day Mediterranean side of the Dinarides (Bulić and Jurišić-Polšak 2009).

Similar applies to the Middle Miocene (the Badenian) correlation between the Central Paratethys and the Mediterranean via the Slovenian Corridor (see Horvat 2004), which points to the completely different morphology of the area at that time.

Although the External Dinarides thrusts (*sensu* Placer et al. 2010) formation in the Paleogene cannot be fully equated with the uplift of more extensive and high mainlands above the sea level, the underthrusting of "Adria" and thus Istria under the External Dinarides from the Miocene on (*sensu* Placer et al. 2010) has been of great significance for karstification and geomorphic modification of the area that can be, to a certain extent, traced until present-day, especially in

Fig. 14.16 A fairly dynamic paleokarst surface in the *bottom part* of the Liburnia Formation (see Fig. 14.16) with a karst pocket

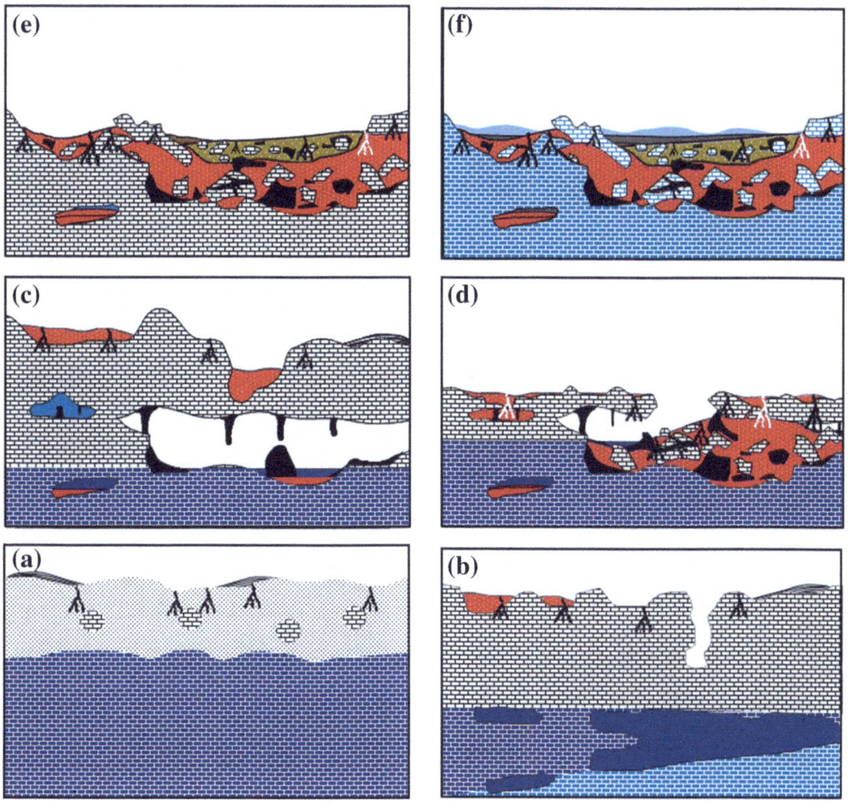

Fig. 14.17 Evolution phases of the paleokarst surface in the area where the ceiling of halocline paleokarst caves was denuded (unroofed caves); the evolution is explained in the text

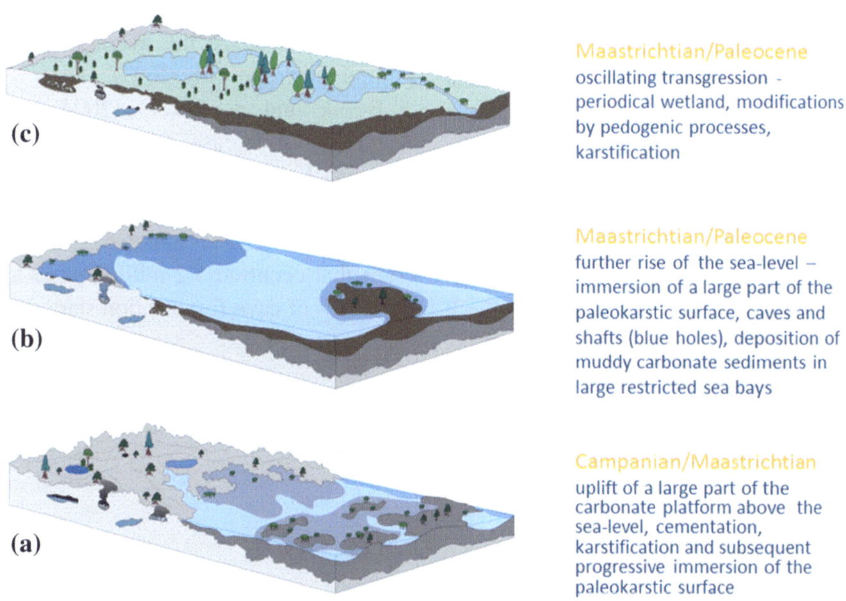

Fig. 14.18 Evolution of the paleoenvironment between the Campanian/Maastrichtian and the Paleocene in the area of the present-day Kozina (Matarsko podolje). The paleokarst surface that developed on the peripheral bulge is covered by palustrine carbonates of the synorogenic carbonate platform

Fig. 14.19 In the upper part of stone teeth measuring over 1 m in height, rain flutes were carved out by rainwater

Fig. 14.20 Rain flutes and solution pans were carved out by rainwater (app. image width 0.75 m)

the SW part of Slovenia. The period of filling caves, which began around 4 Ma ago (proven age of sediments: see above) and which points to the tectonically calmer period and the phase of levelling the karst surface in the area of underground water fluctuation as well as to a milder relief (Mihevc 2001, 2007), was followed by the final tectonic phase that is related to the underthrusting of Istria under the External Dinarides. This area uplifting phase can also be inferred on the basis of the water animal called *Marifugia cavatica* that lived on the cave walls of the Črnotiče II fossil cave around 3 million years ago (Bosák et al. 2004) and is now located, together with the filled cave, at least 370 m above the average level of the underground water (Mihevc 2007).

Bosák and coworkers (2004) suggest that the beginning of intense speleogenesis in the entire of Classical Karst is related to the inflow of high-energy sinking streams in the karst system due to the drop of the sea water level or the erosion base during the Messinian crisis that occurred 5.2 million years ago and that the processes of cave fossilization started after this event.

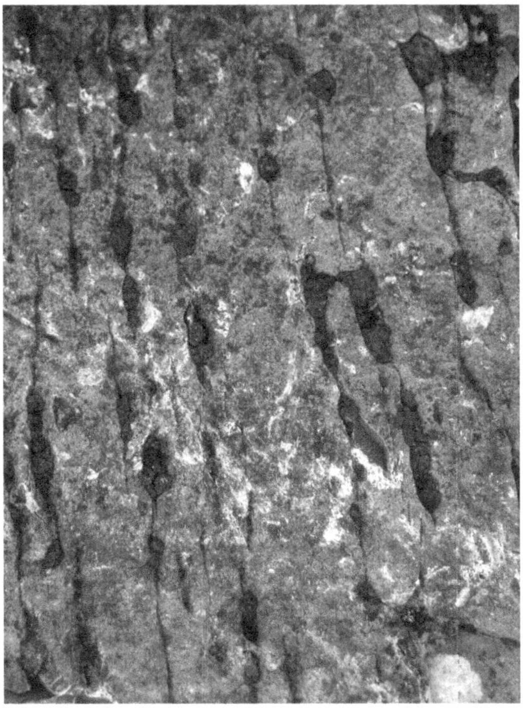

Fig. 14.21 Subsoil hollows on the freshly uncovered rock surface (app. image width 0.75 m)

14.4 Conclusion

Although the described karst systems developed in conditions that differed in terms of geotectonics, geochemistry, hydrology and climate, their evolution bears certain similarities. In both cases, the processes of long lasting relative lowering or stagnation of the underground water level and simultaneous denudation have exposed the relict and fossil phreatic caves to the surface, where they were modified through surface processes and became a morphologically typical part of the karst surface.

Even though the fluctuation of the underground water level is significantly influenced by climate and time in terms of karst aquifer development from "protochannels" to large-scale channels, it seems that the regional character or appearance of the karst profile in the cases described is largely dependent on the geotectonic area where karstification takes place.

The extent to which a certain area undergoing karstification will be relatively uplifted above the regional erosion base, as well as its boundaries, the nature of the area's structural elements, the sedimentological and diagenetical appearance of carbonates that will undergo karstification, and the preservation potential of the karst are largely dependent on the position of the area in terms of the edges of the geotectonic plates and the related tectonic stresses and geodynamics.

Both described karst areas were formed as a result of deformation occurring in the collision area of two lithospheric plates and, as such, logically follow one another in terms of geodynamics; due to relatively predictable tectonic deformation in such circumstances, the development of karstification and the preservation potential of the karst can be anticipated up to a point.

Part III

Comparison with Low and Covered Karst and Karst in Breccia: Dolenjska region

15 The Dolenjska Karst Area Uncovered on the Bič–Korenitka Motorway Section

During the motorway construction, carbonate rocks were uncovered and the present-day epikarst zone was cut into at several sites, revealing interesting formations typical of this particular part of the karst which is mainly covered in thicker layers of alluvium and soil, while the groundwater in lowlands is not far below the surface. This includes various types of karren, especially subsoil karren, subsoil stone forests, karst uvalas with estavelles and caves (Fig. 15.1).

15.1 Geological Characteristics of the Area

According to Osnovna geološka karta (the Basic Geological Map, hereafter the BGM), the Ribnica sheet (Buser 1969), the north-lying layers of Early Triassic limestone (T_3^{2+3}) come into contact with the south-lying Lower Jurassic layers of limestone in the area south of Medvedjek. The Triassic limestone is heavily crystallized and contains numerous inserted coloured pieces of sedimented breccia. According to the BGM, this is a normal transition from Trias to Jura which runs along the morphologic valley of Dolge dole south of the Mali (409 m) and Veliki Medvedjek (417 m) hills. The layers of both the Triassic and the Jurassic limestone strike from northwest to southeast. The layers of massive and breccia-like Triassic limestone are vertical, while the Jurassic layered limestone dips to northwest and southeast at an angle of 20°–50°. South of the Mali Medvedjek Hill (409 m), darker Jurassic oolitic limestone can be found, dipping to southwest at 20°–40°. Slightly more to the west, south of the settlement called Veliki Gaber, megalodontid shells can be seen in the Jurassic limestone. Here, the limestone dips at an angle of 30°–50° to the northeast. This means that a syncline runs through the Jurassic limestone, its axis oriented from northeast to southwest. Along the limestone bedding planes we can also detect slips; between layers, there are faults which bear the traces of vertical shifts (Fig. 15.2).

Buser (1976) considers the territory undergoing research as part of the Mesozoic blocks of the western Dolenjska region. The Temenica fault runs along the Temenica River valley, past Veliki Gaber and Dobrava near Dobrnič. The Stična fault can be traced through the Stiški potok Stream valley past Rdeči Kal, north of the Šumberk Hill (540 m) and south of Dobrava near Dobrnič. Buser determined the course of the Žužemberk fault differently from Placer (1998a). According to Buser, it runs past Lipoglav, north of Zagradec, past Žužemberk and Dvor along the Krka River valley. According to Placer, however, the Žužemberk fault does not continue along the Krka River valley past the settlement of Krška vas; instead, it runs more to the south, through the village of Plešivica.

Part of the fault in the cross-Dinaric direction called the Orlica fault (Placer 1998a) was named the Straža fault by Premru (1976). Pleničar and Premru (1977) also call it the Krško fault. The Straža tectonic depression is filled with Pleistocene sediments. The vertical shift along the Straža fault is located 350 m to its southern side (Premru 1976).

The territory in question is part of a triangular area between Idrija and the Mid-Hungarian tectonic zone called the Sava compressive wedge. This is an area that was exposed to pressures in the direction from north to south, causing folding and resulting in the formation of the Sava folds in the direction from east to west. The folding shrank the compressive wedge

M. Knez and T. Slabe (eds.), *Cave Exploration in Slovenia*,
Cave and Karst Systems of the World, DOI 10.1007/978-3-319-21203-6_15

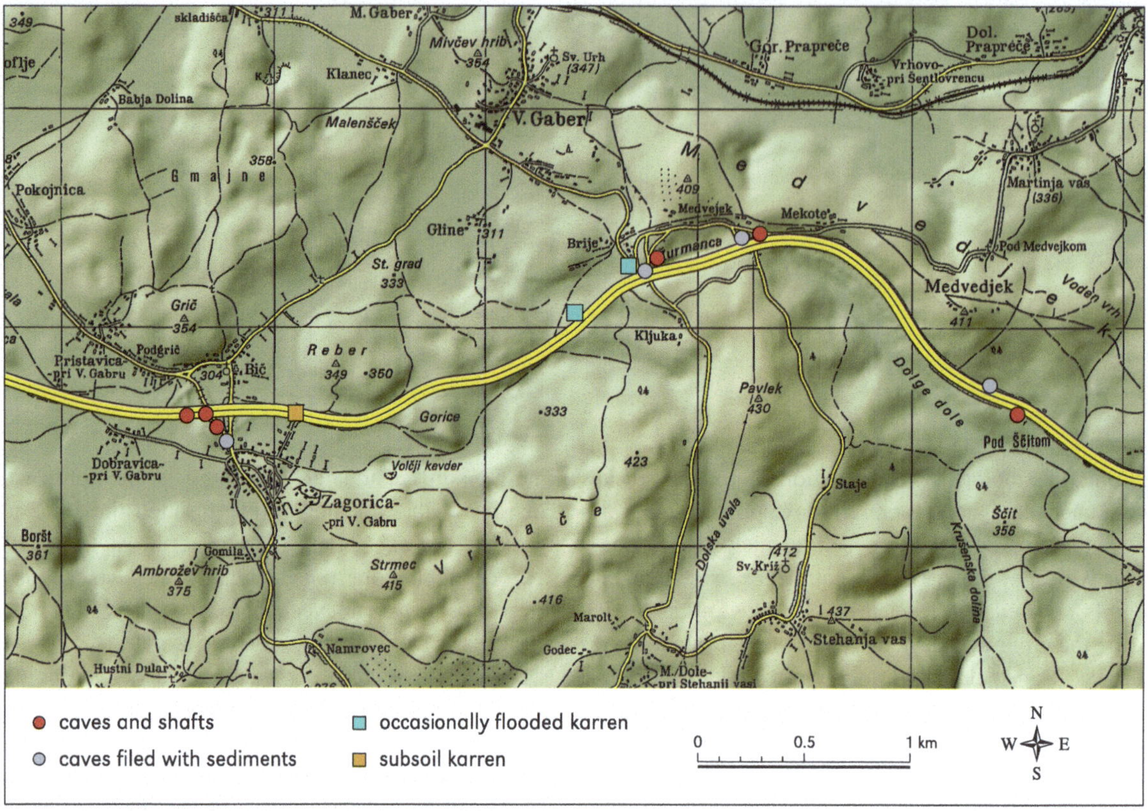

Fig. 15.1 Karst formations uncovered during the motorway construction

from north to south by about 20 km, which led to the shift of the Periadriatic tectonic zone to the south, and its narrowing (Placer 1998a, b).

The Sava folds area probably underwent intense folding in Miocene or Pliocene. The process of compression was also at work in Quaternary and could, hypothetically, be active even at present (Placer 1998a).

We attribute the formation of the Sava compressive wedge to the development of the Idrija–Mid-Hungarian trans-section tectonic zone in which the dominant shear direction cannot be established. This means that the compressive wedge developed in a period of stagnation of intensive shear movements in the inner part of the Idrija and the Mid-Hungarian tectonic zones (Placer 1999a).

The creation of the Sava folds is related to the formation of the Idrija–Mid-Hungarian trans-section tectonic zone and to the change in tension conditions from the dextral shear character, over transpressive to normal ones oriented perpendicular to the Periadriatic tectonic zone in the direction of north to south, which is probably connected with the rotation of wider proportions.

Although the BGM, the Ribnica sheet (Buser 1969), shows the contact between the Upper Triassic limestone and the Lower Jurassic limestone as a normal lithologic contact running south of the Medvedjek Hill along the Dolge Dole valley, the tectonic mapping indicates that the morphology of the mentioned valley is also tectonically determined by a powerful fault zone running from northwest to southeast (230/60). Parallel fault zones also extend westwards where we have measured a 220/80 right-shear fracture and a 10–40/70–90 left-shear fracture. We believe that the fracture in the northwest to southeast direction involves at least two phases of horizontal tectonic movement (right and left shears). The BGM also shows that this area features a significant fault with a dip direction of 60°, which comes from the Temenica River valley in the north, runs east of the Veliki Gaber settlement, crosses the western edge of the Mali

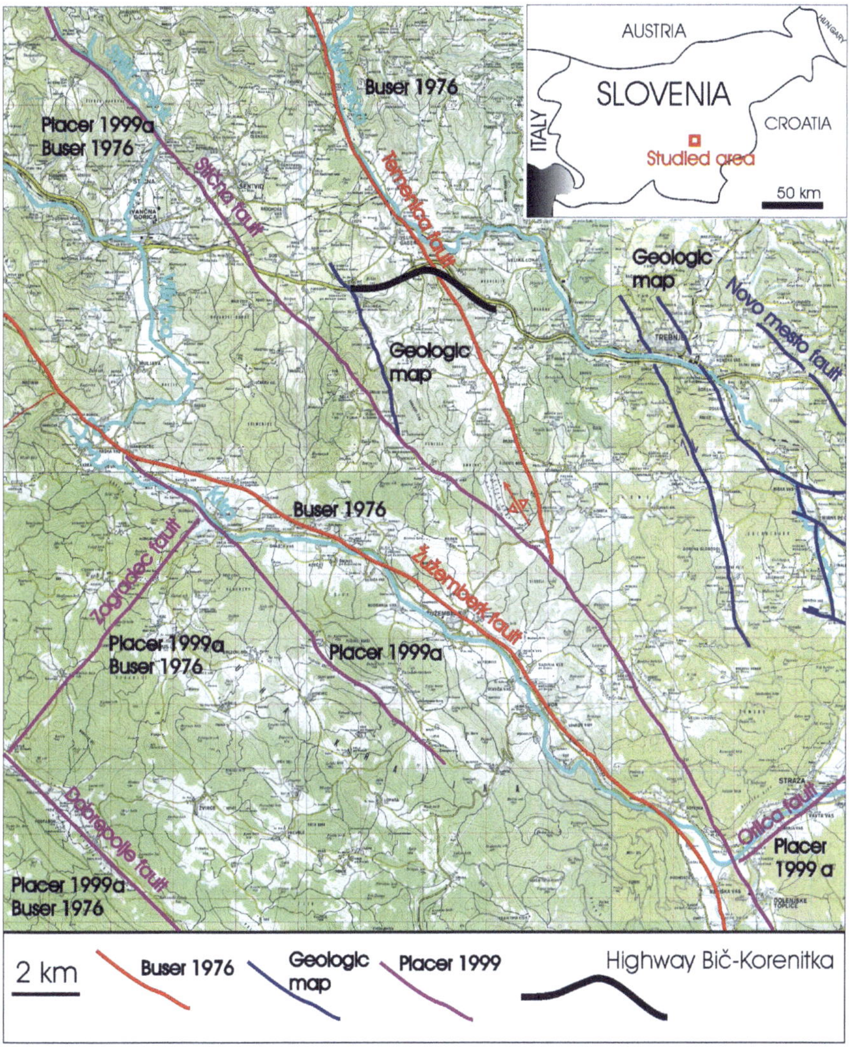

Fig. 15.2 Fault zones in the broader studied area

Medvedjek Hill (409 m) and continues to the southwest of Dobrnič. In 1976, this fault was named the Temenica fault by Buser. The fault running through the Dolge Dole valley is the fracture oriented to the northeast and accompanies the Temenica fault.

The fault zone located before Korenitka, with its dip direction of 310/70, is classified among the fault zones with tectonically powerfully crushed rock. It is probably an older fault zone measuring around 4 m in width, intersected by the fault zone oriented from northwest to southeast that runs along the Dolge Dole valley. We believe that the oldest tectonic deformations of the zone are those in the direction of northeast to southwest, followed by the zones from northwest to southeast (horizontal movement, right and left shear), while the zones from east to west are the youngest (horizontal movement, right shear). Thrust deformations are probably older as well.

15.2 Uncovered Karst Formations

15.2.1 Karst Uvalas with Estavelles

South of Bič, at the bottom of a morphologic flatland that is reminiscent of a periodically flooded karst field (polje), there are many estavelles (sinkholes). The entrances are open in the sediment. The Dob uvala, on which the western part of the research route runs, is one of the many non-typical depressions along the Dolenjska lowland, with fluviokarst being the predominating landscape type. It is located at the contact of the Upper Triassic dolomite on the northern side of the valley, and the Jurassic oolitic limestone on the southern side. In view of its diversified geological composition, locally varying water permeability or generally poorer water permeability can be expected. The uvala is the lowest part of the Šentvid area. Given the dimensions of its bottom, the depression could be defined as a karst polje. Due to the absence of a larger sedimentary flatland at the bottom, and since most of the rim has not been closed and therefore does not have the form of a typical Dinaric karst polje, the depression is termed uvala (Gams 1987). Surface streams (the Dobovščica, Šentviški potok and Glogovnica streams) flowing from dolomite sink in oolitic limestone. During heavier rainfall, the water also flows into the uvala from the west and the east. When the waters are higher, smaller springs increase considerably and extend the surface water course towards the southeastern edge of the uvala. Prior to the motorway construction, floods in this area occasionally hindered the traffic on the old main road (Gams 2004). Quicker drainage of the surface water and smaller scope of the floods, along with adequately permeable sinkholes, indicate a widespread underground network of channels located right below the surface. This is also confirmed by the frequent subsidences and formations on the rock surface, which are the result of fluctuating water.

Other morphologic depressions—uvalas such as Kljuka and Žurmanca can also be found along the motorway route. The Kljuka uvala runs from east to west, the same as the fault zone.

The Dolge Dole morphologic depression has developed at the lithologic and tectonic contact.

15.2.2 Karren

The subsoil karren located in this area can be divided into two types. The first type has formed due to the steady vertical percolation of water through the more or less thick layer of soil covering the fissured carbonate rock. Under the thicker layer of sediment and soil, there are wider or narrower V-shaped fissures formed on considerably crushed rock. Their depth can

Fig. 15.3 Subcutaneous stone teeth

reach several metres and the rock shape between is reminiscent of subsoil teeth developed in a heavily crushed rock (Fig. 15.3). In places where fissures are more thinly scattered and the rock is more corrosion resistant, genuine stone teeth with pointed or blade-like tops have developed, while some places even feature smaller stone forests (Knez et al. 2003) (Fig. 15.4). They are dissected by subsoil channels. Underground, their surface is typically corroded; it is relatively smooth on the uniform and fine-grained rock, and rougher on diversely structured or crushed rock. Prior to the removal of the soil, only smaller karren or pieces of rock were jutting out to the surface which was overgrown and often covered in moss. There were also solution pans, indentations and partly reshaped subsoil formations (Fig. 15.5), channels and notches (Slabe 1998a). The latter are indications of slow soil erosion.

The second type of subsoil karren was discovered during archaeological excavations at the bottom of the

Fig. 15.4 Wider area of subcutaneous stone teeth

Fig. 15.5 Subsoil indentations on karren

Fig. 15.6 Karren near Bič that were uncovered during archaeological excavations

lowland near Bič (Figs. 15.6, 15.7). They were formed entirely underground. In some places, the layer of sediment and soil is relatively thick; in other places, the karren almost reach the surface. They feature sharpened tops and an interesting rock relief that conveys a lot of information. The upper part prevails in a relatively smooth rock typical of subsoil formation and fine-grained sediment. The most distinct feature of the lower part of the karren are subsoil notches. Larger and horizontal notches reach one metre in diameter, while the smaller ones are placed on top of one another. Semi-cylindrical notches represent the ends of vertical subsoil channels formed along the most conductive ways. Some of the tops of the subsoil teeth located above the most prominent notches are mushroom-like. The subsoil channels on these karren can be divided into vertical and horizontal ones. The former are the conduits of fluctuating water flows along the most conductive ways. The latter criss-cross the more gently sloping rocks, including larger ones, and are co-shaped with moisture which stays in them for the longest time, even after the level of the underground water decreases. Formed along flaws in the rock, which mostly occur in the form of small fissures, are subsoil solution cups which can develop into tubes. Between solution cups and channels are the subsoil tubes that criss-cross the rock at different inclinations.

Similar formation of subsoil karren was illustrated by an experiment in which gypsum columns were covered with soil and then exposed to artificial rain (Slabe 2005). The water drained from the model at the bottom. The upper part of the columns was formed by water penetrating the soil in a dispersed manner, and the lower part was formed in the locally flooded zone. The drainage was too slow, which explains why the water filled the bottom of the model.

To sum up, two prevailing ways of karren formation can be inferred from their shape and their rock relief. The rock shapes, which attest to the frequent fluctuation in the level of underground water that floods karren from underneath, left a special mark on them. During low levels of underground streams, karren are formed by water that occasionally flows in a dispersed manner from the surface through the soil, gliding evenly down the rock. It stays longer in subsoil solution cups and gently sloping channels as well as along less permeable contacts of rock and the surrounding sediment.

Fig. 15.7 Occasionally flooded subsoil karren near Bič

15.2.3 Caves

There are only a few larger caves in the explored area. More of them can be found on the higher-lying edges of uvalas. While the karst water level was closer to the surface, thick Plio-Quaternary sediments undoubtedly hindered in-depth karstification (Gams 2004). Because the level of underground water in the explored lowland was high and erosion was modest, thick layers of weathering material, which covered a large part of the Slovenian karst in the Triassic period, remained preserved.

Off the motorway route, several sinkholes had been discovered prior to the construction. The Dobravska Jama sinkhole is located between Dobravica pri Velikem Gabru and Bič. East of Bič lies Špaja jama Cave (Reg. No. 3464), a gradient shaft measuring 68 m in length and 44 m in depth; east of Zagorica pri Velikem Gabru, there are the Kovačeva Rupa and the Volčji Kevder (7–8 m deep) sinkholes.

Ten new karst caves were discovered on the 5-km section of the Bič–Korenitka motorway, most of them in a deep cutting of the Mali Medvedjek Hill. The majority of them were filled with fine-grained sediment (Fig. 15.8), while some shafts were hollow (Fig. 15.9). The largest passage, the walls of which were covered with a heap of flowstone, measured 5 m in diameter, while the diameter of others ranged from 1 to 2 m. The rocky circumference of passages was transformed at the contacts with the sediment. The fracturedness of the rock is most often reflected in the position and shape of passages. Fissure caves represent a special type of caves which were formed along the more prominent upright fissures and faults, and at intersections of layers and tectonic deformations. In the cross-section, they are more or less vertical, sometimes also meandering. In all cases, they conform to the lithotectonic state of the rock. The width and volume of filled spaces in the rock correspond to the local fracturedness of the rock. Such caves often follow narrower fault zones from which the interior fault zone material was transported underground and was replaced with superficial sediment. On the walls of these caves, whose width does not exceed 1–2 m, there are still visible traces of the last movement of rock blocks. Vertical circulation of water did not take place at one point alone, but alongside the entire spread of the fault or alongside the equivalent fracturedness of

Fig. 15.8 A cave filled with fine-grained alluvium

Fig. 15.9 Entrance to the shaft in the cutting of the Mali Medvedjek hill

Fig. 15.10 A cave in brecciated rock

the rock. Their shape indicates a somewhat specific evolution pattern. It seems that they formed as sub-sedimentary caves. Smaller caves also developed in a markedly fractured and partly brecciated rock (Fig. 15.10), which explains why their circumference is very rough.

15.3 Conclusion

The fact of the matter is that limestone along the Bič–Korenitka motorway route is tectonically heavily fractured, forming broad collapsed and crushed zones, where the limestone is often fractured to the level of tectonic breccia. This is understandable, as the area in question is part of the Sava compressive wedge (Placer 1999a) which has undergone several phases of tectonic processes. Unlike southwestern Slovenia, where Cretaceous limestone prevails, this karst developed in the Triassic and Jurassic limestone which is covered with thicker weathered material. The most prominent karst formations are sinkholes which can be found at the bottom of uvalas. In the course of construction work, subsoil karren and caves were also uncovered. Those that developed at the bottom of the lowlands and are often flooded by underground streams are of exceptional shapes. They feature a unique rock relief which, to the best of our knowledge, is now being described for the very first time. The old caves, through which streams were once flowing, attest to the lowering of the groundwater level, which reaches the surface only in the lowest-lying lowlands. A good part of the caves is filled with fine-grained sediment and has been transformed underneath it. The shaping of fissure caves located predominantly along vertical cracks could be defined as having occurred subterraneously. The water widened the cracks into fissure caves and at the same time filled them with fine-grained sediment. The characteristics of the shallow (underground water is close to the surface), sediment-covered, specific type of Slovenian karst are in the process of being brought to light. These characteristics should be fully taken into consideration in future interventions in the karst. Although not visible at first sight, they are revealed by each intervention into the landscape. Many of them (caves, subsoil karren or stone forests) are worth being protected and preserved.

Karst Formations Uncovered on the Pluska–Ponikve Motorway Section 16

16.1 Subsoil Karst Formations

In the lower-lying parts of the route and the lower part of the slope surface, in places covered with thicker layers of alluvium and soil, there are subsoil stone forests (Fig. 16.1), while the steep parts of the slope and the dolomitic surface feature subsoil karren (Fig. 16.2).

Stone forests are comprised of stone columns that reach up to 8 m in height along the more prominent fissures. In view of the network of fissures along which they were formed, the stone columns are either thick-set or narrow, with typical subsoil rock formations developed on them. The traces of water gliding down along the contact between the alluvium and the rock can be seen in the funnel-shaped notches which most frequently dissect the tops of subsoil rock columns and represent the openings of upright subsoil channels. Under the soil, which partly covers the larger, mostly denuded tops, subsoil indentations and channels were formed on relatively horizontal surfaces.

The stone columns developed on gently sloping rock beds, while interbedded anastomoses formed along the bedding planes. The larger ones are paragenetically enlarged. They indicate poorer local permeability of the contact and the formation of locally saturated areas. Their tips were frequently uncovered already prior to the earthmoving work and were partially reshaped by rainwater and biocorrosion. The subsoil and uncovered tops are typically pointy (Fig. 16.3).

Near Ponikve some of the subsoil tops are relatively horizontal or, rather, only slightly rounded and formed along a bedding plane. This area is criss-crossed with a thinly scattered network of upright fissures, which is why the stone columns are large and expansive, and their peaks are dissected by a network of funnel-shaped notches (Fig. 16.4). The cracks between the stone columns are also relatively narrow, giving the stone forest its typical shape. It would seem that the pointy tops were removed while the field was cultivated, and the stones were used to form a path. During construction work, the tops along the bedding planes were also the first to be removed.

In places covered with a thinner layer of alluvium and soil, usually on steep sections of the slope as well as on the dolomitic rock, the rock surface is less prominently dissected, mostly into subsoil karren. The transverse strips of the karren tops measuring up to 10 m in diameter dissected the construction surface, while other parts were covered in alluvium and soil, therefore featuring rounded shapes (Fig. 16.5). The latter were frequently uncovered in the form of stone teeth and forests in the course of construction works. The tops of karren were mostly located in the forest and therefore bear a typical shape. The denuded

M. Knez and T. Slabe (eds.), *Cave Exploration in Slovenia*,
Cave and Karst Systems of the World, DOI 10.1007/978-3-319-21203-6_16

subsoil rock relief was covered by moss, lichen and weathered material, finely rounding and reshaping the rock in a typical manner. Channels and indentations were also formed under the weathered material. Parts of karren that were directly exposed to the rain were dissected by rain flutes. They, too, bear visible evidence of the reshaping that took place in the subsoil rock relief. The tops, in places stone teeth which are pointy and feature several tips, or rock crests reach the height of up to a few metres.

The subsoil karren on dolomite are relatively less prominently dissected. Smaller tops were formed between fissures, usually not exceeding 1–2 m in height. Larger subsoil notches measuring one to several metres developed only along the most prominent fissures. They usually include larger rock masses measuring one to several metres in diameter, which mostly don't feature distinctive subsoil stone teeth shapes—the latter can only be observed in individual spots. The fractured dolomite rock under the alluvium disintegrates more quickly, which is why the rock is only finely and less prominently dissected into karren.

16.2 Caves

As many as 22 caves were opened in this motorway section during construction (Fig. 16.6). Four of them were formed due to the flow of water streams or flood waters, while two of them were filled with alluvium. Others are shafts which developed on account of water percolating from the surface.

The large cave, measuring 18.5 m in length and 15 m in depth, reached the underground water level (Fig. 16.7). The entrance lies already outside the outer edge of the roadway in the direction of Novo mesto–Ljubljana. Located at the altitude of 283 m, the entrance measured 3 × 2.5 m at the time of our visit,

Fig. 16.1 Subsoil stone forest on the Pluska–Ponikve motorway section

Fig. 16.2 Subsoil karren on the Pluska–Ponikve motorway section

with the longer axis in the azimuth direction of 90°. The upper 2 m of the entrance shaft (depth 8.5 m) were formed in loam, while the shaft walls beneath were formed in bedrock. At the bottom of the entrance shaft, there is disintegrated rock from the surface: some of it fell 4.5 m lower into a smaller cave hall called the Zgornja dvoranica hall (point 1). The passage from the entrance shaft to the Zgornja dvoranica hall is up to 0.5 m wide (point 3). Under the entrance shaft, the hall measures 4 × 3 m, with an average height of app. 3 m. The ceiling above the hall is at least 7 m high. In the hall, it is possible to observe distinct water level fluctuation, even by 2 m when the waters are very high. Behind the slanting passage measuring 0.5 × 0.8 m, there lies the Spodnja dvoranica hall (point 2) with a little lake at the bottom. The water level is at the altitude of 268 m, around 3 m above the Temenica River valley near Trebnje. At the time of our visit, around 0.5 l of water drained out of the lake per second in the direction of azimuth 40°, into an impassable fissure. The flow's further direction is unknown. Since this little hall is 1.5 m lower than the Zgornja dvoranica hall, traces of flood water fluctuation were seen everywhere along the walls, reaching the top of the hall, i.e. 3.5 m high. We estimate that, when the water levels are high, a few litres of water can drain through the bottom of the shaft in a second. The water carries the alluvium away, depositing some

Fig. 16.3 Pointed tops of stone teeth

of it on the walls. In the Spodnja dvoranica hall, 1.3 m above the water, the fissures contain quartz sand, a sample of which we took for further research purposes.

We proposed filling the cave with coarse-grained material that would stop at the entrance shaft. At the top of this heap, a concrete plate should be leaned against the bedrock, thus ensuring uninterrupted water flow without the possibility of further terrain subsidence. Above the entrance shaft, there is thus a 1 m high concrete space measuring 2 × 3 m, connected with the northern slope of the motorway embankment by means of a 14-metre concrete tube measuring 60 cm in diameter.

In the road-cut, at the altitude of 333 m, an entrance opened into a cave that is 65 m long and 25 m deep (Fig. 16.8). It is a network of smaller passages connected by fissured shafts that are wider in their lower part. The floor of the bottom part of the cave is covered with alluvium and flowstone (Fig. 16.9). The alluvium indicates the occasional fluctuation in the underground water level or flooding in the bottom part of the cave. The cave is connected with the northern slope of the motorway embankment by means of a concrete tube (Fig. 16.10).

The old Velika jama nad Trebnjem Cave, the entrance of which is located at the altitude of 434 m, i.e. 85 m above the motorway, is a 120 m long, fairly

Fig. 16.4 Larger stone columns

horizontal passage with widened sections, the bottom of which is covered with loam. The walls and the upper part of the vault feature a circular cross-section, with only a part of it reshaped due to disintegration. The rock relief with the prevailing dome pits and large scallops uncovers the process of cave formation by means of a slow water flow beneath the underground water level.

In the course of excavating the tunnel in the middle of the hill's slope, at the altitude of 302 m, a cave opened up, measuring 5 m in diameter and completely filled with fine-grained silicate alluvium (Fig. 16.11). The surrounding surface was formed subcutaneously.

A similar cave measuring 3 m in diameter and also completely filled with alluvium opened up on the slope at the altitude of 330 m.

Shafts developed along prominent fissures and are therefore mostly creviced, some even dissected. Up to 10 m deep and up to 5 m wide, they were formed under the alluvium and soil that covered the karst surface. It seems that, initially, at the time of formation, subsoil shafts were filled with alluvium which the percolating water carried in, but were mostly emptied later on, leaving only a few of the shafts filled. During construction, the earthmoving works uncovered several crevices in the cross-section, which were

Fig. 16.5 Stone karren

completely filled with alluvium. Along the entire longitudinal cross-section of the route, shafts can be found both in limestone and dolomite. The bottom of one of the shafts is covered with water at the altitude of 305 m. This shaft is located on a slope, with water flowing into it along the bedding plane.

On average, three caves were opened up on 1 km of the motorway route section. The cave above the road, its entrance located at the altitude of 434 m, was formed by a slow water flow in the saturated zone of a karst aquifer. It was occasionally flooded in the final period of its formation, but has been dry ever since. The underground water level lowered. Here, a relatively fast dissection of the karst aquifer and of the surface took place. The lower-lying cave with its entrance located at the altitude of 302 m, which seems to have been a more or less horizontal passage that developed on account of a slow water flow, was completely filled with alluvium by flood water. Its origin is therefore quite similar. The remaining two larger caves with entrances at the altitudes of 333 and 283 m respectively, were mostly completely formed in

Fig. 16.6 Caves uncovered on the Pluska–Ponikve motorway route

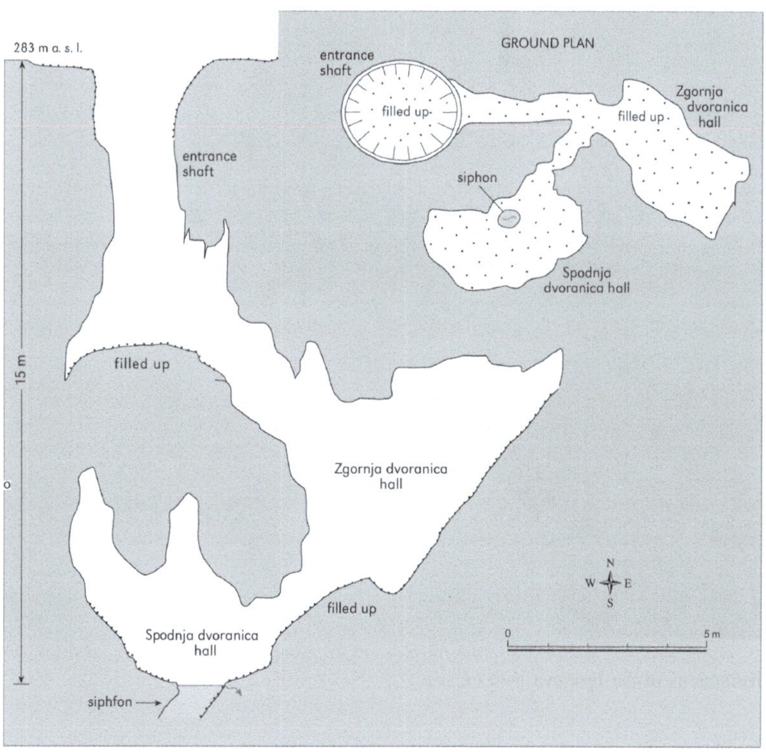

Fig. 16.7 The P292 Cave: ground plan and transverse profile

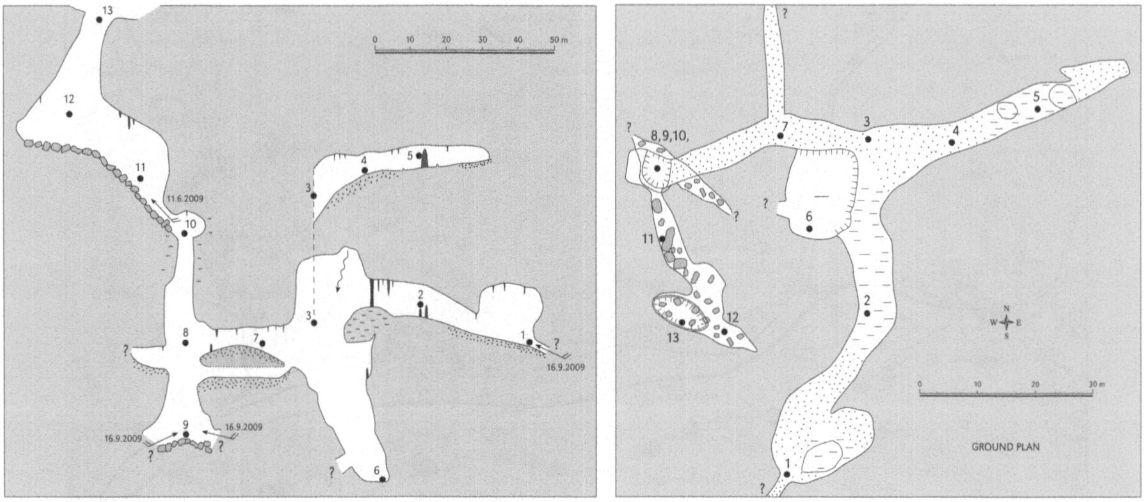

Fig. 16.8 The Igorjeva groza Cave: transverse profile and ground plan

Fig. 16.9 Flowstone formations in the Igorjeva groza Cave

Fig. 16.10 A concrete tube connects the Igorjeva groza Cave to the motorway embankment

the area where the underground water fluctuated, as evidenced by their shapes, alluvium and the underground water level in one of them. This means that they are of later origin. Smaller streams flowed through them, and were flooded when the water levels were high. In the lowest-lying cave, the underground water level is reached at the altitude of 268 m.

Shafts were formed subcutaneously. Today, most of them are hollow and have been reshaped by means of water which trickles along their circumference from the surface or flows into them from the sides.

16.3 Conclusion

The construction works uncovered the characteristics of subsoil formation of the karst area of the Dolenjska region, which is covered with more or less thick layers of fine-grained alluvium and soil. The karst surface is dissected by stone forests and stone teeth, the exception being strips of karren, i.e. stone formations on the surface, which mostly represent the tops of subsoil stone teeth and forests. They develop in an

environment abundant in forest vegetation. Under these formations, shafts develop, usually underground and creviced. Individual shafts reach the underground water level which co-shapes their lower part into a cave that is occasionally flooded when the water levels are high. Traces of older periods of co-shaping karst aquifers can be seen in caves that are completely filled with fine-grained flood alluvium, which means they are beneath the underground water level and finally in the epiphreatic zone.

17 Stone Forest Near Trebnje

17.1 Morphological and Geological Features of the Area

The Triassic dolomites of the Dolenjska lowland witnessed a development of incomplete carbonate karst; it prevails in surface drainage, although the cracked and porous rock absorbs a considerable amount of precipitation as well (Habič 1982). The infiltrated water feeds small but permanent springs. Although the surface and underground karst formations are few and far in between, they are typical of dolomitic karst, which is why the latter can be designated as a specific type of fluviokarst.

An almost normal relief with valleys and a sporadic surface hydrographic network has developed on a predominantly dolomitic base. The prevalent torrent drainage in the high areas of the Notranjska and Dolenjska regions influences the formation of steep and deep ravines cut predominantly into broken and crushed fault zones. On the dolomite of the northern Dolenjska region, a rather lower and milder relief was formed. Karstic features of dolomite are manifested in a mildly bumpy surface, characteristic dry valleys, infrequent dolines and shallow uvalas. Relatively smooth, undissected slopes of crested elevations provide evidence of underground drainage. Another obvious sign of karstification are subsidences opening into thicker strata of weathered dolomite at the bottom of hollows.

The Plio-Quaternary alluvium covers a relatively large part of the stone base in the Dolenjska lowlands. During and after the depositing stage, they were, in some places, uplifted as much as up to 450 m above the present-day Sava riverbed, especially in the Sava compressive wedge area, while the area outside the wedge was not uplifted particularly high at that time (Placer 1998a).

As the soil on Triassic dolomites is thicker and less porous than on limestone, and as it was partly impermeable to water due to seasonal freezing in the cold Pleistocene climate, it shows several signs of former tributary streams in the border limestone area (Gams 1998).

The area of Trebnje comprises Triassic and Jurassic shallow-marine carbonate rocks which are often covered by up to a few metres thick Pliocene-Quaternary sediments and soil (Pleničar and Premru 1977). In places, patches of Upper Cretaceous deep-marine marl and sandstone are preserved, lying discordantly over the Triassic and Jurassic carbonate sequences.

In the tectonic sense, the area belongs to the eastern-most part of the Hrušica thrust sheet which represents the most extensive tectonic unit of the northeastern part of Outer Dinarides. The area is also part of the post-Miocene Sava folds (Placer 1998a; Vrabec and Fodor 2006).

In terms of lithology and composition, two different types of rock can be observed on the relevant outcrop near Trebnje. The larger part of the outcrop consists of the grey, coarse-grained dolomite breccia which transitions laterally into a light-grey, massive, internally layered dolomitized limestone or dolomite dipping app. 30° towards the south. In places, the carbonate rock is slightly silicated.

Dolomite belongs to the Upper Triassic lithostratigraphic unit of the main dolomite. This unit is largely comprised of inner-platform peritidal cycles that are shallower in the upward direction, where the subtidal grey dolomitized micrites are exchanged with intratidal

M. Knez and T. Slabe (eds.), *Cave Exploration in Slovenia*,
Cave and Karst Systems of the World, DOI 10.1007/978-3-319-21203-6_17

and supratidal light-grey, locally reddish-coloured laminated and fenestral dolomitized micrites.

Breccia is internally non-homogeneous and exhibits no apparent internal stratification, while clast orientation is chaotic (Figs. 17.1 and 17.2). The clasts are poorly sorted and either touch one another or float in the breccia matrix. The size of clasts varies greatly, ranging from a few millimetres to several decimetres, but they most frequently measure a few centimetres to a little over 1 decimetre in diameter. They come in various shapes, but the slightly oblong clasts are the most frequent. Clasts are mostly coarse and poorly rounded, with only smaller grains featuring slightly less sharp edges. The clast structure reflects the lithology of the surrounding dolomites or dolomitized limestone of inner-shelf sedimentary facies.

The clasts are immersed in a medium to dark-grey, locally reddish and greenish, grainy dolomitic matrix with no visible internal textures. Up to a certain extent, the matrix resembles dolomitized micrite that forms the subtidal parts of upwardly shallower main dolomite parasequences.

The clasts frequently reveal multi-phase fissures filled with matrix and sparry calcite. Some calcite veins intersect both the clasts and the matrix, while stylolites intersect calcite veins as well.

Breccia and internally stratified dolomite are covered by up to a few metres thick clay top that underwent pedogenetic reshaping in the upper part.

17.2 The Shape of the Stone Forest and Its Columns, and Their Rock Relief

Although only 1000 m^2 of the stone forest (Fig. 17.3) have been uncovered, it is possible to infer its main characteristics in terms of form. The subsoil columns, both larger and smaller, are relatively close to one another and are positioned into strips. Their peaks are mostly at the same level.

In terms of shape, columns can be divided into two types. The larger ones are prevalent, their surface measuring several square metres in their lower

Fig. 17.1 Breccia

Fig. 17.2 Rocky surface on breccia

Fig. 17.3 Subsoil stone forest

Fig. 17.4 A stone tooth

part. They are further dissected by subsoil rock formations. They formed from larger rock masses which were not prominently and densely fissured. The tops of such columns are blade-like and have several sides; they are further dissected by funnel-shaped notches beneath which large subsoil channels can be found.

The second type of columns is represented by individual pointy and less massive columns (Fig. 17.4). Their walls are dissected by subsoil rock formations as well (Fig. 17.5).

Both types taper markedly towards the top, which is typical of subsoil rock formation by water percolating from the surface in a dispersed manner. This was also confirmed by the plaster models of the columns (Slabe 2005).

The most prominent rock formations dissecting columns are funnel-shaped notches (Fig. 17.6) and subsoil channels. The funnel-shaped notches at the top of larger columns measure 10 cm–2 m in diameter, feature a semi-circular shape and most frequently represent the openings of subsoil channels beneath them.

The smaller subsoil channels (Slabe 1999) measure up to 10 cm in diameter, while the larger ones reach 1 m or even more. They are relatively shallow on vertical

Fig. 17.5 Stone pillars with large flutes

Fig. 17.6 A funnel-shaped opening of an underground flute

surfaces, and deep only beneath the funnel-shaped notches. Between their vertical parts, there are also less steep sections that are widened into subsoil cups.

The lower parts of columns feature larger pan-like notches which were formed due to the weathering of the rock. It seems that, in some places, mostly already a few metres beneath the surface, a larger quantity of moisture is being accumulated and preserved for a longer period of time, resulting in a more pronounced corrosion.

Before the columns were artificially denuded, only their tops protruded from the soil covering the rock, reaching 0.5–1 m in height. Solution pans formed on the largest ones. Flutes, i.e. smaller rock formations, don't usually develop on this type of rock.

The rock surface can therefore be divided into two types, the first having developed above ground, and the other beneath it. The upper part of the first type is dissected, with slowly dissolving parts of rock (Fig. 17.7) protruding from the surface; it would seem this is due to the faster washing away of the solution. Several centimetres of the rock's bottom surface is weathered, soft and disintegrating rapidly. The surface of subsoil notches underwent the most intense weathering. The rock surface that was above ground

Fig. 17.7 Surface of a subcutaneously corroded stone pillar

has been washed away but shows "dolomitic" dissection typical of this type of rock. It is criss-crossed by a network of indentations that follows the calcite veins. It is similar with the rims of solution pans. A smoother surface formed only beneath the mosses.

17.3 Conclusion

Despite the spatially restricted outcrop and deficient information from the surrounding areas, the described characteristics are distinct enough as to enable defining breccia as a shallow-marine sinsedimentary deposit formed by the gravity-driven mass flow and/or collapses along faults or smaller tectonic ditches in the period of changing geotectonic and palaeogeographic conditions in the area of the previously relatively monotonous Upper Triassic or Lower Jurassic shallow-marine carbonate platform.

Since no guide fossils were found in clasts and the breccia matrix, we could only infer its stratigraphic position based on the stratigraphic superposition of breccia and the lithological characteristics of clasts and the breccia matrix. In view of these criteria, we believe that breccia was formed in the Upper Triassic or Lower Jurassic period.

Even though the rock is heterogeneous, this is not reflected in the shape of the columns of the stone forest. They are of regular shapes and taper evenly towards the top. The rock affects the rock relief of stone columns. Only larger rock formations develop on their surface, along with funnel-shaped notches, subsoil channels and, on the surface, solution pans. Flutes can't develop on this type of rock.

Subsoil stone forests represent a significant karst feature that reveals the subsoil formation of carbonate rocks and the development of the covered karst areas. What is more, they are a uniquely appealing sight for people interested in karst phenomena. Such a discovery should therefore be taken into consideration and adequately protected in further planning of interventions on the karst surface. These features may namely also occur elsewhere where the above-referenced conditions are met, i.e. compact rock criss-crossed with a network of vertical fissures and a thick top of alluvium or soil above it. An extensive part of the karst in southeastern Slovenia is of precisely such a type.

18 Karst Formations Uncovered on the Ponikve–Hrastje Motorway Section

18.1 Karst Surface

The karst surface in this area is most distinctly characterized by its subsoil formation. The lower parts of the surface, dolines and the slope notches in particular were covered with relatively thick layers of alluvium and soil, which is why the rock beneath these features displays typical signs of subsoil formation. The steeper slopes and rock ribs were covered in thinner layers of alluvium, and were dissected by strips of karren, which represented the tops of more or less uncovered subsoil teeth (Fig. 18.1). Their rock relief revealed subsoil formation followed by a typical re-shaping underneath dense forest vegetation (moss and lichen, criss-crossed with roots and accelerated corrosion beneath the weathered material).

The earthmoving works have uncovered the typical subsoil formation of the karst surface on various types of rock. In particular, it is possible to discern the subsoil karst surface developed on a limestone bed that is only slightly inclined and has no cracks as opposed to the karst surface formed in comparable conditions and on similarly structured rock beds which, however, feature a more or less dense network of fissures.

The first type of subsoil surface, which develops relatively rarely and the likes of which has not been presented in expert literature thus far, is formed on a slightly inclined and rather flat upper plane of a non-cracked rock bed. Here and there, up to 10-cm high peaks rise from it, located among the subsoil indentations which prevail (Fig. 18.2). Between the large, well-developed and densely scattered indentations, the walls developed into a network of decimetre bumps. Their walls are dissected by horseshoe-shaped shelves, i.e. the remains of a subsoil indentation level. Upon simultaneous flowing of water down the rock, these deepened the fastest in their central part, while the edges remained hanging on several levels. The walls of some bumps are dissected by centimetre-long subsoil rain flutes, down which the water flows towards subsoil indentations and channels. The subsoil indentations either feature circular cross-sections measuring one to several decimetres, or are heavily dissected, consisting of several indentations combined into one, with diameters amounting to 1 m or even more. They are frequently connected in the direction of the rock surface slope. Larger subsoil channels, which feature semi-circular cross-sections and diameters measuring one to several decimetres, also follow the rock surface slope direction. Their bottom is usually dissected by subsoil indentations. This is actually a composite rock formation combining subsoil channels and connected subsoil indentations. Sometimes, the shape of a channel with indentations is prevalent, while other times, the formation features indentations connected so as to form a channel. The channels are prevalent on the somewhat steeper rocks, while indentations are more frequent on more gentle slopes. The rock relief undoubtedly bears traces of relatively even water percolation through alluvium and soil that

M. Knez and T. Slabe (eds.), *Cave Exploration in Slovenia*,
Cave and Karst Systems of the World, DOI 10.1007/978-3-319-21203-6_18

Fig. 18.1 A slope dissected by strips of karren

covered the rock, and of its flowing away down said rock.

Under the thicker layers of alluvium and soil, subsoil stone teeth and stone forests were formed on cracked rock (Fig. 18.3). The most extensive and prominent continuous surfaces of stone forests can be found under the thickest layers of alluvium, which prevail in notches and the lower parts of slopes. Stone columns reach the height of up to 8 m, but most of them are lower, i.e. stone teeth. The shape of stone forests and stone columns as well as the extensiveness of the latter are directed by the network of fissures that criss-crosses the rock in the vertical direction. A dense network leads to the formation of narrow and pointy stone columns, while a thinly scattered network would result in the formation of more extensive and dissected columns. They can develop both on limestone and dolomite. When formed on dolomite, they are often rough and dissected due to the structure and fine cracks in the rock. The pointy tops are another typical feature of the subsoil rock formation when water percolates evenly on it through a fine-grained top. The rock relief of more massive stone columns reveals the method of their formation in greater detail. The vertical gliding of water is attested by subsoil channels that reach 1 m in diameter. Usually relatively shallow, they become deep only along fissures. Where the fissures are more prominent, small underground shafts may develop. In the upper part, they usually begin with funnel-shaped notches which often dissect the

Fig. 18.2 Subsoil stone teeth

tops into several peaks and crests. The structure of larger subsoil funnel-shaped notches can be composite, their walls frequently dissected by smaller notches. The notches are also dissected by crests, as channels can develop on both sides. On larger tops, they can be connected by slightly inclined subsoil channels.

Along the gently sloping bedding planes, anastomosis networks featuring channels measuring a few centimetres developed on the lower planes of beds. They bear evidence of smaller quantities of water percolating in locally saturated zones which are formed under the alluvium top with different flow-through. The water that flows into the channels deposits alluvium, causing the channels to develop upwards. Individual flutes that developed where the rock was most porous are of similar origin.

Smaller, up to 3 m high stone teeth covered in thinner layers of alluvium and soil that predominantly fills the fissures between them, extend over greater surfaces.

18.2 Caves

As many as 9 caves were opened in the course of constructing the motorway; one of them is a passage that is completely filled with alluvium, while 8 of them are shafts (Fig. 18.4).

Fig. 18.3 Subsoil stone teeth within a thicker layer of soil and alluvium

In the route cross-section a passage was uncovered, measuring 5 m in diameter and completely filled with fine-grained alluvium and quartz sand. It contains preserved traces of flood waters which brought alluvium to the cave. Today, the passage is located on a slope, several tens of metres above the underground water level.

The rest of the caverns are shafts (Fig. 18.5). The largest shaft is 30 m deep and 55 m long. It is a system of vertical shafts and larger fissured areas (measuring up to 8 m in length) located between them. Ten metres below the surface (the entrance lies at the altitude of 313 m), the walls show traces of the levels of water that stood under the fine-grained alluvium that filled the shaft.

Shafts mostly developed under fissures that dissect the (subsoil) karst surface in the shape of a letter V of varying width, and are filled with fine-grained alluvium and soil. The fissured shafts are usually up to 5 m deep and 4–5 m wide. They were mostly formed as underground shafts, which means they were filled with alluvium at least in the initial development

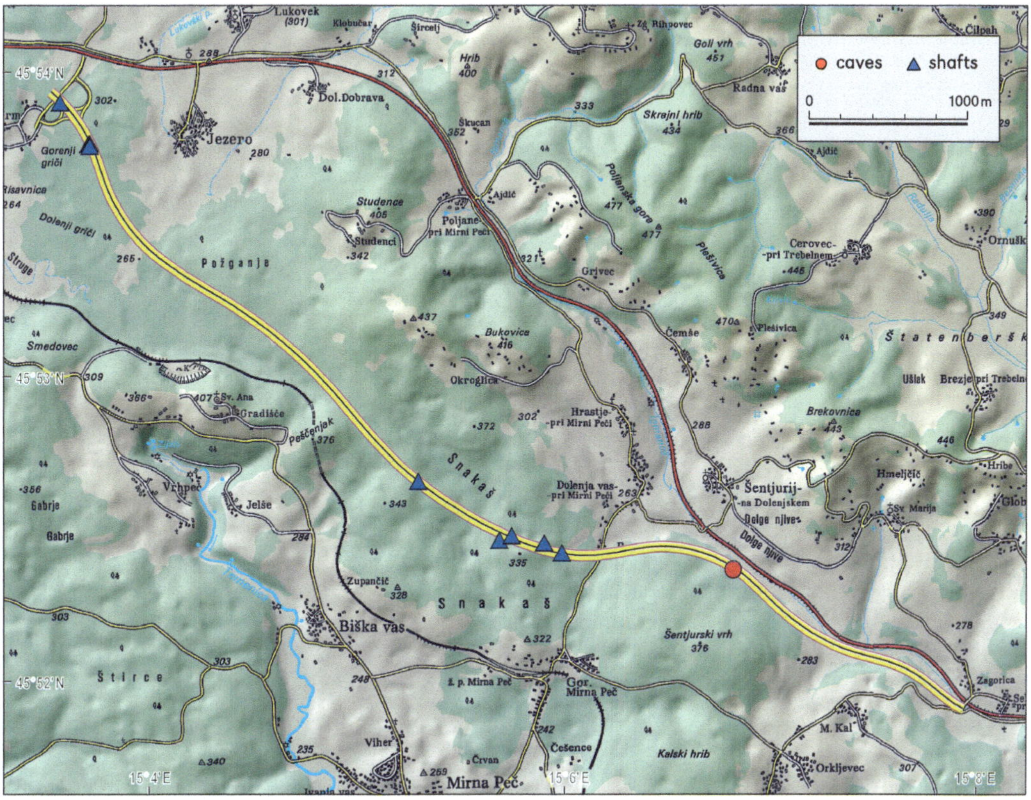

Fig. 18.4 Caves uncovered on the Ponikve–Hrastje motorway route

periods. The water percolating from the covered karst surface and dissolving the most porous parts of the rock into shafts carries fine-grained fragments and deposits them in the cavern. The cavern can be either completely or partially filled with alluvium (Fig. 18.6). Parts or entire amounts of alluvium were carried from some shafts by water. When only the lower parts of shafts are filled with alluvium, the water stagnates above the poorly permeable alluvium when the water flows are greater, partly flooding the cavern.

18.3 Conclusion

The construction works uncovered the characteristics of the subsoil formation of the karst area of the Dolenjska region, which is covered with more or less thick layers of fine-grained alluvium and soil. The karst surface is dissected by stone forests and stone teeth, featuring a typical rock relief and the uniquely formed surface of the slightly inclined and

Fig. 18.5 The entrance to the shaft on the road route

Fig. 18.6 Fissured shafts filled with fine-grained alluvium

non-cracked rock bed that is described here for the very first time. Strips of karren, which mostly represent the tops of subsoil stone teeth and forests, are denuded. They develop in an environment abundant in forest vegetation. Underneath these formations, shafts develop, usually subsoil and fissured. Certain shafts are occasionally partially flooded after a copious amount of water flowed onto the alluvium that fills its lower parts. The cave that is completely filled with fine-grained alluvium bears evidence of aquifer development from the time when this part of it was still beneath or at the underground water level.

19 Subsoil Stone Forests and Other Karst Formations Between Hrastje and Lešnica

19.1 Morphological and Geological Characteristics of the Region

The northwestern part of the motorway section starts in the vicinity of the ponor of the Igmanica Stream near the villages of Hrastje, Dolenja vas and Šentjurij, and runs past the village Selo on a typical Dolenjska lowland. The landscape there is covered with laterally transforming and relatively thick layers of alluvia and soil. The soil water is located near the surface, with fluviokarst prevailing. Here we can find individual karst formations, including smaller sink caves, estavelles, and to a small extent also the outcrops of carbonate rocks. The terrain towards the hills Strmec and Dobrava is slightly sloped and the road there has a steeper ascent, and a short distance after the pass the terrain lowers towards the Krka River. Along the pass, we noticed some thinner layers of sediments and several outcrops of a highly karstified rock which is mostly crumbled into tiny fragments. Due to the incline of the surface, thinner layers of alluvia, and open soil top, there is not much surface water since it runs into the underground relatively quickly. The stone forests and karren, which reach into various depths, are made of both compact and tectonically crushed rock. In places where the rock was not crushed or where fissures were predominantly filled with limestone blocks, the surface was noticeably abound with stone teeth (Fig. 19.1). The excavations revealed that many of these stone teeth were actually stone pillars. On the other hand, in places where the rock was tectonically crumbled or crushed, there was no evidence of karren on the surface; at some locations, there were some early formations of stone teeth which, however, instantly crumbled during earthworks. Between the Brezovica hill and Lešnica, the typical Dolenjska karst area reappears, with soil water near the surface and characteristic subsidences and hollows. On the surface, a thinly scattered river network is forming; on the other hand, a considerable portion of drainage-basin ravines and lateral ravines tend to have no permanent flow. Smaller sources of the underground waters flowing over the rocks with moderately widened corrosion-provoked fissures are a frequent occurrence. Surface and subsoil karst formations are rare. Nearby valleys are dry for the greater part of the year, with streams and floods appearing after heavy rainfalls. Infiltrated water feeds small but permanent springs. Flow oscillation of the springs is small and only occasionally more abundant, which is an indication of greater permeability and karstification. Underground water flows close to the surface; the phenomenon with such characteristics is known as "shallow karst".

Thick layers of Plio-Quaternary sediments on carbonate rocks, especially on a wet terrain, tend to be acidic. It is not entirely clear whether these sediments originated here or were brought from the nearby dolomite areas (Gams 2004). After several comparisons with the conditions in the tropical karst, the general opinion was that a thick layer of weathered debris was preserved as a result of the high level of

M. Knez and T. Slabe (eds.), *Cave Exploration in Slovenia*,
Cave and Karst Systems of the World, DOI 10.1007/978-3-319-21203-6_19

Fig. 19.1 A subsoil stone forest

underground water and poor erosion; underneath it, subsoil karst has developed.

The motorway route Hrastje–Lešnica runs, with a few exceptions, on Upper Jurassic (Lower Malm) rocks. In some instances, it crosses the Plio-Quaternary alluvium and the alluvial river sediments. From the geotectonic point of view, the area falls under Outer Dinarides which are known for their block fragmentation and the Dinaric direction of the faults and the fold axes. A covered fault runs along the western part of the road section. The anticlinal fold of the Upper Jurassic strata runs in the Dinaric direction from Dolenja Nemška vas in the northwest to Novo mesto in the southeast. The road runs mostly on its northern side. The dip direction is variable; in general, however, the strata tend to dip towards the northeast. Along the route, the rock shows various forms of micro- and macrofauna, and in the middle section of the route also forms of macroflora.

In the north, the Upper Jurassic strata border the Upper Cretaceous brown and green marl, sandy marl, marl limestone, and grey and reddish platy limestone with interbeds of limestone breccia. In the south, they border the Plio-Quaternary brown loam weathered debris. North of Prečna there is a smaller area of Upper Triassic grey dolomites, both layered and massive.

According to the geotectonic map BGM (Pleničar and Premru 1977), the area of the motorway section lies on the so-called Novo mesto horst. Its northern part is a transition between the Sava folds and the Dolenjska-Notranjska blocks. The oldest rocks here are Middle and Upper Triassic dolomites with Jurassic limestone discordantly deposited over them and Upper Cretaceous pelagic rocks as the upper stratum. The terrain shows characteristic Dinaric-directed synclines and anticlines which, in some places, divert from their original direction.

The Jurassic strata in the vicinity of Novo mesto are generally made up mostly of light grey limestone which lays on the Cordevolian Upper Triassic dolomite and is the substratum for the Upper Cretaceous pelagic sediments discordantly deposited over them (Pleničar and Premru 1977).

The Lower Malm rocks that run along the motorway section north of Novo mesto (Pleničar et al. 1976) are highly diverse. The so-called northern and southern development is an alternation of white to grey limestone, oolitic limestone, reef limestone with hydrozoa, and bedded limestone with cherts.

In a broader sense, the northern evolution can be traced between Poljane and Mali Slatnik east of Novo mesto (Pleničar and Premru 1977), i.e. mostly along the northeastern parts of the motorway section. Prevalent here are the bright grey, massive reef limestone and the coarse-grained reef breccia. Here and there among these strata there are also traces of the dark grey and nearly massive limestone. The massive limestone contains a rich hydrozoan fauna. At some locations, platy limestone with cherts can be found as well.

Along the southwestern sections we can follow what is probably the middle development, which is lithologically similar to the northern development; the hydrozoa in the rock are less common. In between the reef limestone, oolitic limestone can also be found (Pleničar and Premru 1977). Corals can also be found at certain locations. According to the BGM

Fig. 19.2 A biocorroded surface of karren

information and the information provided by the expositor, the rocks of the southern development, which are characteristic of grey, thick and oolitic bedded limestone (Pleničar et al. 1976; Pleničar and Premru 1977), cannot be found along the route.

In the Plio-Quaternary alluvia southwest of Mačkovec there are traces of bentonite clays. The clay is deposited in pockets of Triassic dolomites and Jurassic limestone. At some places, the clay is deposited in strata up to 12 m thick.

19.2 Karren Surface

The majority of the higher-area surfaces—the valleys are covered with layers of alluvia—are intersected by karren. They can be categorized into two distinct types. In areas that are not covered with alluvia, but rather partly with soil, karren can be found in the true sense of the word. They cover the majority of the surface. The carbonate rock is intersected along the cracks and its surface shows traces of its past subsoil formation

Fig. 19.3 Large surfaces of the subsoil stone forests

(Slabe 1999; Slabe and Knez 2004), which are relatively small in number, then traces of indirect transformation due to precipitation, and traces of fine intersection due to biocorrosion processes. This is because the surface was overgrown mostly with forest. The karren of this type were found on the surface of a heap in the valley at the initial section of the motorway.

In particular in the area of the hills Strmec and Dobrava, highly intensive biocorrosion processes were discovered and observed on numerous karren outcrops of carbonate rocks. Signs of intensive biocorrosion melting of the rock were evident in areas with more shade. These areas are predominantly covered with moss, whereas lichen are more common in more sunny areas. The biocorrosion processes do not take place evenly across the entire surface, but rather selectively (Fig. 19.2). The clasts (which are probably lithologically slightly different) in the crushed and recemented rock are melted to different depths or there are various organisms on various neighbouring clasts. Especially melted due to corrosion are the contact points between individual clasts, at certain spots even several millimetres deep.

A greater part of the surface overgrown with forest was intersected mostly by individual rocks of various sizes that showed traces of formation which is, at least

Fig. 19.4 A doline in the subsoil stone forest

in part, similar to that of the karren described above. Often, they are segmented with funnel-shaped openings of larger vertical subsoil flutes. They reached 1–2 m high and were narrower towards the top, with larger areas of soil in between.

The earthworks revealed them as the tops of a larger area of stone forests.

19.3 Subsoil Stone Forests

The relatively large areas of subsoil stone forests (Fig. 19.3) bear witness to the long-lasting underground formation of this part of the karst surface which is covered with fine-grained deposits and soil, and how this formation took place. The stone pillars are entirely covered with alluvia and soil or only their tops stick out.

The surface is intersected by smaller and bigger dolines (Fig. 19.4). The biggest of them are several tens of metres in diameter. Some are filled with grey loam, the origin of which is still being studied. On the other hand, both types of dolines are intersected by subsoil stone forests.

The subsoil stone forests are made of a thick network of more or less stout and pointy pillars (Fig. 19.5) reaching 8 m, some seldom even 10 m high, while the majority tend to be lower. Narrower pillars with 1–2 m in diameter have one sharp or rounded tip at the top, whereas the stout pillars with up to 5 m or more in diameter have the tops shaped more or less like sharpened and winding ridges. Among them, there are predominantly funnel-shaped openings of vertical subsoil flutes or more or less horizontal subsoil channels.

Fig. 19.5 A subsoil stone pillar with subsoil rock forms

Prevalent in the rocky relief of the pillars are subsoil rock forms (Slabe 1999; Slabe and Knez 2004) which demonstrate the gravity flow of the water from the surface. These are mostly subsoil flutes (Fig. 19.6), most commonly vertical channels 1 m or more in diameter; as determined below, the largest among these can be called underground shafts which broaden into funnel-shaped openings at the top. They accumulate water from the surface which then percolates down the rocks and flows from the soil. A cross-section of the funnel-shaped openings reveals that they come in different shapes. They can be open, semi-circular, or even nearly circular. Their shape is often the result of permeability at the contact point between the rock and the alluvium across which the water flows down. A long-lasting shaping of the opening, coupled with a poorly permeable contact point, makes the rock form dip deeper into the rock. On some rare occasions, the water managed to penetrate through the rock which led to the formation of actual funnels. If the contact point is less permeable or when a smaller amount of water is gathering on the surface—which is, of course, common for narrower pillars whose tips protrude from the soil—smaller and winding flutes tend to form. Subsoil flutes, which generally demonstrate that the contact point between the rock and the alluvium around it is permeable, are a rare occurrence; moreover, the walls of the pillars can have elongated indentations which are an indication of waterlogging if the contact point is partly less permeable, and of an accelerated notching of the rock next to it.

The rocky area of the stone pillars just below the surface, where the rock is covered with soil, is relatively smooth; however, deeper down where it comes in contact with alluvia which cover the surface, it is rough, coarse, and intersected by smaller and rounded protrusions. This is because the rock is weathered. The weathered layer of the soil is up to 1 cm thick. When wet, it is soft; when dry, however, after being exposed on the surface for a longer period of time and the water

Fig. 19.6 A subsoil flute

evaporates, it solidifies. The weathering of the rock's upper layer is the result of the contact with the alluvium which is wet most of the time. The contact point is relatively impermeable, meaning that the water, which percolates down the contact point, only slowly washes the melting away. The contact point with more porous soil is more permeable.

19.4 Karst Cavities

The cavities that have been discovered are mostly the result of vertical penetration of the water through the epikarst and vadose part of the aquifer. The shafts can be categorized into hollow and alluvium-filled. The latter bear the name "underground" due to the formation which is similar to that of underground rock forms, i.e. the vertical underground flutes.

Fifteen of the hollow shafts (Figs. 19.7 and 19.8) are vertical and mostly plain, and only one was distinctly gradient. The deepest shaft was 24 m deep, and another three were over 10 m deep, while others were not as deep. Their diameter reached up to 5 m, but mostly less. They indicate permeability that was, in some places, greater, which makes the water percolate in a continuously vertical fashion due to the vertical fracturedness or vertical rock stratification. The shafts are located between subsoil karren and forests. They do not reach the surface. Their walls are intersected by both bigger and smaller vertical flutes, and are often cowered with a thin layer of alluvium which causes their fine dissectedness (Slabe and Knez 2004). The

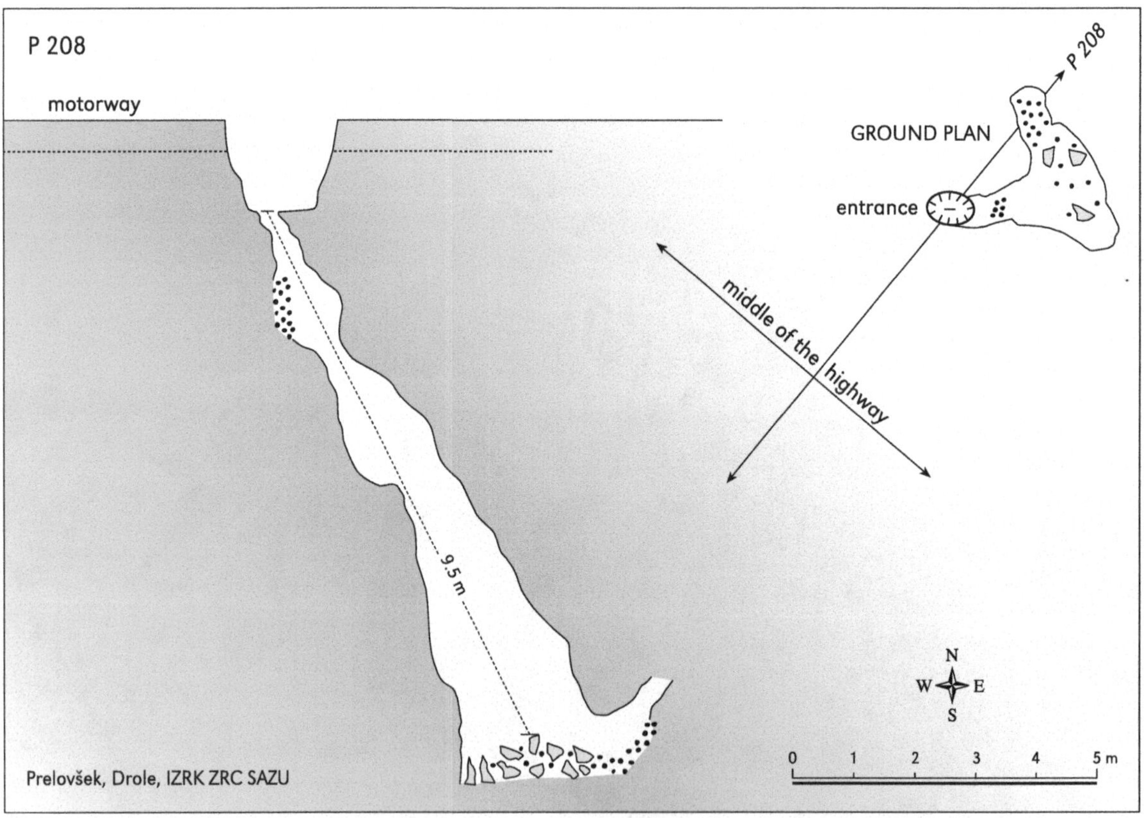

Fig. 19.7 One of the hollow shafts in the motorway route

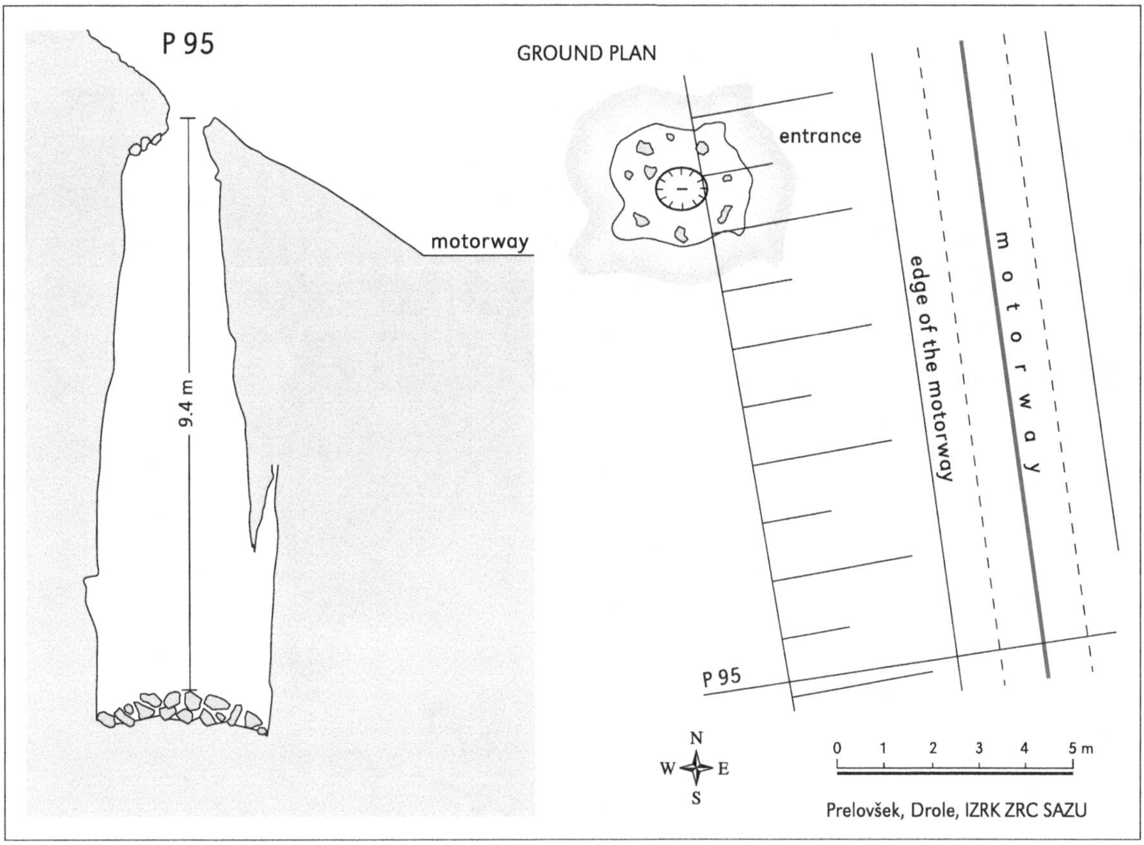

Fig. 19.8 A shaft uncovered during motorway construction

shaft floors are often covered with alluvium or the alluvium fills their lower section.

Underground shafts (Fig. 19.9) are, for the most part, vertical cavities similar to plain shafts, over which also percolates the water from the karst surface; however, they are almost entirely filled with alluvia, with only a few vertical sections that may be hollow. The water flowing through them brings in and deposits alluvium which covers the surface. Their cross-sections are more or less circular or elongated along the cracks and bedding planes. Their diameter reaches up to 2 m. Because they are filled with alluvia, they can be intersected around the rim, whereas less permeable contact points can cause the formation of notches. Underground shafts are formed along a localized continuous flow of larger quantities of water. They can develop from subsoil flutes. Their walls have rock forms which were formed due to the contact with a fine-grained alluvium. During higher conductivity within the karst interior the underground shafts can empty out.

Alongside horizontal bedding planes, there are often supra-alluvial flutes or networks of anastomoses which are the result of paragenetic stratification. Thus, localized areas develop which are occasionally flooded, and the water carrying and depositing the fine-grained alluvium cuts upward.

Fig. 19.9 An underground shaft filled with fine-grained alluvium

19.5 Conclusion

The peculiar natural heritage and the evolution of the covered Dolenjska karst area are brought to light in an increasing fashion.

On this occasion, we were provided with the opportunity to follow the percolating water into the epikarst and the upper section of the vadose zone; both were formed under relatively thick layers of alluvia and soil, where over large areas subsoil stone forests, shafts, and underground shafts can form.

The uniqueness of stone forests, these special geomorphological karst forms which are common especially for this part of the karst (Knez et al. 2003), indicates the starting points for the further planning of human intervention into the karst relief.

20 Palaeomagnetic Results from the Filled Karst Depression on the Motorway Section Hrastje–Lešnica

A sediment profile was taken in filled depression at Profile No. 207 (45° 50′ 46.02″N; 15° 08′ 59.01″E; 360 m a.s.l.) of highway construction in the section Hrastje–Lešnica, on the pass between Karteljevo and Novo mesto in the Dolenjska region (Fig. 20.1). The depression was filled with greyish laminated to thin bedded clayey deposits (Fig. 20.2), rather uncommon in the area, where reddish-brown homogenous clayey material usually covers karstified Mesozoic carbonates.

The carbonates in the area belong to the Upper Jurassic white to grey limestone, oolitic limestone and thin-bedded limestone with chert, with stratal dip to the east/northeast (Pleničar et al. 1976). Karstified limestone is covered by several metres of grey, yellowish-brown to red laminated sediments.

20.1 Profile

20.1.1 Lithology

The profile of laminated sediments 383 cm thick (Fig. 20.3) was composed mostly of clays and silty clays with interbeds and bands to laminas of clayey-sandy and clayey silts. The colour of the sediments was dominantly grey, sometimes brown and beige mottled and with yellowish brown lamination (Fig. 20.4). Finds of gastropod shells, plant remains and plant roots were concentrated in the lower half of the profile (Fig. 20.5). Structures of protopalaeosols (mottled zone with plant roots and iron rich geodes—ferruginized plant roots) occurred in the same section.

20.1.2 Palaeomagnetic Results

A total of 34 samples from Hrastje profile were studied for their palaeomagnetic properties (Table 20.1). They are characterized by NRM intensities between 3 and 27 mA. m^{-1} and MS values from 101 to 494 × 10^{-6} SI units. Based on both sets of values, the profile may be divided into two parts and categories. Samples are characterized by low up to intermediate J_n and k_n magnetic values. Only one sample with R polarization is characterized by a high NRM (1973 mA. m^{-1}) and MS (5060 × 10^{-6} SI units).

Three components were isolated after AF demagnetization. The A-component is undoubtedly of viscous origin and can be demagnetized in the AF (0–5 up to 10 mT). The characteristic *C-HFC* is stable. It can be demagnetized or isolated in the AF (ca 15–80 up to 100 mT). The stereographic projection of the C-component with N polarity is shown in Fig. 20.6. Table 20.2 summarizes results of the mean direction of samples from this profile.

The mean palaeomagnetic directions of C-components for the N polarity are D = 7°, I = 49°. Palaeomagnetic directions for one sample of R polarity are D = 212°, I = −40°, but statistical parameters of α_{95} and k can be calculated for one sample only. The systematic acquisition of palaeomagnetic data within the studied section allowed the construction of a detailed magnetostratigraphic profile (Fig. 20.7).

M. Knez and T. Slabe (eds.), *Cave Exploration in Slovenia*,
Cave and Karst Systems of the World, DOI 10.1007/978-3-319-21203-6_20

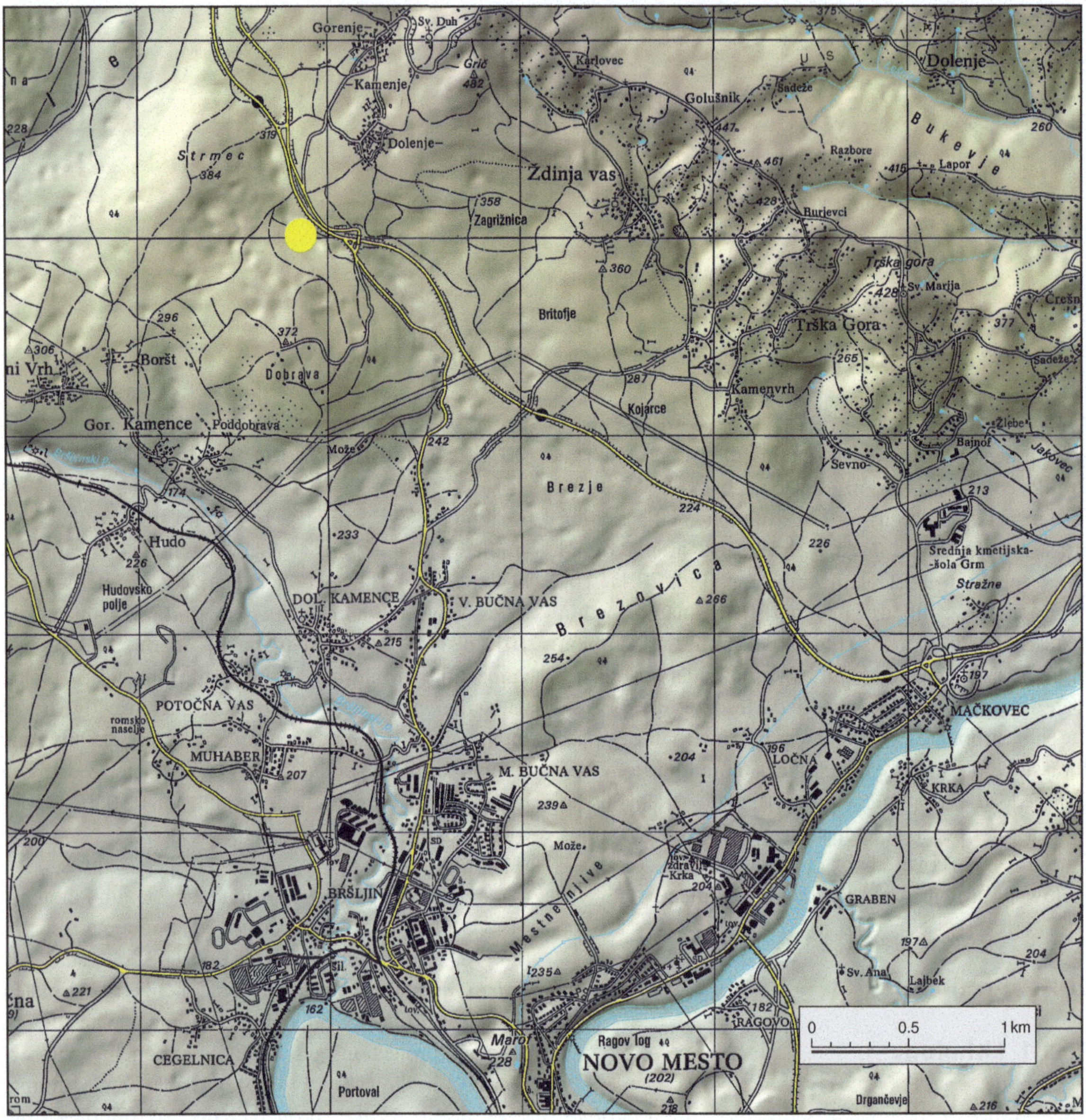

Fig. 20.1 Hrastje (SE Slovenia), site location

Fig. 20.2 Construction of the highway north of Novo mesto has an exposed profile of laminated grey sediments, which were covered by red soil. The profile lies behind the drilling machine in the middle of the photo

Fig. 20.3 Grey laminated sediments of the Hrastje profile

Fig. 20.4 The upper part of the Hrastje profile

Fig. 20.5 The lower part of the Hrastje profile

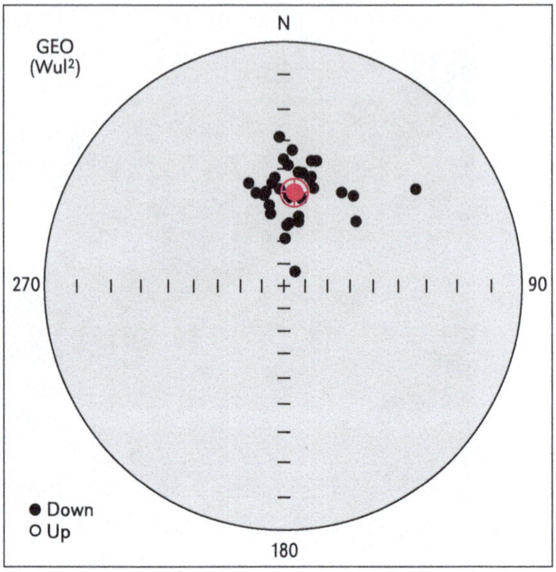

Fig. 20.6 Directions of C-components of remanence of samples with N polarity, Hrastje profile. Stereographic projection, open (*full*) *small circles* represent projection onto the lower (*upper*) hemisphere. The mean direction calculated according to Fisher (1953) is marked by a *crossed circle*, the confidence circle at the 95 % probability level is circumscribed about the mean direction

Table 20.1 Mean palaeomagnetic values and standard deviations, Hrastje

Hrastje	J_n (mA. m^{-1})	$k_n \times 10^{-6}$ (SI)	Interval (m)[a]
Mean value	7745	183.5	0–3707
Standard deviation	5005	77.2	
Number of samples	33	33	
Mean value	1972.87	5060.0	3827
Standard deviation	–	–	
Number of samples	1	1	

[a]From top to base

Table 20.2 Mean palaeomagnetic directions, Hrastje

	Polarity	Mean palaeomagnetic directions		α_{95} (°)	k	n
		D (°)	I (°)			
Hrastje	N	6.93	48.9	5.11	23.46	32
	R	212.4	−39.7	–	–	1

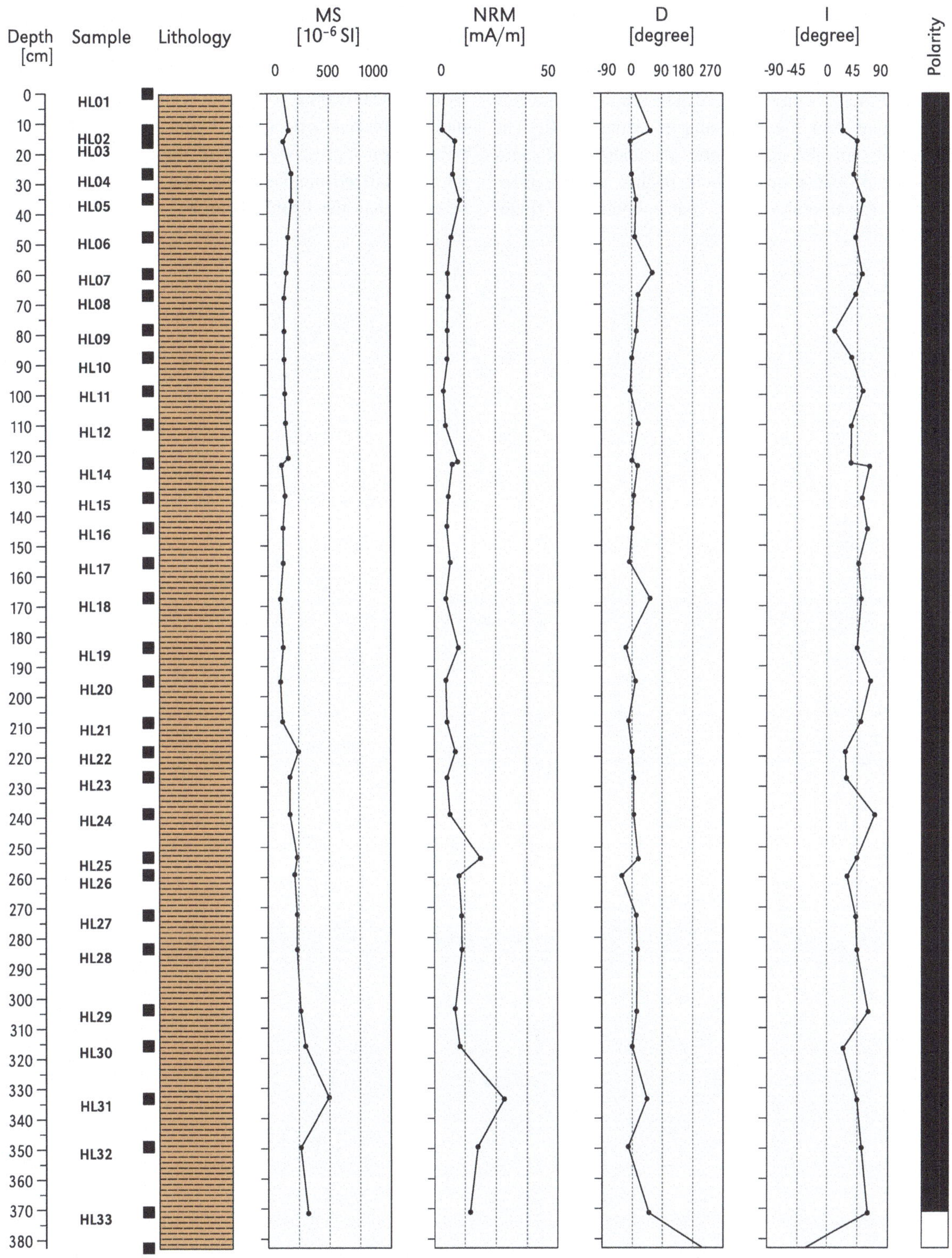

Fig. 20.7 Basic magnetic and palaeomagnetic properties, Hrastje profile. Legend: *MS* magnetic susceptibility, *NMR* natural remanent magnetization, *D* declination, *I* inclination, polarity: *black* normal, *white* reverse

20.2 Conclusion

The laminated clays and silts were deposited in a very calm sedimentary environment. The whole profile is N polarized except the last sample, which is R. The uniformity of palaeomagnetic parameters indicates quite rapid deposition. Without results of palaeozoological (gastropods) and palaeobotanical (plant remains) determinations, there can be three possible interpretations of the age: (1) the deposition took place within the Brunhes chron (<780 ka), or at the Brunhes/Matuyama boundary (780 ka), which then must be placed between 3.71 and 3.8 ma, (2) the R polarization represents some of the excursions within the N polarized magnetozone, and (3) the profile can be older than the Brunhes chron.

21 The Karst Between Lešnica and Kronovo, Revealed During the Motorway Construction

We have been monitoring the construction of the motorway on the low and shallow karst of Dolenjska. The majority of the surface of the motorway route was covered with thicker layers of fine-grained alluvia. The formation of the karst underneath is what gives the surface its uniqueness. The better part of the surface is intersected into subsoil stone forests. The stone forests formed on various types of rock. The cavernosity of the exposed part of the epikarst—the road cut up to 5 m deep—is mostly tied to the formation underneath the alluvia. Larger cavities are, in fact, underground. At places, they are causing the stone pillars to hollow out, while along the crevices between them underground shafts have formed.

21.1 Newly Discovered Karst Phenomena

The earthworks during motorway construction revealed a karst surface and a part of the epikarst. Both were most distinctly characterized by their underground formation, i.e. the development underneath the thicker layers of alluvia and soil. The water percolating through them intersected the carbonate rocks, limestone, and dolomites into subsoil stone forests and carved out underground cavities.

21.1.1 Subsoil Stone Forests

They expand across the entire surface of the route (Figs. 21.1 and 21.2). Apart from time, their shape and size mostly depend on the characteristics of the rock. The rock is, in fact, fractured with a more or less thick network of vertical cracks, alongside which the subsoil crevices were carved out by the water. Between them, stone pillars of various sizes have formed. The biggest among these are 10 m wide at the base and reach up to 8 m high. Most are narrower, though. Their long-lasting formation and development are evidenced by their shape which becomes narrower towards the top. Many of them have their tops either pointy or blade-shaped.

Thickly fissured limestone intersects into the stone forest as well. Their shape is preserved under the surface. In fact, distinct crevices have formed alongside numerous fissures which criss-cross the rock in various directions. The rock next to them is crumbling. The rock is intersected by subsoil rock forms, i.e. flutes, tubes, notches, and niches (Fig. 21.3).

The stone pillars are intersected by subsoil rock forms. The most distinct are the more or less vertical subsoil flutes on the walls of the pillars. For the most part, they do not exceed one metre in diameter. Only a handful of pillars cutting deep into the rock and having a big funnel-shaped notch at the top tend to be bigger. Some of the larger stone pillars thus have semi-circular tops, some even have several tips, which are actually the walls of the openings of funnel-shaped notches. Smaller flutes can be winding. The water runs down the flutes at the contact between the rock and the alluvium surrounding the pillars. The second large group of subsoil rock forms, which can be found on lower sections of the stone pillars, are subsoil notches. These are semispherical notches with a circular top and open downwards. Their diameter reaches up to 1.5 m. They are mostly shallow, individual notches which formed along bedding planes or fissures, and can cut more than 1 m into the rock; some of them can

M. Knez and T. Slabe (eds.), *Cave Exploration in Slovenia*,
Cave and Karst Systems of the World, DOI 10.1007/978-3-319-21203-6_21

Fig. 21.1 Uncovering of a subsoil stone forest

Fig. 21.2 Subcutaneous stone teeth

Fig. 21.3 Vertical subsoil channels on the walls of stone columns and of the shaft

even evolve into cavities. The notches form along the contact points of the rock and the alluvium that are, in parts, less permeable, or at spots where the rock is weaker. The water that is slipping down the walls of the pillars tends to stagnate, making the rock corrosion even more intensive. The surface underneath some of the larger pillars is eroded, and the bottom sections of these pillars are narrower than their upper halves.

The rock is covered with a layer of weathered rock 1–3 cm thick. The contact between the rock and the alluvium is, in fact, very narrow, so the water does not wash away the melting entirely. The weathered rock surface tends to consolidate during dry periods and remains soft when moist.

We tried to analyse such subsoil formation of the rock by conducting laboratory tests on plaster.

The block of plaster (Figs. 21.4 and 21.5) was cut into small pillars 30 cm high with a square cross-section and the length of its side of 6 cm. They were placed closely next to one another, put into a big bucket, and covered with soil. Then small holes were drilled through the bucket's bottom and the bucket was then filled with water. The water started to percolate through the soil and flow out through the small holes in the bottom of the bucket.

The upper two thirds of the pillars' wall were corroded with tiny shapes which indicate the spots where the water was percolating through the alluvium and flowing at the contact between the plaster and the alluvium.

The top sections of the pillars had their entire surface corroded with tiny pockets. The pockets had a diameter of up to 2 cm, but the majority were smaller. This can be explained as the result of steady vertical penetration of water through the most permeable layer of the soil.

Larger pockets that tend to form at the midsection of the pillars have their surface relatively smooth.

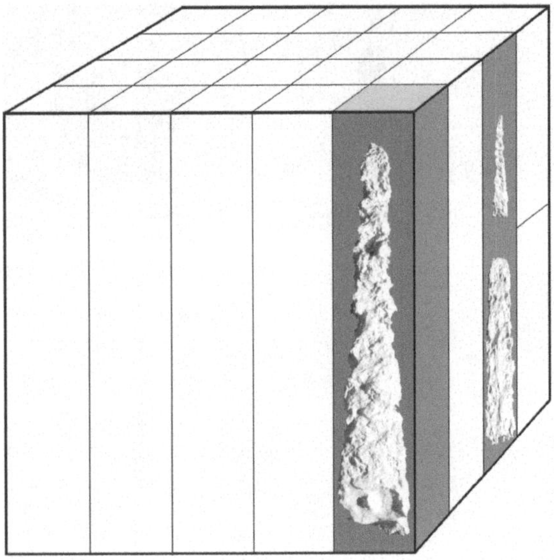

Fig. 21.4 A subsoil formation of stone columns in the course of laboratory testing with gypsum

Such pockets, much larger, of course, were named subsoil scallops (Slabe 1998b, 55).

Along the fissures or other spots where the rock was weaker, individual deeper semispherical or flute-shaped pockets tend to form which can, with time, transform into subsoil tubes.

On the walls of the lower sections of the small pillars flutes have formed, which generally measure 1–3 cm in diameter. The locally flooded zone reached up to the upper level of the flutes; the zone was created due to the inflow of water at the contact point which was larger than the quantity of water that flowed out through the perforated bottom of the bucket. This rim was often marked by notches.

The upper planes of the pillars, including the midsection of cut pillars, had distinct networks of above-sediment anastomoses.

The angular shape and the square cross-section of the pillars were preserved along their entire height.

Fig. 21.5 Uncovered subsoil forest in the course of laboratory testing with gypsum

The typical pointy shape of the upper sections of the pillars is due to the dispersed percolation of the water through the soil-covered plaster.

On the last test, the somewhat larger plaster pillars (the bigger ones were 20 cm in diameter, the smaller ones 15 cm, and both were 25 cm high) were covered with a finer-grained loam which is less permeable for water (Fig. 21.6). The contact between the loam and the rock reflected this. First, subsoil shafts formed between the pillars. As a result, vertical subsoil flutes formed in the upper sections of the pillars; these flutes were a part of the shafts and had funnel-shaped openings on their tops. The openings were up to 3 cm in diameter. It seems that when the contact between the plaster and loam is poorly permeable, the water tends to find paths that are the most conductive and merges them into streams. This assumption was also confirmed by the bubbles on the surface of the water which revealed several distinct ponors. Special flutes, such as those described in the test above, formed in the lower section of the walls of the pillars and were shaped in the locally flooded zone. Anastomoses formed on the lower plane of the pillars.

At the top, the pillars became gradually sharper and tip-shaped. The test was suspended several times in order to observe all stages of the development of subsoil karren.

After 800 h when the test was finally stopped, the larger two pillars were 20 cm high and 10 cm wide. Of the smaller pillars, two 5 cm tall and up to 1.5 cm thick pieces of plaster remained.

The pillars were entirely pointy. The sharp side edges were preserved. With time, the rock relief became increasingly similar to that in the first test, especially on individual planes. The surface of the pillars was generally more flat, however. The new characteristics tend to remain preserved, i.e. the intersection by more or less vertical and more or less winding subsoil flutes which is due to the varying permeability through the contact point between the plaster and the loam. Higher up the walls of the pillars than on the initial tests, horizontal or leaning notches with different inclinations were preserved; this demonstrates that the crevices were almost entirely filled and the contact points between the plaster and the loam were narrower, making the model less permeable. The surface of the plaster was weathered in several places, meaning that the melting was not carried away consistently.

Fig. 21.6 A subcutaneously sharpened top of a gypsum column

The tests have confirmed our findings obtained in nature. Of course, the plaster's ability to melt rapidly and consequently a more finely intersected surface need to be considered as well. Underneath the soil and various fine-grained alluvia, through which the water percolates downwards and runs through the fissured rock, the tops of the rock become sharper, which leads to the formation of tips or blades, fissures transform into crevices which then fill with the alluvium, along which a typical subcutaneous rock relief is forming. At the permeable contact point between the rock and the alluvium, the water is slipping across the entire surface which leads to the formation of subcutaneous scallops; at less permeable contact points, on the other hand, the water finds a more tight path and carves out underground shafts and flutes. A unique rock relief forms in the locally flooded zones. Such are also the subsoil karren in the valleys of the lower Dolenjska karst area.

The groundwater occasionally completely floods them.

The tops of the stone pillars, which were uncovered even before the earthworks and were thus protruding up to a metre (rarely 2 m) from the alluvium and the soil, are intersected by rock forms which was carved out by the rain water, grooves and solution pans, or formed underneath the moss and lichen which—these are common on the forest-covered surface.

21.1.2 Underground Cavities

The cavities in the upper section of the uncovered epikarst zone are underground since, during formation, they are filled with fine-grained alluvium brought from the surface by the water.

They can be categorized into underground shafts and smaller caves (Fig. 21.7). The underground shafts form along vertical fissures, most commonly by widening the crevices between the stone pillars. They reach up to 1.5 m in diameter, have circular cross-sections, and their walls are intersected by subsoil rock forms, especially flutes and pockets. During the digging of roadcuts, they uncovered within their walls or loam soil as smaller subsidences. From some of them, the rain water washed away the alluvium.

Underground cavities which were uncovered during earthworks formed along more or less gently sloping horizontal or along slightly sloped fissures and bedding planes. The permeability of the epikarst zone to the water percolating vertically is relatively small; as a result, the water stagnates at certain points or the

Fig. 21.7 The entrance to an underground shaft

Fig. 21.8 An underground cavern filled with fine-grained alluvium

hanging flooded sections of the epikarst aquifer form. In these, the formation of the cavities with up to 1 m in diameter tends to be the most intensive. In general, the uncovered cavities crisscross the stone pillars. They are entirely filled with a fine-grained alluvium, and above-sediment flutes and anastomoses have formed on their ceiling (Fig. 21.8). The above-sediment anastomoses are common along gently sloping bedding planes as well.

21.2 Conclusion

The research into the karst along other sections of the motorway route, which runs across the Dolenjska karst area, has shown that the surface under the thick layers of alluvia and soil is highly dissected. The better part of it is covered with stone forests. The epikarst is almost entirely characterized by its underground formation. Cavities form underground as well and are filled mostly with alluvia.

The image of the rounded shapes on the surface of the Dolenjska karst area, dotted only with protruding low tips of stone pillars, mostly on slopes, of course, becomes entirely different when the surface is bare. Most common are sharp shapes of carbonate rocks that have been subcutaneously thoroughly dissected. This uniqueness of the karst surface is an important part of the natural heritage which should dictate the appropriate planning and human intervention into the karst relief; last but not least, it should also dictate its proper presentation.

Part IV

Comparison with Low and Covered Karst and Karst in Breccia: The Vipava Valley

22 The Karst in the Breccia of Rebrnice in the Vipava Valley

On the slopes of Rebrnice, we had an opportunity to study a part of the young karst which has been forming in the talus breccias. Breccias form from the rubble under the steep western edge of Mount Nanos. The rubble is covered with flysch which rims around the southern and the western part of the high karst plateau. The surface of the slopes was formed by the collapses which could, on the flysch bedrock, turn into landslides. Water that flowed above the flysch also dissected the slopes. The layer thickness of the rubble or breccia varies from place to place. Rainwater has carved out rock forms on larger rocks protruding from the karst surface. The most distinct are the grooves. They require two thousand years of development to reach a mature form (Gams 1990). Therefore, the surface of breakdowns and landslides on this part of the slope has not changed significantly for a long time. More or less vertical fissures developed in breccia that indicates tensions in the slopes. During construction as the route cut deeper into the slope, the contact between the rubble, breccia, and the flysch bedrock revealed an extremely fragile balance. After abundant precipitation, numerous smaller streams spurted out, revealed by the cuts along the contact between the flysch and breccia. Many of these streams have been captured.

Characteristic and important, but in Slovenia relatively rare, karst phenomena were discovered in breccias that lie on a sloping foundation of impermeable flysch. We identified dolines in their early stages of development and common cave types. The latter are smaller karst cavities within the most compact parts of breccia; the largest cavities formed above the contact with flysch, whereas fissure caves formed along diagonal fissures which crisscrossed the slope in a relatively thick fashion.

The motorway runs across three geomorphologically diverse units, i.e., across the southwestern slopes of Mount Nanos across Rebrnice and Breg, and across the bottom of the Vipava valley.

The slopes of Mount Nanos, Breg, and Rebrnice are distinct geomorphological units. The specific geological thrust structure, transitional climate of the area, and special talus processes are reflected here in the morphology of the slopes and in its botanical peculiarity. These features have led to the proclamation of a landscape park covering the southern and the western slopes of Mount Nanos. The geomorphological mapping encompassed the area along the route and at certain locations also the wider area above and below the route of the planned motorway. It turned out that the road that runs diagonally to the slopes in fact cuts all kinds of relief types and relief forms.

The slopes of Mount Nanos and the range of the Trnovski gozd Plateau were first studied by the geologists during the geological mapping of the area. On the slopes, they established in detail the contact between the flysch and the limestone and the distribution of the Quaternary sheets of rubble and larger breakdowns (Buser et al. 1967; Mlakar 1969; Buser 1968, 1973; Pleničar 1970; Placer 1981). The contact between the limestone and the flysch on the eastern side of Mount Nanos along the edge of the Pivka basin was also studied (Čar and Juren 1980).

From the geomorphological point of view, the number of studies was smaller. The distinct slopes were described by Melik (1960) in his monograph. Habič (1968) studied the slopes of Mount Nanos within the scope of studies of high karst plateau of Nanos and the Trnovski gozd Plateau. He described various slope material, rubble, conglomerates, remains

M. Knez and T. Slabe (eds.), *Cave Exploration in Slovenia*,
Cave and Karst Systems of the World, DOI 10.1007/978-3-319-21203-6_22

of breakdowns, and larger scree and fans. In the Rebrnice area, he observed the inverse relief with thicker layers of rubble and breccias in former ravines carved into the flysch bedrock. The rubble and breccias are more resistant, thus later on the streams cut into former flysch ridges.

Around Črniče, Radinja (1962) observed various talus sediments and slope creep of the rubble on flysch and attempted to determine individual genetic types of gravel.

The predominantly non-carbonate flysch contains thick layers of limestone (limestone sandstones and breccias) which are genetically tied to the flysch. This limestone is karstified. Here and there at the contacts between the low-lying impermeable non-carbonate flysch springs can appear. To a lesser extent, the poorly cemented talus breccias can also get karstified.

22.1 Geological Conditions in the Area of the Road Route

The road runs across two landscape units: across the bottom of the Vipava Valley and the southwestern slopes of Mount Nanos, and across Breg and Rebrnice (Fig. 22.1). The route runs in the northwest to southeast direction and starts turning east at the southernmost point before Razdrto.

Breg and Rebrnice are a part of the Mount Nanos mass which is the southern part of the extensive thrust sheet. It is composed of the Cretaceous limestone in the shape of massive overturned or rather settled, occasionally also partly thrust dips on the Eocene flysch in the Vipava Valley (Pleničar 1970). The lithological contact between the Cretaceous limestone and the Eocene flysch runs mainly high up over the slopes of the Mount Nanos, mostly approx. 200 m above the motorway route. The contact is morphologically distinct since the slopes are steep or even vertical on limestone and more gently sloping on the flysch rock. The geological contact of the rocks is accompanied with a morphologically and vegetation-rich string of still living and sparse scree.

On the southern part of the Mount Nanos, in the area at Razdrto and Rebrnice, the layers of flysch and above them concordantly also the layers of Senonian, Turonian, and Cenomanian limestone dip towards northeast (50/40). At Podboršt, north of Podnanos, they dip in the same direction, but more steeply (50/70). Farther to the north at Vipava, the layers of limestone and also flysch are almost vertical.

The Vipava Valley is a younger, tectonic valley with a wide and flat bottom. The bottom is composed of flysch rocks that have been covered with younger Quaternary sediments, mostly with alluvia from streams and torrents from Mount Nanos or the part of the Vipavska Brda hills that are made of flysch.

A small section of the road route runs across the flat, material-deposited bottom of the valley. Here, the flysch rocks are located at various depths beneath the alluvia brought along by torrents and streams from the slopes of Mount Nanos or Brda. The deposits from the Nanos side contain, alongside flysch, also roughly oblique and sorted limestone pebbles. Limestone outcrops in the vicinity of the motorway can be found in only a handful of places. In the area of Mlake, on the military firing range, the limestone made of nummulitid breccias forms a rather small slope. The nummulitid limestone is inserted into the Eocene flysch layers.

A few smaller areas in the slopes are composed of Cretaceous rudist limestone. Typical examples are the Šembijski zatrep Cave above Podnanos and a smaller patch of Cretaceous limestone at Orešje above Lozice. These cirques demonstrate a bedded Cretaceous limestone; however, judging by the fracturedness and fissuredness of the stone, these are larger chunks of limestone that were ripped off Mount Nanos and crept across the flysch surface.

In the area of Breg and Rebrnice, the road runs across the flysch stones or across Quaternary talus sediments.

The flysch rocks are mostly marls and sandstones with a carbonate binder. These layers are mostly thin, very folded, and otherwise tectonically damaged. Here and there among them the layers of nummulitid breccias can be found as well. These limestone layers

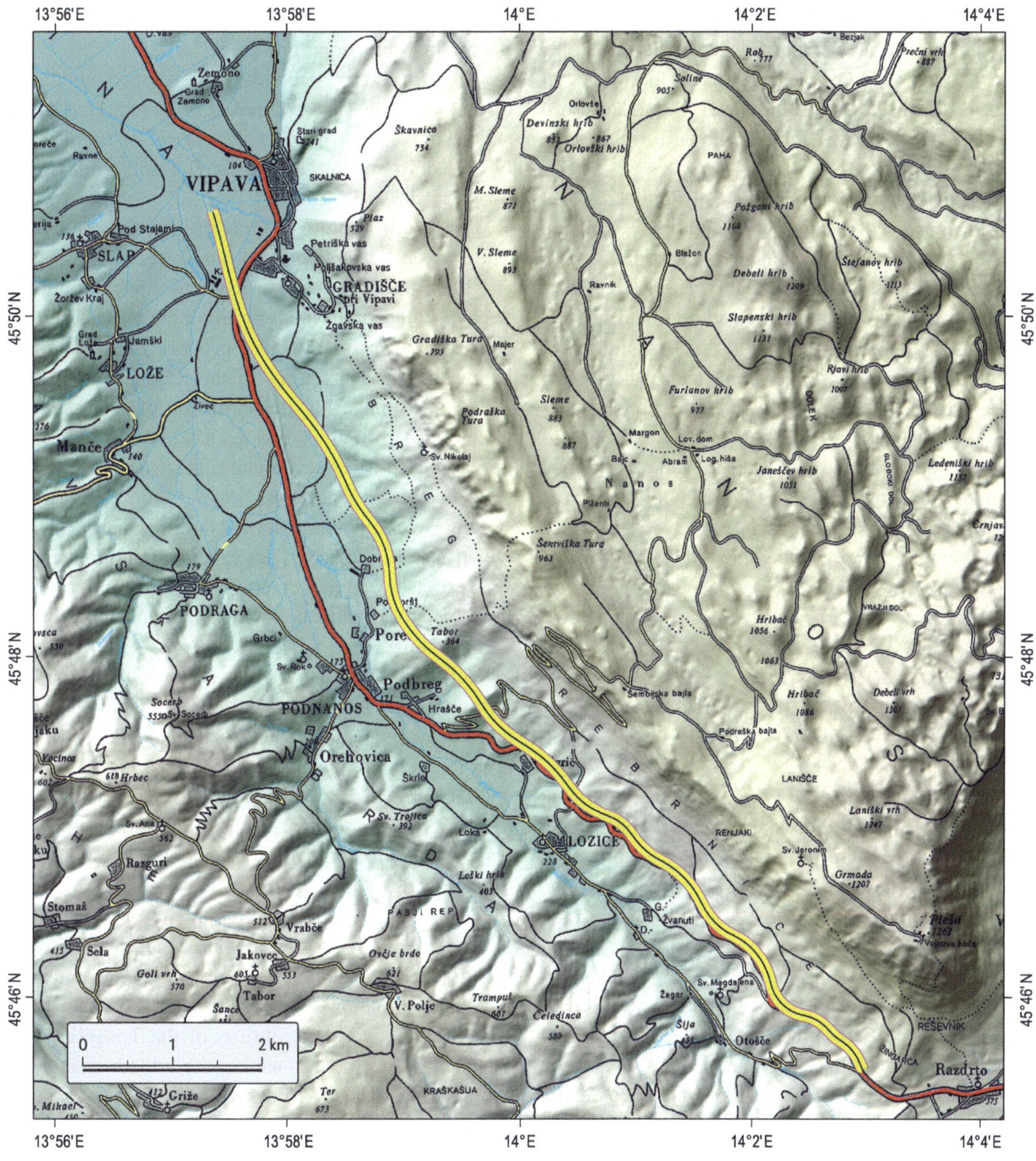

Fig. 22.1 The motorway route along the slopes of Mount Nanos

are more resistant and manifested in the relief. Alongside, smaller folds have formed in the slopes, while in stream beds and erosion trenches smaller plunges or waterfalls appeared as well.

A greater part of Breg and Rebrnice is covered with talus sediments and weathered debris of various thickness. At places, these are composed only of Cretaceous rudist limestone which often contains particles of flysch rocks. The thickness and grain size composition of talus sediments are highly varied. They are composed of square rubble mixed with bigger limestone rocks and blocks. At places, they can be bedded, but mostly they have no visible sedimentation structures. They can also be mixed with small pieces of carbonate material and reddish loam. Here and there, these sheets are cemented with calcite flowstone.

The origin of limestone rocks is in the slopes of Mount Nanos, from where they have expanded down to the bottom of the Močilnik Valley in form of scree or larger fragments.

Talus breccias are morphologically quite distinct. They are composed of talus debris and blocks 1 m and more in size. The breccia binder is carbonate, while the breccias themselves are probably among the oldest preserved talus sediments. They can be found in form of larger sheets or as individual smaller chunks of cemented talus debris. Larger breccia sheets are morphologically manifested with even angles, and at their edges with distinct folds in the slopes, smaller gradients, and walls. Individual breccia chunks formed within the unbound talus limestone rubble at places where local conditions allowed for the release of the calcite binding. Their arrangement has no distinct order.

22.2 Geomorphological Development of the Slopes of Mount Nanos

The southeastern slopes of Mount Nanos, Rebrnice, Breg, the Močilnik Valley, and a part of the Vipava Valley's bottom where the road runs were formed at the contact of three distinct morphostructural units. These units are Mount Nanos, Vipavska Brda hills, and the Vipava Valley. Hypsographic changes, as well as hydrological and morphological development of these three units were the cause of the general development and formation of the slopes, whereas the small relief shapes formed based on local conditions.

Mount Nanos is a high karst plateau where the precipitation water from both the plateau and from the steep, in limestone formed slopes flows down underground and towards the springs of the Vipava River. Therefore, there are no fluvial forms on the surface, no valleys or ravines that would reach over the plateau's edge and cut into the slopes. Since this is a karst outflow and the springs can be found only on the northwestern side of the massif 100 m a.s.l., there are no karst springs higher up the slope. The rough shape of Mount Nanos has thus been determined by the geological and structural conditions, whereas the plateau's edge has been shaped solely by karst, disintegrating, and slope processes.

The surface river network has developed on the part of the slopes that are composed of flysch. Due to the smaller quantities of water and sheets of rubble over flysch, only smaller fluvial relief forms could have formed here.

Vipavska Brda hills are made of the Eocene flysch. They cover the area between Senožeče, the Raša fault and the valley of the Raša River at one side and the Vipava Valley and Mount Nanos at the other. They are covered with a surface river network of streams that belong to the river basins of the Raša River and the Močilnik Springs. For the formation of Rebrnice, the valley of the Močilnik Springs has been of significant importance since it accumulates the water from Brda, as well as from the slopes below Mount Nanos. The stream's valley has a narrow bottom which does not widen before Lozice. The valley's bottom and the speed of its deepening determine the erosion terminant for the small streams below Mount Nanos and for the various slope creep processes as well.

North of Podnanos, the valley of the Močilnik Springs opens into the Vipava Valley which is, however, a tectonic graben with a flat and deposit-filled bottom. Its bottom is a local erosion base for all fluvial and talus processes of the entire area. This place is the merging point of the streams from the slopes of Mount Nanos, the Močilnik Springs, and the streams from Vipavska Brda hills.

The past geological development of these three units is less known or rather studied and its time frame has not been accurately determined. It is connected with the formation of the post-Eocene nappe thrust structure in which large thrusts of the Trnovski gozd Plateau have formed, as well as the folded structure of Mount Nanos.

Later radial movements caused the uplift of Mount Nanos and subsidence of its edge, most intensively in the area of the Vipava Valley. This allowed for the flow of water through the karst system and the formation of the surface of Mount Nanos. The formation of this plateau was not affected by surface waters (Fig. 22.2), while the underground drainage merged into an outflow as a group of springs at Vipava. Due to the low altitude of the springs, sinking streams from the northwest edge of the Pivka Basin started to flow towards the springs.

Fig. 22.2 The formation of this plateau was not affected by surface waters, whereas the underground drainage merged into an outflow as a group of springs at Vipava

On flysch rocks at the southwestern edge of Mount Nanos, a water basin of the Močilnik Springs has emerged. The Močilnik River is characterized by its asymmetric river network which gets only the left tributaries from the flysch area of Vipavska Brda hills. These tributaries have shaped deep valleys and ravines. Tributaries from the slope of Mount Nanos carrying less water have not shaped the most important valleys and have not significantly dissected the slopes (Fig. 22.3).

Therefore, the characteristically shaped relief of the slopes of Mount Nanos is due to the karst manner of water outflow from Nanos, the geological structure, and the formation of the valley of the Močilnik River.

The limestone lying on flysch has caused the formation of a distinct structural level in the slope of Mount Nanos. The slopes that have formed in flysch are more gently sloped, whereas the slopes that have formed in limestone tend to be steeper. The fold in the slopes in the upper section of Rebrnice runs at the altitude of 800 m, above Lozice at around 650 m, and above the springs at Vipava at around 350 m. Thus, the contact of rocks lies high above the bottom of the valley of the Močilnik River and the Vipava Valley, i.e. at about 200 m above the motorway route.

Since the Nanos plateau and its limestone slopes drain into the underground, there are no surface waters on the slopes that would shape the fluvial valleys or intersect the straight, structurally conditioned slopes of Mount Nanos.

The structural level with limestone above the flysch rocks is highly important since it allows the formation of unstable and steep limestone slopes, as well as intensive mechanical limestone corrosion. Crumbling,

Fig. 22.3 Tributaries from the slope of Mount Nanos carrying less water have not shaped the most important valleys and have not significantly dissected the slopes

fragmentation, or collapses of steep slopes have covered the more gently sloped flysch part with large sheets of limestone gravel. Here and there, these sheets are cemented with calcite flowstone (Fig. 22.4).

At certain locations, the sheets of limestone gravel and of bigger rocks tend to be more than 10 m thick. They start out at the contact between limestone and flysch and reach downwards, sometimes even to the bottom of the valley of the Močilnik River. Most of the sheets are confined by distinct folds in the slopes. The folds have formed at their bottom edges, often above the springs or groups of springs. The distinct folds in the slopes confine them at the sides as well, i.e. at places where erosion ravines of streams have formed parallel to them where the flysch is not covered with limestone gravel.

Morphologically distinct are predominantly the sheets at Razdrto, between Gradišče and Žingarca, and two sheets at the top of the Hraščak hill, above Hrašče and on Slatna or at Gradišče above Podnanos.

At current conditions, the sheets are no longer forming. Limestone crumbling and corrosion are limited to the steep slopes of Mount Nanos and are reflected in the form of still living scree reaching down to the contact with flysch. It is therefore obvious that the large limestone sheets have formed in different climate conditions when limestone corrosion was more intensive; these stages of development have no time frame, however. The age of the sheets can be discerned from the way they are cemented into breccias and from the way they are lifted above the slopes in their upper section.

Fig. 22.4 Here and there, the sheets of limestone fragments are cemented with calcite flowstone

The napes are an important morphological and structural unit of the slopes which due to their characteristics affect further shaping of the slopes. The precipitation water effortlessly percolates through these layers all the way down to the contact with flysch in the bed. The water accumulates along the contact, washes away finer particles, and causes the instability and creeping of talus sediments, the emergence of springs in the lower sections of the slopes, as well as the formation of shallow and small valleys or deeper ravines at certain locations. Precipitation water also percolates through them, meaning there are no fluvial forms on them. The latter form underneath, below the springs.

Since the water percolates through the sheets only slowly and these sediments have a high retinence, the outflow from the springs below is suppressed. This is reflected in the greater continuity of springs and lesser erosion ability of their water or rather in a lesser erosion dissection of slopes.

22.3 Characteristic Relief Forms in the Motorway Route

22.3.1 Structural Forms

Structural relief forms have formed due to various physical or chemical properties at the contact among different rocks or rather different geological structures, such as faults, thrusts, or bedded rocks. Apparently, such relief forms show the geological structure of the surface.

Structural relief forms have formed at the contact between limestone and flysch, limestone and talus debris, and talus debris and flysch. Structural forms in flysch rocks are common; they are composed of layers of marl of various resistance and thickness, sandstone, and also limestone. More distinct structural levels in flysch have formed only where thick layers of nummulitid breccia or limestone are stacked in between layers of marl. The most distinct structural levels in flysch rocks have formed on the northern slope of the Slatna Hill.

22.3.2 The Structural Level of Mount Nanos

Structural levels form at places where the more resistant rocks lie on top of the less resistant rocks. Due to talus processes, the more resistant rocks led to the formation of steeper slopes, and the less resistant rocks to more gently sloping slopes. The fold of the slope is at the contact of various rocks.

The structural level of Mount Nanos is the largest structural form on its entire southern slope. The upper section with the incline of about 35° to 45° has formed in limestone and belongs to Mount Nanos in the strict sense. Below it lie the less sloped slopes of Mount Nanos, Rebrnice, and Breg. They have formed in flysch rocks. Their incline is about 15°. The fold is distinct, sharp, and runs along the slopes from

Razdrto to Gradišče at Vipava. The contact of rocks and the fold run about 200 m above the motorway route.

22.3.3 Relief Forms on Carbonate Rubble or Breccia

A large part of Mount Nanos's slopes is covered with carbonate rubble and collapses from its steep, limestone slope. Crumbling, fragmentation, or collapses were intensive predominantly in cold Pleistocene climates. This sheet was later heavily transformed and intersected. The main reasons for the transformation were corrosion, washing of carbonate particles, erosion, and soil creeping. Thus the uniform sheet, especially lower in the slope, segmented into laterally separated or cascade sheets. The route of the road cuts into several sheets of this type.

In addition to the thick layers of rubble and breccia, the slope is full of large areas of thin sheets. These are simply a thin layer of limestone rubble covering the flysch. This layer prevents the formation of erosion relief forms, modifies soil thickness and soil properties, and makes the surface more rocky.

Thin layers are more frequent on upper sections of the sheets or below their steep lower edges. Morphologically, they are manifested as rocky surfaces with rare erosion relief forms.

22.3.4 Breakdown Forms

Along the steep structural level of Mount Nanos, larger parts of the slopes have broken off due to the instability of the surface. A piece that has broken off slowly creeps down over the flysch surface. Because it is moving over the surface, such a large chunk made of Cretaceous limestone becomes heavily fractured; stratification, on the other hand, is largely preserved and therefore such chunks can separate from the talus breccia where the structure is completely different.

Behind the larger fractures, talus debris and blocks have accumulated as well, which is why such forms can be similar to rubble sheets and breccia sheets. The largest forms of this kind are Šembijski zatrep and the fracture above Gladežnica.

22.3.5 Breakdown Blocks

Large limestone blocks are being torn off the steep slopes of Mount Nanos. Such blocks can roll far down onto the flysch. The blocks that lie on the surface are relatively young and emerged after the main breakdown and rubble sheets had already formed.

There are no such larger blocks in the route of the road; they do, however, lie above it, i.e. in the area of Ravenca, Parti, and below Šembijski zatrep.

22.3.6 Erosion Forms

The entire studied surface has formed in flysch rocks and has a surface outflow; however, the flysch rocks in the major part of the slopes have layers of carbonate rubble of various thickness. This rubble has had a major effect in the shaping of the surface.

The flysch rocks are impermeable which has led to the surface outflow and erosion forms. However, a large part of the surface has no surface flows. In fact, the precipitation water flows under the surface through the soil and talus debris. Since the amount of water that accumulates on the short flysch slope is small and since the rubble is suppressing the outflow, the erosion capability of all these streams is poor. Because the sources of the streams are smaller, the majority of the streams dry up during summer.

All this has led to the formation of distinct but small and shallow erosion relief forms which have formed on the slopes intersected by the road. They can be divided into erosion beds, erosion trenches, and erosion ravines. Erosion beds are beds that have a permanent or a periodic water flow. They have formed with erosion processes.

In the area of the motorway, only erosion beds and erosion trenches can be found. Erosion ravines, on the other hand, have formed lower in the slopes, just before the confluences with the Močilnik River.

Especially on the sheets of rubble and breccia there are shallow trenches which are, however, dry or inactive and covered with vegetation. They are most likely remnants of different climate conditions, local changes in water flow, or have formed in different conditions during deforestation. A rapid change of the vegetation sheet could have caused a locally enhanced erosion.

22.3.7 Hummocky Ground

Hummocky ground can differ both in shape and in formation. On flysch slopes where the layer of weathered debris is thicker and where there is plenty of underground water, a part of the flysch weathered debris can start creeping. They can form by washing of finer particles in non-homogeneous scree, similar to glacial hummocks, or due to the creeping of the entire chunk of the talus material down the flysch surface.

Such hummocks can be found on all rubble sheets. They tend to be uneven and are predominantly dependant on the size of rubble or rocks of which such a sheet is composed of. They are particularly large at the edges of sheets and at places where large rocks or blocks of breccia are nested in between the rubble.

22.4 Anthropogenic Forms

22.4.1 Paths Reshaped Due to Erosion

Since the slopes are erosion-sensitive, numerous forest roads have sunk into actual erosion trenches due to erosion. Such roads can be found both on flysch rocks and in places where flysch is covered with thin sheets of rubble.

22.4.2 Escarpments, Piles and Dry Walls

Permanent settlement, which also means land use for agriculture in the studied area, reaches all the way back to prehistory. It is evidenced by archaeological discoveries and remnants of the fort above Žganica at Razdrto and at Gradišče nad Hraščami in close proximity to the route. Their shape was affected by the agricultural use of the land in Rebrnice, Breg, and the Vipava Valley. The most noticeable traces of surface use, however, are the cleared rocky surfaces with stones folded into escarpments, piles, and dry walls.

Escarpments can be found in the area of the road's route mostly on the flysch rocks, where they have been used to level the ground in the slopes into the terraces for fields. Such escarpments can be found above the roadhouse above Lozice, on the slopes of Gradišče nad Hraščami, on the slopes of the Slatna Hill, and to a lesser extent above the villages of Podboršt and Dobrova.

Piles can be found everywhere on rubble and breccia. The largest can be found on larger sheets and flat cirques. They indicate that pastures and meadows have been cleared and fixed up. The largest piles in terms of quantity, however, can be found on the slopes between the villages of Podboršt and Dobrova.

22.5 Breccia Formation

In cross-sections of the slope exposed by cuts for the motorway, it was frequently possible to observe many layers of rubble or breccia (Fig. 22.5). Their total thickness mostly reached up to 10 m, in individual places that were exposed during the digging of the wells for the viaduct pillars even up to 25 m. Generally, the layers can be differentiated relative to the degree of cementation of rubble into breccia. Only individual layers were relatively well-cemented. The carbonate binder binds the rubble into breccia; the process first takes place around bigger rocks. In fact, the earthworks often revealed chunks of breccia several m^3 in size, which were surrounded by unconsolidated or poorly unconsolidated rubble. Between the larger and smaller pieces of rubble and rock forming breccia were hollow spaces partly filled with flowstone (Fig. 22.6). The degree of cementation of breccia is therefore the consequence of the age of breakdowns, a deeper or shallow position below the surface, and characteristics of waters that percolate regularly from the surface, which is covered with a thin layer of soil and vegetation. In certain places, breccia is already uniquely karstified.

The water that has been percolating through the rubble and has stuck it together has been carrying with it dissolved calcium carbonate. The breccia clasts binding on the slope of Mount Nanos is therefore almost exclusively carbonate in origin. The breccias are predominantly already bound solid, except on the surface where they tend to be less solid. Pieces in

Fig. 22.5 Layers of breccia

breccia are fragments of Senonian, Turonian, and Cenomanian limestone and most likely of Lower Cretaceous limestone as well. The Lower Cretaceous layers in the upper half of the Nanos slope have developed similarly to the Upper Cretaceous Cenomanian limestone (Pleničar 1970; Buser 1973) and are thus not shown separately on the maps (Buser et al. 1967; Buser 1968).

During works along the route, we came across several different types of breccia. The material building the breccias is of different size, from tiny sand to blocks of 1 m or more in size. The majority of breccias is still very porous. Most of the time, their clasts do not exceed 5–10 cm in diameter. The clasts contain no fine material and are also not covered with flowstone. A similar structure has been observed in seemingly similar breccias, only their clasts are completely covered with 1–2 mm of layered, mostly porous or white flowstone. There is also no fine material nested in between (Fig. 22.7). This type of breccia is very porous. Only partially porous breccias have formed where water between the clasts carried in weathered debris and fewer fine limestone clasts from the surface. In places where fine limestone fragments with the size of sand accumulated between clasts that are a few centimetres in diameter, the areas with breccia are substantially less porous. They also contain weathered debris that is bound into flowstone (Fig. 22.8). In some cases, the clasts have been bound together by the calcium carbonate in which the water deposited major quantities of weathered debris. Even in this case, breccias are poorly or almost non-porous, and the binding is of characteristic red-brown colour. In places, flowstone-covered cavities have formed between the clasts; here, the beginnings of larger calcite crystals can be observed.

Fig. 22.6 Parts of breccia

22.6 Breccia Karstification and Cavities

The carbonate blocks on the surface have rain flutes which indicate a long-lasting shaping of the rock with rainwater. In close proximity of the route solution pans appear as well (Fig. 22.9). None of them is significantly big, although the longer axis of some solution pans is 0.8–1 m long. The majority are smaller, though.

In some locations, dolines have begun to form; they were formed with water that mostly dispersedly percolates from the permeable surface (Fig. 22.10). They are smaller and covered with soil which makes them less distinct on the surface. Their diameter reaches up to 3 m. Underneath, due to earthworks, of course, there are vertical, 1–2 m wide belts of uniquely washed out rubble and breccia that are clearly visible in the cross-section (Fig. 22.11). These are therefore traces of already consolidated water percolation which is also due to the rubble coagulating into breccia.

During construction works several cavities have opened. They can be divided into three types; however, these do not include small cavities between chunks of rubble and rocks as described above.

The first type are networks of smaller cavities (Fig. 22.12) which have formed in the most consolidated layers of breccia. Their cross-section is more or less circular. Their diameter did not exceed 0.5 m.

Fig. 22.7 There is almost no fine material in between breccia clasts

Most often, they can be found along the cracks and at the contact with layers of breccia where they tend to be the biggest as well. They are often filled with alluvium. These are true karst cavities.

Others can be found at the contact between flysch and breccia above them. The biggest among them could be accessed. They had branched-out, at places also three-dimensional networks of smaller passages with domed expansions (Figs. 22.13, 22.14 and 22.15). The domes reached up to 3 m in diameter. Particularly along the cracks, the domes tend to be higher and narrower. Their walls are dissected and rough indicating that they are composed of breccia. It can be concluded that these caves have formed due to the water percolating at the contact with flysch and due to the rock particles that have been washed out. The shape of the caves is reflected in different degrees of cementation of rubble into breccia. Larger stacks of flowstone often formed under the domes, provided that a sizeable amount of rocks has accumulated on the

Fig. 22.8 Some breccia also contain weathered debris bound into flowstone

flysch along smaller collapses and was not carried away by water.

During the motorway construction at the section Vipava–Selo, smaller cavities with the diameter of up to 0.5 m have opened in breccia and at the contact with flysch. Smaller water flows were running through them.

Along the contact, flows of smaller or larger amounts of water can be traced. These become increasingly prominent after more abundant precipitation. Flysch turned out to be a relatively poorly permeable rock, particularly in places where its upper layer was weathered. Along larger continuous streams, which are particularly prominent in talus notches, conditions for the formation of caves are favourable. At several locations, breccia has already been carried away and thus smaller valleys have emerged above the streams; the process is that more prominent at locations where the layer of breccia and rubble over the layer of flysch is thin.

The third type are fissure cavities (Fig. 22.16). They form along fissures that have formed in breccia. They tend to cross the dip of the slope diagonally. The biggest among them can be accessed. However, they are mostly narrower since their diameter does not exceed several decimetres in size, while their depth depends on the thickness of the breccia layer and the characteristics of the fissure. Their walls are often cowered in flowstone, and smaller walls can be filled with flowstone.

To provide a comparison, Remškar (2006) compiled data about caves in breccias of the Vipava Valley. He provided definitions of individual cave types

Fig. 22.9 Solution pan

and their formation. He divided them into those that developed along fissures, those formed by streams of water, and abris.

Flowstone often fills smaller spaces between the rock fragments that compose breccia. Where there is enough room, straws tend to form, and when breccia is composed of rocks, smaller flowstone domes form whose stratification in places indicates that this was originally a filled cavity (Fig. 22.17). A ribbed flowstone sheets occur on the walls of talus fissures. Larger domes of flowstone were indeed discovered in caves originating along the contact with flysch.

22.7 Karstification of Flysch and of the Contact with Carbonates

Carbonate sedimentation which resulted in the formation of limestone stopped approximately 50 million years ago after the alveolinid-nummulitid limestone had deposited in Cuisian. This was followed with the deposition of flysch and flysch-like sediments. After flysch sediments had deposited, the sea in southwestern Slovenia finally receded.

Fig. 22.10 Small dolines

The flysch lies in direct contact below the limestone; at places, alveolinid-nummulitid breccia have been observed as an intercalation in flysch. At the contacts with carbonate (permeable) and non-carbonate (impermeable) rocks in other locations throughout Slovenia a significant hydrologic activity has been observed. It is known that the impermeable rock is not only a water barrier, but also an area where water accumulates; at the same time, the water level can change drastically, thus washing and carrying away the rock; pressure changes can also occur here, resulting in larger channels shaped by water. Both limestone and flysch particles are transported along these paths. Despite the fact that flysch rocks, which are, in the karst, in a constant direct contact with carbonates, are often presented as exclusively impermeable strata, it can be pointed out that flysch (often of lesser thickness) is merely an isolated lenticular body on the permeable carbonate rocks. Furthermore, as has been determined on several occasions, underground conductive channels are also forming in flysch rocks, though in smaller numbers, and the precipitation water that has accumulated on the flysch can flow into the karst.

To illustrate this, it is worth mentioning an interesting flysch section between two covered entrenchments, i.e. Šumljak and Boršt (Fig. 22.18). Predominant there are marl layers in grey-olive and yellow-brown colour. The sandstone is not as abundant. The layers are mostly very thin, ranging from several cm up to 15 cm, only in the southwestern part of the exposed profile they can be up to 1 m thick.

Fig. 22.11 Belt of breccia underneath a doline

Fig. 22.12 A cavity filled with fine-grained alluvium

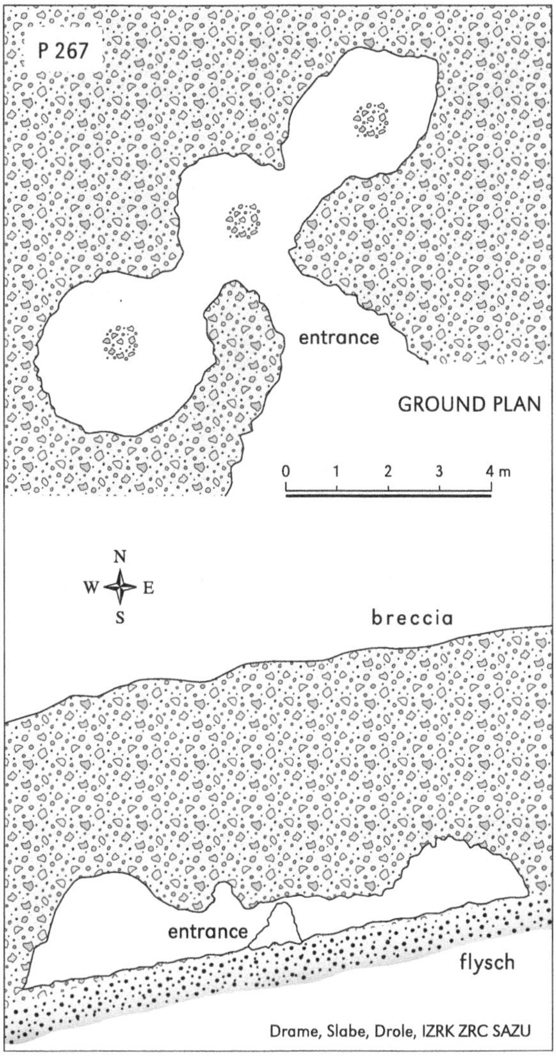

Fig. 22.13 A cave above flysch

Flysch layers dip steeply towards SW and progressively take a sub-vertical direction. Thicker layers are tectonically highly fractured, whereas thinner, a few centimetres thick layers have a splinter-like shape. In the SW part of the profile, syngenetic flysch clasts are quite prominent; perpendicular to the dip direction, they indicate a distinctly characteristic concentric weathering. The layers are not only leaning to an almost vertical position, but are also folded. Everywhere where the layers are either fractured or folded, the water flows down the fissures or down the clearance between layers which are due to the folding. There is a strong flow of water along the interbedded contacts, most likely due to the almost vertical layers. The water flowing along these contacts carries away flysch material, widens the fissures and, at the same time, periodically and laterally deposits calcium carbonate in different ways. Frequently, several centimetres thick calcite fillings of fissures have been observed. In places, the fissures are completely filled, whereas elsewhere up to 1 cm large scalenoedric calcite crystals formed in the fissures; a great deal of fissures is covered with a thin (a few millimetres) coat of flowstone. In marls which tend to fracture in a distinct conchoidal manner, many a fissure has already been filled with thick calcite crystals. It must be pointed out that the binder is carbonate and that many a layer can contain significantly more than 10 % of particles of carbonate origin. Therefore both erosion and corrosion occur when water flows through the fissures and along the faults. However, there is no doubt that karstification takes place in a very small extent.

Heavy weathering of the rock in the interior of the tectonically undeformed block of rock occurs along fissures and faults where precipitation and surface waters flow. Calcium carbonate (flowstone) is deposited across the majority of such contacts.

In thinner layers that have a splinter-like shape, the karstification in flysch is virtually unnoticeable; there are also fewer calcite veins. As soon as such a layer is exposed to the atmosphere, it begins to disintegrate rapidly.

According to the analysis of the samples taken in the vicinity of the road's route, it can be expected that marl and particularly sandstone contain, beside the calcite binder, a considerable amount of carbonate particles. This is why flysch is not only a "water impermeable barrier", but also the cause for karstification, although in a substantially lesser and somewhat specific extent than is the case in carbonates.

Fig. 22.14 A network of dome-shaped passages

Fig. 22.15 A passage above flysch

Fig. 22.16 A fissure cave

Fig. 22.17 Flowstone in breccia

22.8 Conclusion

The geological, geomorphological, speleological, and hydrological diversity of the Slovene karst has been demonstrated also by the study of karstification of breccia that have been forming beneath the western slope of Mount Nanos. Water, in most cases percolating diffusely through the permeable surface of rubble or breccia and into the more or less impermeable flysch bedrock, is forming young karst phenomena.

Rainwater covers large blocks on the karst surface with rain flutes and solution pans. Essentially, there has not been significant surface movement for a long time. Fissures crossing the rock alongside the slope indicate tensions in the rock mass and its exposure to sliding. Breccia and rubble lie on the slanting flysch and the majority of water flows along the contact causing its instability.

The motorway route Razdrto–Vipava has cut into the entire southern slope of Mount Nanos in the area of Rebrnice, Breg, and a portion of the Vipava Valley's bottom. Since the motorway runs diagonally to the slopes, at the same time it cuts many and various relief forms in the upper and lower sections of the slopes.

The morphological mapping has shown the distribution and the size of various relief forms in the area of the route and its near vicinity. The new road has changed many relief forms, but has not substantially changed the biggest among them. The route cuts into them diagonally, so that the larger sheets of rubble and breccia remained unchanged. Tiny forms in the route have been completely destroyed and changed in its neighbouring areas. Road construction has also indirectly affected many, particularly fluvial relief forms. Due to the changes in the outflow from the slopes, changes in the old and the formation of the new relief forms are likely.

Fig. 22.18 Flysch

The percolating water accumulates where breccia is most consolidated. Earthworks have revealed the early stages in the formation of unique dolines.

Characteristic types of cavities developed in the young and very porous breccia which is consolidated only in places that lie on the more or less slanting flysch, i.e. on an impermeable bedrock. The true karst cavities are small and their development was influenced by the alluvium that has been depositing on their bottom and mainly fills them. They formed in a locally and periodically flooded zone or they were paragenetically enlarged. The largest cavities formed above the contact with the impermeable flysch bedrock where the big streams joined together. Their shape reflects the varying degrees of consolidation of breccia. In areas where breccia is less solid and along fissures they rise into domes. Along fissures which emerged due to the sliding of breccia and rubble down the slanting bedrock of frequently soggy flysch, fissure caves formed diagonally to the dip direction of the slope; some of them are very long and wide enough in places to make them accessible. Their walls are mainly covered with flowstone.

Although the described karst is relatively young, discovered in its early development stages, it still reveals all the characteristics of the karstification of breccia in characteristic geological, geomorphological, and hydrological conditions. The knowledge about it broadens our knowledge of the diverse natural heritage of the karst and is the basis for further planning of human intervention into the karst relief.

23 Karst Phenomena Between Vipava and Selo

23.1 Geological Characteristics of the Region

The bottom of the Vipava Valley is made mostly of clastic rocks and sediments of different age and origin. The motorway between Vipava and Selo initially runs across the agriculturally changed Quaternary alluvium sediments, then it crosses the slightly undulating landscape made of Lower Eocene non-carbonate and carbonate turbidites (Fig. 23.1), and in its last section it cuts across Quaternary, locally cemented talus debris (Fig. 23.2).

Beds of siliciclastic sandstone, claystone and marl, with megabeds of resedimented carbonates in between, alternate in the sequence of flysch beds.

Beds of carbonate flysch are a few decimetres to over 10 metres thick and are made of grey limestone sandstone (calcarenite) and organogeneous limestone breccia (calcirudite) of different grain sizes. The thick carbonate beds are often fissured and crisscrossed with thick calcite veins. Previous research using bores has shown that the limestone is karstified in certain places, whereas the cavities are filled with greasy and bare clay, as well as flysch rubble and gravel.

The Quaternary coarse-grained carbonate talus debris which, at places, features large limestone blocks, is locally cemented which is why breccia formations (Figs. 23.3 and 23.4) in between fine rubble appear "spongy". To a lesser extent, the talus breccia of this type can be karstified.

Springs can emerge at the contacts with Eocene resedimented carbonate and Quaternary carbonate rubble, as well as at the contact with the low-lying impermeable flysch. Smaller permanent springs also emerge at the contacts of predominantly pure carbonate rubble and breccia with larger blocks of rock and clay.

23.2 Karst Cavernosity of the Rock

The cavities along the route can be divided into smaller shafts which formed in flysch-bedded limestone, i.e. due to the contact or due to the accumulation of water on impermeable rock, and into caves which formed in breccia that overlays the impermeable flysch.

Past Dobravlje at the Koševec Stream, an entrance to a 3.7 m deep shaft has opened during excavation works for the underpass (Fig. 23.5). The entrance to this small shaft was oval in shape and 0.7 × 0.5 m in size. Up to 2 m down its diameter reaches 1 m, past this point it widens to the size of 3.5 × 2 m. The main axis of the shaft's ground plan runs in the direction of 320°. At the bottom of this small shaft there is a small fissure reaching 2 m deep. The shaft was most likely formed by the water from the periodic Koševec Stream which permeates into the rocks 4 m northwest along the fault (320/90).

M. Knez and T. Slabe (eds.), *Cave Exploration in Slovenia*,
Cave and Karst Systems of the World, DOI 10.1007/978-3-319-21203-6_23

Fig. 23.1 Contact area between limestone and flysch

The second shaft (Fig. 23.6) is 4 m deep and up to 1.5 m wide. The bottom section of the shaft was buried due to the breakdown of rocks into it. The shaft lies in a thick-bedded limestone which lies in flysch.

During roadcut excavation works, small cavities have formed at the contact between the upper, at places several metres thick bed of carbonate breccia, and the low-lying flysch; these cavities have a diameter of up to 0.5 m. Smaller water streams flow out from the contact (Fig. 23.7). The water that percolates diffusely through the permeable breccia therefore flows across the intertwined network of smaller cavities which formed along the contact. Moreover, two slightly larger cavities have opened as well. The larger of the two consists of several passages 1 m in diameter. The juncture of the passages is a cave 3 m wide and 1.5 m high. The circumference of the cave is crumbling since breccia is not well cemented.

The motorway route in the section between the villages of Cesta and Potoče south of Skrilje has cut into a narrow ridge which is composed of Palaeogenic, alveolinid-nummulitid limestone. In the cross-section, the limestone was clearly visible, but no significant karst formations or phenomena were observed in it.

23.3 Along the Protective Archaeological Excavations on the Motorway Route Between Log and Ajdovščina

Underneath the layer of soil which contained the remnants of prehistoric ceramics, the archaeologists have stumbled upon a sterile rubble layer. The upper surface of this layer was not even, but showed shallow sags and elongated trenches. Since it was possible that

Fig. 23.2 Layers of breccia

Fig. 23.3 Breccia above flysch

these uneven surfaces were human-made, we took a closer look at profiles that have been dug out and tried to determine the origin of this layer and estimate when these uneven surfaces in its upper section were created.

Although we were not able to pinpoint when exactly the hollows were formed, there is a strong indication that they are of natural origin. The Slovenian literature on geomorphology describes similar surface hummocks which formed due to the corrosion in moraine rubble of various thickness.

The layer of rubble was composed of limestone rubble mixed with yellow-brown loam. The limestone that makes up the rubble contains numerous fossils from the Jurassic era. The rubble is rough but slightly corroded at the surface, which is why it is slightly rounded at the edges. The clasts size differs, with the biggest up to 20 cm in size, and with larger limestone rocks here and there among them. The rubble is unsorted and without any fluvial sediment structures. The layer of rubble is deposited onto the bedrock. The bedrock in this part of the Vipava Valley is Eocene flysch sandstone and marl.

Fig. 23.4 Impermeable flysch bedrock underneath breccia

About 0.5 km to the north, the slopes begin that are steeper and have formed in Eocene flysch marl and sandstone. Several smaller streams originate here as well; however, in the more flat part of the valley, these streams are ameliorated and heavily transformed, carrying along only fine gravel or sand.

The bed of rubble mixed with loam underneath the plough layer undoubtedly originates from the rubble which can be found higher at the outskirts of the Vipava Valley. The even size of the rubble indicates that it was formed in the cold Pleistocene conditions on the steep edges of the Trnovski gozd Plateau. Later on, the rubble either slid or was carried by the waters and deposited onto the lower part of the valley. The rubble is not rounded which indicates a short-distance transport.

The ground in the area of the excavation works is slightly sloped (1° to 2°) and is descending towards the lowest part of the valley along the Vipava River. On such a low incline, the streams can only carry small flowing matter. They cannot, however, carry the rubble made of larger clasts. The rubble was most likely resedimented onto the more gently sloping part of the valley from the scree, either during heavy rainfalls or by solifluction. Coarse rubble, low incline of deposited rubble mixed with loam, and unsortedness indicate solifluction in the cold Pleistocene climate.

The upper plane of the rubble bed was not even. It has spherical or elongated sags that are several tens of cm deep and several tens of cm to over 1 m wide. The elongated sags are probably the beds of smaller streams which were later on flowing across this dyke. Due to the low incline they have not had the strength to carry rubble, but they have been able to wash away finer particles of clay.

The spherical sags could have formed in several ways. They could have formed in places around the tree roots or bushes when the rubble was sedimenting. They could have formed later when the trees began to

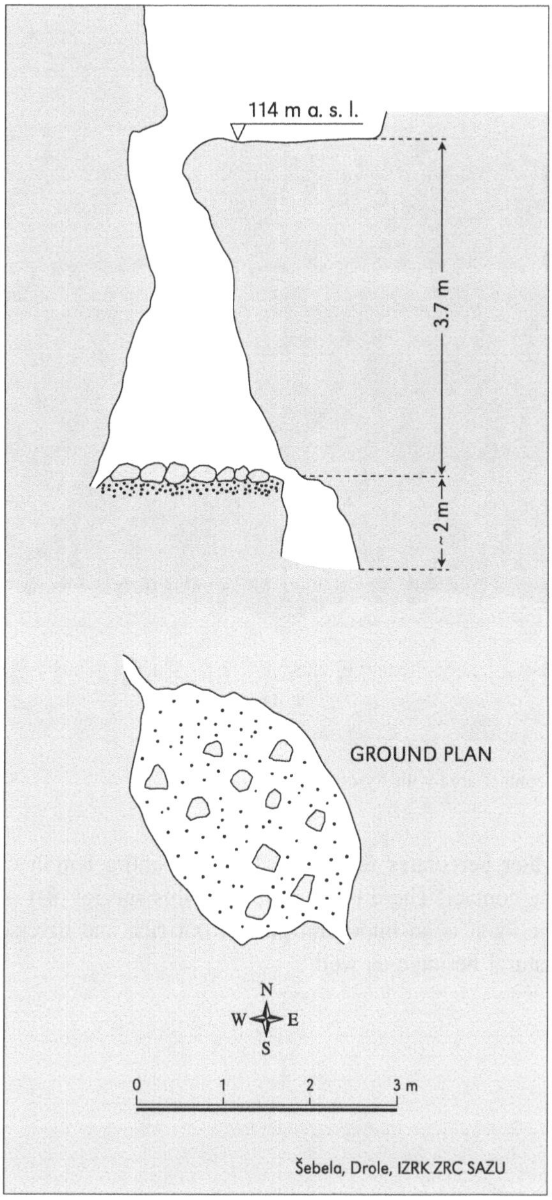

Fig. 23.5 A plan of a cave that was uncovered during excavation works for the overpass

grow on this surface. The tree roots can push the rubble apart; when a tree dies off, it leaves behind a hole which is then filled with finer surrounding

Fig. 23.6 A shaft which opened during excavation works for the roadcut

material. Lastly, the sags could have formed in places where rubble corrosion was locally more intensive, e.g. in places with plenty of finer rubble.

23.4 Conclusion

Despite the fact that not many karst phenomena have been revealed in this section, the ones that have been are unique and interesting. The caves in the carbonates in the flysch (and the breccia that covers it) have unveiled a way of formation and development of the karst in these special geological conditions, in different rocks, and in the contact of carbonates with

Fig. 23.7 A small passage with a water flow in breccia above the contact area with flysch

non-carbonate and impermeable flysch. In the breccia which lies above the flysch, a true system of caves with cavities has formed; the cavities reached several decimetres in diameter. This system indicates that the water percolates regularly along the entire length of the contact. There is no doubt that this special part of the karst is an important part of our rich and diverse natural heritage as well.

Part V

Planning

24 Karstologic Research for the Engineering of a Preliminary Design for the Motorway Section Bič–Hrastje, Subsection Ponikve–Hrastje, Sv. Ana Variant

The entire area of the motorway subsection Ponikve–Hrastje is made of impermeable carbonate rocks which do not outcrop with the same intensity throughout the area. In many places on the east side of the Temenica River Valley, the large relief forms are of erosion type, whereas the rock base, the hydrographic function, and the tiny fluvial forms are of karst origin—therefore, what we have here is fluviokarst or rather fluviokarst formations which have developed here more beautifully than anywhere else in Slovenia.

Karst waters, which sink into the area in question, flow across a broader area of SE Slovenia and supply certain karst springs that are important from the water supply point of view; at the same time, they supply the common underground karst aquifer which is an important source of drinking water in that part of the country.

The issue of a potentially negative impact on the underground karst water reservoirs as one of the most important questions in studying the karst terrain, has been singled out in respect of the motorways more than anywhere else.

Therefore, the motorway was under construction on an ecologically very vulnerable karst terrain.

24.1 Geomorphological Conditions

The motorway section Ponikve–Hrastje runs across the landscape that belongs to the central Dolenjska region. It lies on the area where major landscape or geomorphological units come into contact: the Sava Hills with the predominating erosion (normal) relief and the Dolenjska part of the Slovenian Dinaric Alps with the predominating karst relief.

The contact area between these two landscape units is not striking, but rather gradual which is why the studied area is not merely a "contact area" or, in other words, it cannot be said that one portion of this landscape belongs to the Sava Hills (Posavsko hribovje) and the other one to the Dolenjska karst area, but it is a special, inherently unified landscape which is a merging area and made of morphological elements of both neighbouring units. Because of this, a special type of landscape has formed where the predominant or rather characteristic special forms are fluviokarst formations. Such transition between the two major units does not take place only in the area of the motorway section Ponikve–Hrastje, of course (Fig. 24.1), but in a wider part of this region of Dolenjska. The common name for this region is the Dolenjska Lowland. Due to its internal integrity, the studied small landscape unit includes the Temenica Lowland as well.

Slovenian geographers and geomorphologists have proposed various ways of regionalization for this part of Slovenia; they agree, however, that this is a contact area, and within it an independent landscape and geomorphological unit having been named differently by different authors: Temenica (Melik 1961, 1962), Lower Dolenjska karst (Ilešič 1958), Temenica Lowland, Dolenjska Lowland, Middle Dolenjska fluviokarst (Gams 1959, 1962a, c, 1987; Kladnik 1996).

Regardless of the name, there is a belt of lower terrain between the Sava Hills to the north and northeast and the Dinaric Karst to the south and southwest—a valley that stretches out to the Ljubljana Marshes on one end and the Krško Basin on the other.

The Temenica Valley is both absolutely and relatively lower terrain, the bottoms of the valleys or depressions are generally below 300 m above the sea

M. Knez and T. Slabe (eds.), *Cave Exploration in Slovenia*,
Cave and Karst Systems of the World, DOI 10.1007/978-3-319-21203-6_24

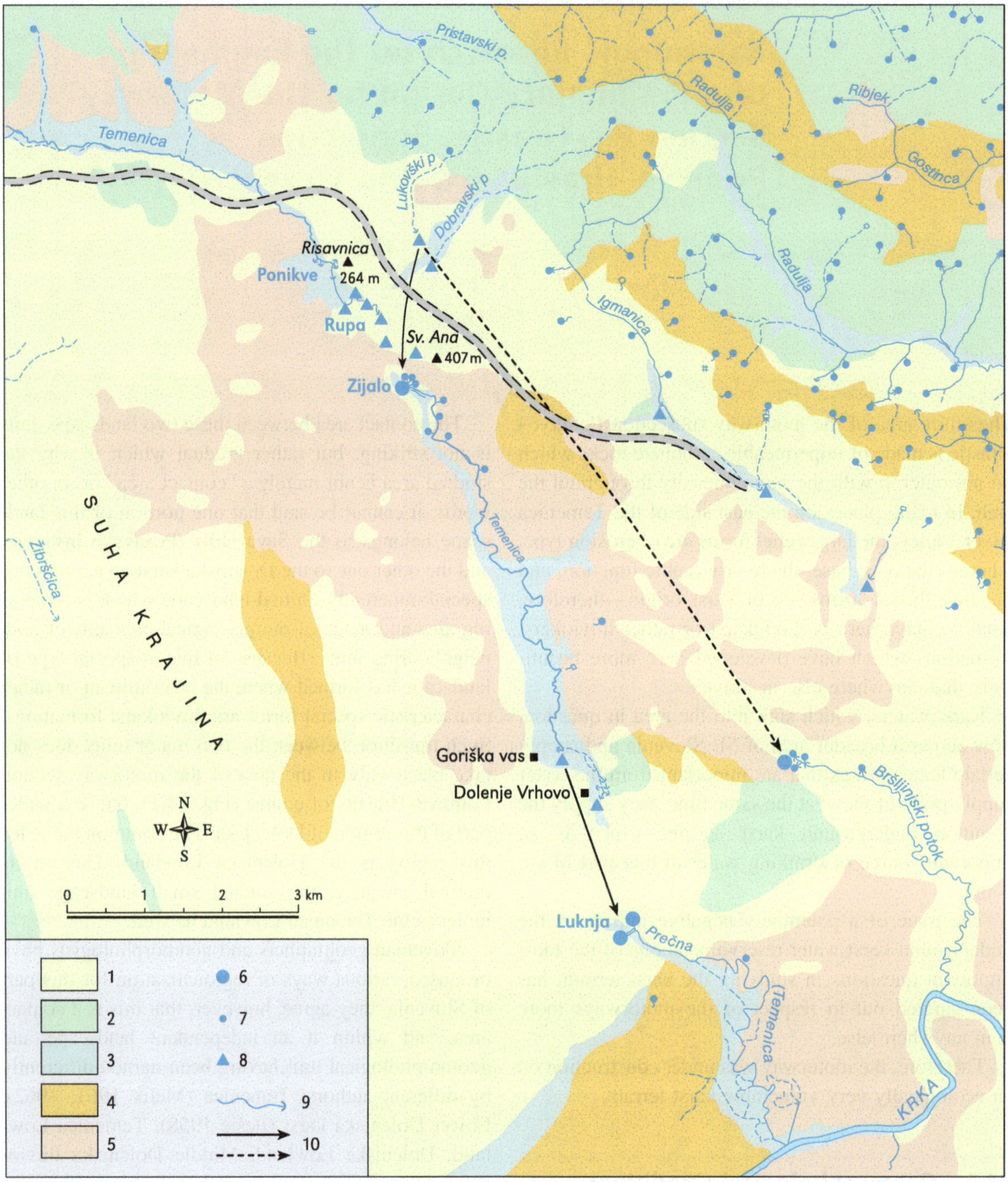

Fig. 24.1 Hydrogeological map of the area planned for the motorway section. Legend: *1* karst aquifer, *2* fissure aquifer, *3* porous aquifer, *4* very poorly permeable rocks—hydrogeological barrier, *5* very poorly permeable rocks—hanging hydrogeological barrier, *6* major karst spring, *7* spring, *8* ponor, *9* surface flow, *10* confirmed and unreliable underground water link; the route is marked with thicker grey line

level, whereas the hills and ascents in the lowland do not exceed the relative altitude of 100 m. In terms of neighbouring relief units, the lowland does not get significantly deeper either since the relative altitude differences do not exceed 300 m. The lowland is limited to the north and northeast by the hills Mirnsko gričevje (Rihpovec 492 m, Poljanska gora 477 m) and by the Radulja River, and to the south and southwest by the eastern Suha krajina region (Ajdovska planota—Srobotnik 593 m or Golobinjek 461 m above Mirna peč).

The relief forms from the north and northeast, characteristic of the Sava Hills, reach into the Temenica Lowland while from the other side, i.e. from the south and southwest, the karst relief is expanding into the lowland. The first are outgrowths—nicely rounded and flat side ridges of Mirnsko gričevje (Jagodnik 374 m, Hrib 400 m)—and intermediate valleys with relatively steep slopes and somewhat narrow but flat bottom (alluvial plain) with a more or less thick alluvial deposit. Moreover, with regard to the deposition or inundation of sediments, the transition of the landscape is clearly visible: the waters that have been flowing (in the geological past, the river basins changed several times) from the Sava Hills have been depositing quartz gravel and sand (Šifrer 1970). Good examples of this are small valleys of the streams Lukovški potok, Dobravski potok, and Igmanica. These are characteristic, nicely developed forms of the normal, erosion relief that are typical of the Sava Hills and of the "non-karstic" part of Dolenjska in general, i.e. for the relief that is typical of Dolenjska and is often called "Dolenjsko gričevje hills".

However, these relief elements do not end with a larger valley as one might expect; instead, the relief on carbonate rocks transforms into the karst relief. Moreover, streams that flow across the valleys leading from the Mirnsko gričevje hills do not flow into a bigger flow, stream, or river, but instead sooner or later sink into the karst underground, depending on the geological base, the sediments on the valley's bottom, and on the quantity of water. Generally, a "normal" valley continues into a dry valley (which is a karst formation) which has some water flow during floods and high water rises that ponors cannot swallow consistently. For example, the streams Lukovški potok and Dobravski potok usually sink instantly after flowing onto the carbonate bedrock, whereas during floods their flow extends by 1–1.5 km, reaching the village of Jezero where they finally sink below the surface. An even better example is the Igmanica Stream which usually sinks below Šentjurij; however, during floods it continues to flow 2 km across the valley and sinks below Selo. In many places on the east side of the Temenica Lowland, the large relief forms are of erosion type, whereas the rock base, the hydrographic function, and the tiny fluvial forms are of karst origin—therefore, what we have here is fluviokarst or rather fluviokarst formations (Melik 1959) which have developed here more beautifully than anywhere else in Slovenia. The karst relief that lies on the contact area of non-carbonate and carbonate rock or that has developed on carbonate rock in the vicinity of non-carbonate rock has its own name: contact karst. The lower part of the Temenica Lowland or rather its blind valley below Ponikve is also an example of such karst relief or a distinct form of contact karst. The Temenica River flows from Trebnje across the valley that cuts into the non-carbonate rock base. When it crossed over the carbonate rock, it cut into them a blind valley, into which it then finally sinks.

The western edge of the Temenica Lowland is formed by the outgrowths of the karst plateaus belonging to the Suha krajina region. Here, the plateau lies relatively low and is not very distinctly developed, in particular there is no clear distinction between the valley, the steep slopes, and the plateau. Prevalent are the rounded forms and slow transitions between individual relief units and the remnants of fluvial elements: dry valleys opening into the Temenica Lowland (at Dolenje Ponikve, at Kurja vas, and below Hrastje) and the dry valley southwest of Sv. Ana.

The central part of the Temenica Lowland is of karst origin. Flat sections alternate here, like for example the blind valley with a flat bottom before the ponors of the Temenica River below Sv. Ana (the triangle of Sv. Ana–Jezero–Gorenje Ponikve is the largest blind valley in Dolenjska) where, in addition to the Temenica River, the waters of the already mentioned Lukovški potok and Dobravski potok streams sporadically sink as well. There is a wide valley along the lower section of the Temenica River (between Vrhpeč and Vrhovo) and at the lower section of the Igmanica Stream. The midway part of the Temenica

Lowland is a lower karstified section with gently sloping hills. The exception here is a series of domed hilltops under which the Temenica River disappears underground between the ponors below Sv. Ana and then resurfaces from under the Zijalo cliff at the village of Vrhpeč. This is a series of four hills, from Smedovec (335 m) in the northwest to Peščenjak (376 m) in the southeast, with Sv. Ana (407 m) as the tallest among them. The Temenica Lowland also counts as the eastern border of the Slovenian centralized karst.

Smaller fluvial karst forms (mesoforms) are smaller dolines that can be found particularly on flat parts of land and can amount to several 10 per km^2. Bigger dolines or closed depressions and valleys are not as frequent. A typical example is the valley above the Zijalo cliff. It is about 500 m long, 300 m wide, and 20 m deep.

The karst springs and ponors are both morphological and hydrographic forms, most typical and common specifically for fluviokarst. The Temenica Lowland is mostly full of ponors that come in various forms, from ponors in alluvium and in the form of rock crevices, to true sink caves. According to their function, ponors can be either permanent or periodic. Special consideration should be given to the karst spring below the Zijalo abri below Sv. Ana where the Temenica River resurfaces as a spring. It is therefore not surprising that the spring, together with the abri, and the nearby surroundings are protected as natural monuments (Ordinance 1992; Skoberne and Peterlin 1991).

There are many places in the Temenica valley that exhibit shallow karst or rather relatively shallow piezometric level, i.e. "the level of the underground karst water" that is relatively close to the surface. This is also often one of the characteristics of fluviokarst. For this reason, the morphological changes are more apparent and demonstrate that a relatively strong karstification is taking place in the Temenica Lowland, i.e. the karst processes are prevailing over the fluvial and erosion processes, and that the hollowing-out of the karst (corrosion) is prevailing over the sedimentation processes. These are relatively young morphological changes (that can be directly recognized and observed) which, at the same time, also indicate hydrographic changes.

There are two typical examples of this: in 1959, a 3 m deep subsidence has suddenly opened 30 m further on from the ponor in Ušivk. The second example is the Igmanica Stream which used to flow a great distance away from Šentjurij; however, in the middle of the last century, a hole has opened at the church in Šentjurij which is now swallowing all of its water (Savnik 1962).

Such changes are not exclusive to the Temenica Lowland, but are common in other parts of the shallow or transitional Dolenjska karst area, as demonstrated by the examples from Škocjan pri Turjaku and from Žalnsko polje. This is the main reason why special care needs to be taken when carrying out construction works on this type of terrain; at the same time, this is yet another confirmation that karst soil water on such landscape is particularly vulnerable and at risk due to contamination. The traffic and the motorway are indeed two very significant risk factors.

24.2 Caves

There are two caves in the route of the motorway or in its vicinity. The first of the two is the Zgončarica Cave (Reg. No. 2362). It is one of the larger caves in this part of the karst. The cave has a 20 m deep entry shaft and a passage at its bottom. The flowstone in the cave is being corroded by the percolating water. The second cave is the Zgonuha Cave which is located 25 m south of the southern variant of the route (Reg. No. 2187). There is a small passage at the bottom of the large, 5 m wide entry shaft. The passage that runs in the northwest direction and contains deposits of flowstone was shaped by the water flow. The walls of the shaft were shaped by the water dripping down.

There are another 13 caves in the vicinity of the route. The knowledge of the caves in the vicinity of the route helps us expand our knowledge about the cavernosity of this region of the Karst. This is an important factor to consider during road construction, as numerous caves tend to open when carrying out the earthworks.

With 150 m, the longest of all caves is the Risanica Water Cave. The Velban kevder Cave which is located in its vicinity is over 100 m long as well. Other caves are all shorter than 50 m. The Risanica Cave is 30 m deep, the Zgončarica Cave 23 m, the Velban kevder Cave and the Zgonuha Cave 17 m, while others do not reach more than 10 m deep. The cave passages are relatively small, their diameters reaching 1–3 m. The water caves are located at the altitude between 255 and 264 m, with the exception of the Gabrovška jama

Cave which is a ponor in the middle of the alluvia-covered flatland at the village of Jezero and is located higher, at the altitude of 273 m. The caves at higher altitudes are old and dry.

The caves can be divided into water caves and old caves; the latter are dry now, but their shape and alluvia indicate that water was flowing through in the past. Small shafts lead to some of the old, in the past water caves.

The majority of the caves are located in the bed of the Temenica River which sinks into the caves Rupa (1 and 2) na Zemljančevem travniku (Reg. Nos. 2365, 2366), and only during longer droughts it sinks before it reaches these two caves. The ponors have been artificially widened and covered with bars which hold back small flowing matter. Below the shaft that is enclosed with a wall, there are two horizontal water passages that are intersected by crevices. The cross-sections of the passages are intersected in the same way. There are two more water caves at the end of the old bed of the Temenica River; these two are the biggest. High waters sink into a collapse doline near the Velban kevder Cave (Reg. No. 2189), and only the highest surface waters are supposed to reach the Risanica cave (Reg. No. 2348); allegedly, the last time this occurred was in 1932.

There is a lake at the bottom of the Velban kevder Cave. Its level varies and the cave itself is located at the altitude of app. 244 m. During the highest waters, the entrance to the cave is flooded. The passages which formed along bedding planes and fissures have been shaped by water flows. There are cups in the ceiling of the passages. The Risanica Cave is the largest cave in this region of the Karst. It is an intersected network of vertical, narrow, and horizontal passages. There are three lakes at its bottom. The walls of the passages are intersected by scallops which show traces of water flowing through the passages during high waters. The floor of the cave is covered with loam. The cave is known for its characteristic fluctuations of the underground water table. In the loam of the Požganjska jama Cave (Reg. No. 5147), which is already covered with flowstone, there are traces of periodic flooding.

Somewhat higher (Table 24.1), at the edge of the bed, old caves can be found. The majority of the caves show traces of water flowing through them in the past. These caves are Jama pod Rupami (Reg. No. 6579), Zgonuha (Reg. No. 2187) with its old passage at the

Table 24.1 Caves in the motorway route and its vicinity

Cadastral number	Name	Cave type	X	Y	Altitude (m a.s.l.)	Length (m)	Depth (m)
393	Jama pri desnem kamnu	Old cave	5,084,125	5,504,256	268	14	5
1764	Klopušna jama	Collapsed cave	5,080,628	5,508,262	355	10	5
2187	Zgonuha	Old cave	5,082,662	5,505,435	277	38	17
2188	Zemljančeva jama	Old cave	5,083,841	5,504,936	285	13	5
2189	Velban kevder	Water cave	5,082,635	5,505,130	260	113	17
2343	Luknja v Cerkvenem talu	Old cave	5,083,821	5,504,853	290	46	14
2344	Ponor of the Temenica River on Požganje	Old cave	5,082,495	5,505,285	256	12	7
2348	Risanica	Water cave	5,082,300	5,505,340	258	150	30
2362	Zgončarica	Old cave	5,081,134	5,507,442	310	40	23
2365	Rupa 1 na Zemljančevem travniku	Water cave, permanent ponor	5,083,138	5,504,250	260	31	7
2366	Rupa 2 na Zemljančevem travniku	Water cave, permanent ponor	5,083,177	5,504,270	260	31	7
5147	Požganjska jama	Old cave	5,082,507	5,505,300	257	22	7
5238	Mala Risanica	Abri	5,082,285	5,505,315	255	5	1
6233	Gabrovška jama	Water cave, permanent ponor	5,083,500	5,505,390	273	6	6
6579	Jama pod Rupami	Old cave	5,082,900	5,504,450	260	13	0

bottom of a larger shaft, and Ponor of the Temenica River on Požganje (Reg. No. 2344), while Mala Risanica is an abri.

Gabrovška jama Cave (Reg. No. 6233) is also a water cave. It is a smaller ponor at the village of Jezero. The Lukovški potok Stream which flows down from the north sinks into it. Its alluvium that covers the carbonate rocks keeps it at the surface. The rocks are uncovered at the ponor.

Other caves have been shaped by past water flows as well. There are two such caves, both located on the karstified hills west and northwest of the village of Jezero; the first is the Zemljančeva jama Cave (Reg. No. 2188), while the other is a network-shaped and creviced set of passages shaped by the water flowing through the Luknja v Cerkvenem talu Cave. This group of caves also includes the Jama pri desnem kamnu Cave and the Zgončarica Cave (Reg. No. 2362) which are both caves on the karstified hills southwest of the village Dolenja vas near Mirna Peč. These caves are dry due to the lowered level of the underground water. The cave Klopušna jama is a smaller collapse doline NE of Mirna peč.

The entire area on the carbonate rock is perforated. The waters flow even under the areas that are covered with alluvia and overflown with surface flows. The latter sink along uncovered carbonates. On the karstified and mostly higher-lying surface where more alluvium has been preserved only in dolines, the precipitation waters sink directly into the ground.

Prevalent among the familiar caves are the caves that have been shaped by waters flowing through them. On the hilly, elevated area, on the other hand, there are indeed more shafts that are nested among karren; these shafts could have been revealed during earthworks as well.

For the most part, the caves are a part of branched-out network of passages which formed along bedding planes and fissures. The passages, which are rather small, are thus either horizontal, vertical, or inclined. They have mostly retained their fissure-like shape which can also be said for the passages that have been shaped by water flows. The entire area is known for its substantial fluctuation of the groundwater table which is, although it depends on the climate, short-lived; at the same time, the area is slowly getting karstified which means that the water table is getting lower. The highest-lying caves are thus dry, the ones at the surface of the underground water are mostly flooded, while others are below the water table, of course.

24.3 Hydrogeological and Hydrological Conditions

The area of the motorway section is bound by two surface rivers: Temenica and Igmanica (Fig. 24.1). Both are typical sinking rivers accumulating water in the upper surface flow from numerous smaller tributaries, which then sink down when they reach the karstified ground and continue flowing underground towards the karst springs. Based on the determined hydrogeological characteristics, this landscape can be defined as a transitional area between the non-karst and karst flow regime. The somewhat less permeable Triassic dolomites and very poorly permeable Palaeozoic clastites in the upper course of the Temenica and Igmanica rivers, as well as their surface tributaries (Buser 1969), allow the water to accumulate on the surface. The conditions typically change at the transition onto karstified, well-permeable and predominantly Jurassic limestone. Although these are in places covered with the Pliocene and Quaternary alluvia (Pleničar et al. 1976), they do have the character of a hanging hydrogeological barrier, which is due to their small thickness, and have but a local effect on the hydrogeological conditions. With the transition onto limestone comes the typical karst terrain with numerous ponors and karst springs, together with karst caves and shafts that have formed alongside them. The surface and underground flows merge in the valleys of the Temenica and Igmanica rivers, which is why the area between these two rivers has no surface water, making it clearly different from the non-karst area with many smaller springs and streams. The part of the Suha krajina region on Ajdovska planota Plateau, which feeds the Temenica River with water from the southwest, is also a typical karst landscape.

The upstream part of the Temenica River sinks at Ponikve on the karstified Jurassic limestone through numerous ponors. The activity of these ponors depends on the hydrological conditions that reflect in the river's water level. When the water level is low, the river sinks entirely into the ponor Rupa I at the altitude of 260 m (Fig. 24.2); however, when the water reaches its lowest level, the river sinks through the ponors on the right-hand side of the riverbed approx. 150 m

Fig. 24.2 When the water level is low, the Rupa ponor at Ponikve represents the uppermost point of the surface flow of the Temenica river

upstream. As the water level rises, the ponors are unable to swallow the entire quantity of water and the river floods the regulated part of the riverbed as well, which is otherwise dry when the water level is low. The surface flow reaches more and more ponors which become active. Some of them even alter their hydrological function during high water level and start acting as springs. The collapse doline under the Mačkova jama Cave is the uppermost part of the surface flow of the Temenica River when the water level is at its highest. The collapse doline and the Risanica Cave are linked by another 600 m of fossil riverbed.

After approx. 2 km of flowing underground, the Temenica River resurfaces at the southern outskirts of Sv. Ana at the altitude of 245 m as a true headwaters with several permanent and intermittent springs, also known as the Zijalo springs (Fig. 24.3).

From the springs onwards, the river flows at the surface until it reaches the ponor below the village of Goriška vas at the altitude of 230 m where it sinks entirely (Fig. 24.4). Only during the high water level it continues its path another 1 km down the riverbed and then sinks through the swallet below Dolenje Vrhovo. At its midsection, the river has no surface tributaries.

The flight distance between the ponor below the village of Goriška vas to the springs in the Luknja Cave where the Temenica River resurfaces at the altitude of 180 m, is approx. 2.5 km long. Here, the river has a new name: Prečna (Fig. 24.5). The Prečna River has one permanent spring below the Luknja Castle and a few intermittent springs, like for example the spring coming from the Jama Cave below the Luknja Castle. At Zalog, 4 km east of Novo mesto, the Prečna River flows into the Krka River.

The smaller Igmanica River flows on the surface approx. 10 km east of the Temenica River. Afterwards, it sinks into the ponors between Dolenja vas and Zagorica, then continues its path through the underground section of the flow, and finally resurfaces, most likely in the springs of Bezgavška voda. The latter feed the Bršljinski potok Stream which flows into the Krka River at Bršljin.

Fig. 24.3 The Zijalo springs are several permanent and intermittent springs flowing into the Temenica River

Fig. 24.4 The main ponor of the midsection of the surface flow of the Temenica River above Goriška vas

Besides smaller tributaries feeding the Temenica and Igmanica rivers at the surface, the streams Lukovški potok and Dobravski potok are also worth mentioning. They have no direct surface link with both bigger rivers, but sink at the village of Jezero and feed the rivers subterraneously. The link has been confirmed by means of a tracer test, as described below.

Major karst springs emerge in the area in question only as the outermost points of the underground flow of the Temenica (the Zijalo springs and the springs in the Luknja Cave) and Igmanica rivers (the springs of Bezgavška voda). There is another spring in this area, i.e. the spring below Sv. Ana. Many smaller springs northeast of the Igmanica River accumulate the water from the mountainous, predominantly dolomite terrain, and feed the aforementioned surface tributaries. Some of these springs are captured in order to provide water supply for some of the local villages.

24.3.1 Hydrological Measurements

Measuring the flow rate of the Temenica River is carried out by the Slovenian Environment Agency

Fig. 24.5 A permanent spring of the Prečna River in the Luknja Cave

(Agencija Republike Slovenije za okolje—ARSO). The flow rate in the upper course is measured at Rožni Vrh before Trebnje, while the lower course of the Temenica River (or the Prečna River, as is its alternative name) is measured at Prečna (Fig. 24.1). The data about the minimum and the maximum flow rates for the 30-year period (1961–1990) were taken from the Hydrological Yearbooks for the years between 1990 and 1995 (Hydrometeorological Institute of the Republic of Slovenia, 1995–1997) and compared with the measured extreme values for the years between 1990 and 1995. In addition, the mean monthly flow rates for both locations were calculated on the basis of the yearbook data for the 1990–1995 period. The calculated values are shown in Tables 24.2 and 24.3 and in Fig. 24.6.

In the 1961–1990 period, the lowest measured flow rate of the Temenica River was at Rožni vrh with 60 and 560 L/s for the Prečna River at Prečna, whereas the highest flow rate for both of them was 14 and

Table 24.2 Minimum and maximum flow rates of the Temenica and Prečna rivers

	The Temenica River, Q_{min} (m^3/s)	Rožni vrh, Q_{max} (m^3/s)	The Prečna River, Q_{min} (m^3/s)	The Prečna River, Q_{max} (m^3/s)
1961–1990	0.06	14.00	0.56	21.8
1990	0.07	4.18	1.28	17.9
1991	0.09	4.18	0.86	18.9
1992	0.03	4.13	1.22	18.7
1993	0.001	4.44	1.12	18.4
1994	0.06	5.11	1.23	16.4
1995	0.06	7.44	1.35	16.7

Table 24.3 Mean monthly flow rates for the period between 1990 and 1995

	The Temenica River, Q (m^3/s)	The Prečna River, Q (m^3/s)
January	0.73	4.65
February	0.42	3.15
March	0.80	3.84
April	0.68	4.45
May	0.56	3.81
June	0.75	4.01
July	0.40	2.14
August	0.31	1.85
September	0.49	2.72
October	0.86	3.99
November	1.16	6.29
December	1.05	5.72

21.8 m^3/s, respectively. Considering these extreme values, the Temenica River in the period from 1990 to 1995 can be rated as having the lowest flow rates that are below the long-term extreme value, and as having high water levels that are quite below its maximum during the 1961–1990 period. In case of the Prečna River, the circumstances are somewhat different since extremely low water levels in the years between 1990 and 1995 were not recorded.

The diagram at the top (Fig. 24.6) shows the flow rates of the Temenica and Prečna rivers having the same scale, therefore the differences between the flow rates for the upper and the lower river course are quite clear. The comparison validates the finding that the Temenica River accumulates in its entire course both the surface and the underground tributaries. An inflow from the major karst aquifer of the Suha krajina region contributes significantly to this. Direct merging of the waters from this area into the Krka is hindered by the dolomite belt along the left bank of the Krka River from Zagradec to Dvor, which is why a big part of the eastern Suha krajina is being drained into the Temenica River (Novak 1970).

The diagram at the bottom (Fig. 24.6) shows the flow rates on a different scale in order to provide a comparison of the regime characteristics for both parts of the river. There are some deviations which are due to the differences in the feeding of individual parts of the hinterland; in general, however, both curves show similar hydrological regimes. The maximum flow rates

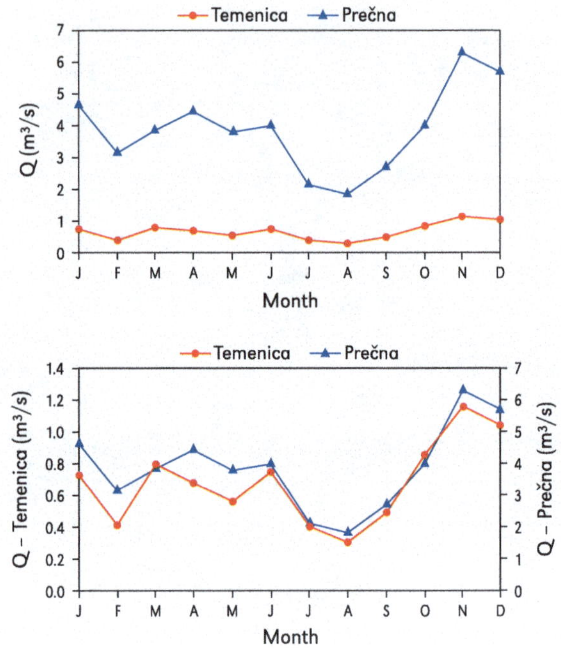

Fig. 24.6 Mean monthly flow rates of the Temenica river in the 1990–1995 period

measured were in November; however, there is no noticeable spring peak in this six year interval. The highest water levels are in August, while the lowest winter flow rate of the river is in February.

24.3.2 Directions and Velocities of the Underground Flow of Water

According to the morphology and the location of the ponors and the springs, an underground link of the Temenica River at the Rupa ponors and at the Zijalo spring is very likely. The link has also been confirmed by chemical analyses which have shown virtually identical values of water hardness (Ladišić 1981). Tracing in this area was not carried out.

Within the scope of inventorying the Temenica River, led by the Institute for the protection of cultural heritage Novo mesto, dye was added into the Lukovški potok Stream on 9 August 1994 in order to initiate the tracing. 5 kg of the fluorescent tracer called uranine was injected before the ponor leading into the Gabrovška jama Cave at the village of Jezero, i.e. during a summer drought when the water level was low (Fig. 24.7) and when the increase in the flow rate

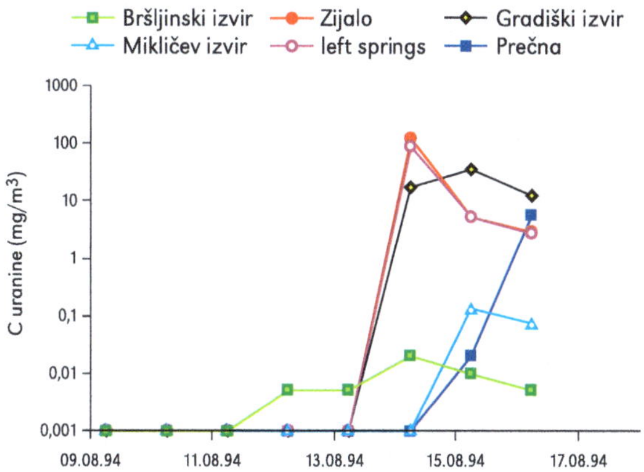

Fig. 24.7 Tracer test carried out in August 1994: detection of the tracer in individual springs (Hudoklin 1995 and the results of the fluorescence analyses carried out by the Karst Research Institute ZRC SAZU)

and the velocity of the water flow are affected mostly only by thunderstorms. We observed the Zijalo springs (the main spring and the springs Gradiški, Mikličev, Žagarski, and Jelševski izvir), the spring in the Luknja Cave, and the spring of the Bršljinski potok Stream. Sampling was carried out once a day. The fluorescence of the water samples was analysed using the Perkin Elmer LS 30 Luminescence Spectrometer and carried out in the laboratory of the Karst Research Institute ZRC SAZU.

In the Zijalo spring, the tracer did not show until the fifth day when, following a somewhat dry period of weather, a two-day rainfall provided a rise in the flow rates and thus increased the velocities of the water flow (Hudoklin 1995). The drop at this section of the water flow was as high as 17.5 ‰. The tracer showed in a high concentration (unfortunately, sampling was carried out only once a day), but then dropped rather quickly. The calculated apparent flow velocity was low (0.45 cm/s) which is mostly due to the slow travel velocity during the time before a precipitation water wave.

The tracer showed intensively in both springs from the left-hand side under the lake and in the lower-lying Gradiški izvir spring; the latter most likely also contains the water from beyond the dam. The lowest tracer concentrations were measured in the Mikličev izvir spring; the tracer in this spring appeared one day after it had appeared in other springs (Hudoklin 1995). These differences cannot be explained with the aid of the existing literature and based on research already carried out.

The tracer then travelled along the Temenica River and reappeared in the spring of the Prečna River in the Luknja Cave. The tracer's travel velocity between the Zijalo spring and the Prečna River was 4.1 cm/s which is due to the high permeability of the surface and the underground water path with the drop of 10.3 ‰ and due to the increased flow rate.

The calculated mean flow velocity from the injection point to the Prečna River, with the mean drop of 11.9 ‰, was 1.3 cm/s (Table 24.4), which is slightly lower than the calculations for the direction of the Bršljinski potok Stream (2 cm/s) with a somewhat lower drop. This indicates that the link with the Bršljinski potok Stream is questionable.

The fact of the matter is that, after heavy rainfall during a thunderstorm, the traces of uranine appeared in the Bršljinski potok Stream which was quite surprising (Hudoklin 1995). Since the appearance of the tracer was rather poor (low concentration, shorter time), it was assumed that precipitation was not abundant enough, thus only a small amount of the tracer reached the Bršljinski potok Stream. However, it is possible that the signal increased for a short period of time due to the washing-away of accumulated matter and waste from the riverbed after a longer

Table 24.4 Flight distance (l), difference in altitude (h), drop, tracer travel time (t), and apparent flow velocity (v_{dom}) in individual sections

Distance	l (m)	h (m)	Drop (‰)	t (h)	v_{dom} (m/h)	v_{dom} (cm/s)
Lukovški potok Stream—Zijalo springs	1820	32	17.5	113	16.1	0.45
Lukovški potok Stream—Mikličev izvir spring	1820	36	19.7	144	12.6	0.35
Lukovški potok Stream—Gradiški izvir spring	1950	37	18.9	144	13.5	0.40
Lukovški potok Stream—Bršljinski izvir spring	8000	87	10.9	110	73	2.00
Lukovški potok Stream—Prečna River	8500	101	11.9	176	48	1.3
Zijalo springs—Prečna River	6700	69	10.3	45	150	4.1

period of drought. We know from experience that such rises are possible when the waste is being washed away, which is why such cases require longer observation prior to injection.

Since the calculated apparent flow velocity from the ponor of the Lukovški potok Stream to the Bršljinski potok Stream was relatively high (2 cm/s), especially since the water level was low at the time and the water flow in the karst underground tends to be slower, the karst landscape between the ponor and the spring should be highly permeable (the flow in the bigger karst channel) in order for the flow to be faster. This link could have been confirmed or rejected only by repeating the tracer test at a high water level and with more frequent sampling. The Igmanica River, which sinks at Dolenja vas near Mirna Peč, most likely flows underground towards the Bršljinski potok Stream (Novak 1994); however, the link was never confirmed with a tracer test.

Based on the tracing and the geological structure it can be concluded that the karst rocks between the ponors of the Temenica River at Ponikve and the ponors of nearby streams (Lukovški and Dobravski potok) and between the spring of the Bršljinski potok Stream on the southeast probably allow for the underground flow of high karst waters. The Igmanica River sinks into these rocks as well. However, only an additional tracer test at a high water level could show the extent to which the underground water flows into the Bršljinski potok Stream as well. Such tracing would provide those highest flow velocities that occur during periods of more abundant precipitation and during periods when the karst terrain has accumulated more water (the August tracing was carried out at lower water levels). On the other hand, these velocities play a very significant role when determining the transfer of contamination through the karst terrain.

24.3.3 Physicochemical Properties and Water Quality

The literature provides little data regarding the pH and the content of carbonates, calcium, and magnesium in the water of the area in question. The data obtained are from the years 1961 (Gams 1962c), 1980 (Ladišić 1981), and 1984 (Plut 1984). However, the same literature contains no data regarding water quality.

The data regarding the composition of waters in springtime (in February) have shown that the water of the streams Lukovški potok and Dobravski potok flow from the dolomite area, as was indicated by the Ca/Mg ratio which was 1.3 and 1.0, respectively. Water hardness of the Temenica River at Ponikve was overall higher than water hardness of both streams; however, its calcium content was also higher, giving the Ca/Mg ratio of 1.9. The Temenica River which resurfaces as the Zijalo springs had a somewhat lower carbonate hardness and calcium content, but a slightly higher magnesium content, giving the Ca/Mg ratio of 1.7. This is in favour of the inflow of the more dolomite water into the Temenica River between Ponikve and the Zijalo springs, which could be explained with the underground inflow of water from the Lukovški potok Stream and the Dobravski potok stream. Compared to the Zijalo spring, the water composition of the spring of the Prečna River in the Luknja Cave showed a lower carbonate content and a particularly lower magnesium content (Ca/Mg = 3.9). This indicates that, along the way, the water is fed with inflows having a lower magnesium content or with the water containing more limestone.

Moreover, the flow rate of the Prečna River in the Luknja Cave is higher than that of the ponor, which can be explained with the mixing of the underground water with the waters of the Suha krajina region that have a high limestone content (Ladišić 1981). June measurements of the Prečna River showed higher water hardness than the February measurements (Plut 1984); however, the water had a similar Ca/Mg ratio which indicates a seasonal water hardness fluctuation. The Bršljinski potok Stream has a similar water composition. Despite high water hardness and the mixing of the dolomite and limestone waters, the Temenica River is not discharging tufa unlike the Krka River (Gams 1962c).

The literature contains no data about the water quality of the Temenica River and its tributaries. On the other hand, the settlement along its current, including Trebnje, numerous small settlements, and rail and road links, have an impact on water quality. In the past, it was already established (Plut 1984) that the spring of the Prečna River could not have been used for the water supply of the population due to the ecologically sensitive hinterland.

A possible cause for the deteriorating quality of the Temenica River is increased human activity (settlements, industry, agriculture etc.) in its hinterland. The waste material on the karst surface is being washed directly into the karst and then into the Temenica River and the Bršljinski potok Stream underground. Road traffic pollutes the area even in normal circumstances, not to mention traffic accidents resulting in the spillage of hazardous and toxic substances which provide that much greater of a threat to the karst water, when the roads are not waterproofed and such accidents cannot be brought under control (Kogovšek 1995a, b). Another cause for the deteriorating quality of the Temenica River are contaminated surface streams flowing into this region of the Karst and into the Temenica River.

24.3.4 The Importance of the Karst Water in the Area of the Temenica River

The water supply is provided by a handful of captured smaller springs on the dolomite area northeast of the Igmanica River. The captured springs are important locally and have relatively small hinterlands.

The karst springs along the Temenica River are not being utilized as the sources of drinking water; however, the water from the spring of the Prečna River is routed to a large fish farm in the Luknja Cave.

It should be pointed out that the Temenica River and the springs along it are an important natural monument. With the Ordinance designating natural sites of special interest and immovable cultural and historical monuments in Novo mesto Municipality, which was published in the Official Journal of the Republic of Slovenia No. 38/92, the Temenica River, the Zijalo springs, and the Luknja Cave were protected.

24.4 Conclusion

The entire landscape in question belongs to the Krka River basin. Both surface and underground waters flow toward the Temenica and Igmanica rivers which act as the local base level. Based on the existing data, the watershed between both rivers on the karst terrain could not have been determined precisely; on the other hand, one needs to consider the possibility of separating a single stream into two or more separate streams (bifurcation), as indicated by the tracer test carried out in the Lukovški potok Stream. Nevertheless, it can be assumed that the larger part of the motorway section of the Sv. Ana variant is draining towards the Temenica River. The study of the hydrological conditions has shown that the landscape in question can be defined as a transitional area between the non-karst and karst flow regime. The surface water flows sink down when they reach the karstified ground, flow underground, and then resurface as karst springs.

The variant of the motorway route Sv. Ana runs predominantly on the karst ground. The surface waters infiltrate into the underground very rapidly, either locally through the ponors or diffusely through the fissured and karstified carbonate rock; in the underground, they flow rapidly (in some cases also over great distances) through karst channels and expanded fissures towards karst springs. The downside to this is a poor self-cleaning ability which increases the danger of rapid spreading of contamination. The vulnerability of the karst aquifer to various means of contamination is high.

The waters in the area in question flow towards the Temenica River. An outflow in the direction of the Bršljinski potok Stream is also likely, but the link would need to be confirmed first. Contamination from the motorway (permanent outflow of contaminated water or a single spillage of hazardous and toxic substances caused by a traffic accident) would very quickly sink into the karst underground and flow towards the springs along the Temenica River, provided that the road would not have been waterproofed (Kogovšek 1995b, c). The studies of the flow through the vadose zone carried out thus far (Kogovšek 1995a) have shown that the flow rate during precipitation is very high (at about 2 cm/s), meaning that the contamination would transfer rapidly. On the other hand, the transfer of contamination is very slow during dry periods or the washing-away does not take place before the next more abundant precipitation; what is more, prolonged retaining of water and longer delays need to be taken into account as well. The tracer test has shown that the velocity of the water flow from the Zijalo spring to the Prečna River is very high. It can be expected that these velocities and with them also a more rapid transfer of potential contamination can increase even further during high water levels caused by the abundant spring and autumn rainfall.

Although the Temenica River and the springs alongside the river are not a source of water supply, they are protected as natural monuments, meaning that suitable measures need to be taken in order to preserve or further increase their quality. There is a large fish farm in the Luknja Cave that has water routed from the spring of the Prečna River. Potential contamination in the hinterlands would have a negative impact on the living resources in it. The wider area has a few small springs that have been captured for local water supply; fortunately, they have not been compromised by the construction of the motorway.

In general, from the geomorphological point of view, a more suitable motorway route would be the one that would run across the non-carbonate surface, i.e. across normal relief, as much as possible.

Since the chosen Sv. Ana route runs predominantly over the fluviokarst surface and over the shallow karst along the contact area with contact Karst, special attention and care were required, especially in places where the motorway runs across the contact point. These places could see rapid morphological changes together with hydrographic changes. Because of this, the stability of the terrain can change rapidly (both in terms of space and time), particularly due to sudden changes (each human intervention is "sudden" compared to the progression of geological and geomorphological processes); at the same time, this is a transitional terrain (with nearby ponors, a shallow unprotected zone or epikarst zone, and the surface of the underground karst water) which is particularly sensitive to contamination.

The route has one cave which is the second deepest in this area, and another one in its near vicinity. It can also be assumed that the cavernosity of the karst terrain is also good.

The area of the dry Temenica riverbed and its edge are a unique piece of karst terrain which should be preserved. It boasts with one-of-a-kind karst surface, caves, and hydrological characteristics, while the Temenica River is a protected natural monument. Our suggestion was to take into consideration all stated facts, both during the construction of the motorway and during its operation, and to avoid unnecessary degradation of this piece of unique and typical Dolenjska karst area as much as possible.

25 Planning of the Motorway in the Pivka and Reka Rivers Catchment Areas, the Evaluation and Reduction of Impacts on Known Caves

Slovenia and Croatia are connected with only one motorway (A2, E61/E70) that connects capitals (Ljubljana, Zagreb). Motorway A4 (E59; Maribor–Zagreb) and semi-motorway H6 (E751; Koper–Pula) are currently under construction. Beside their bilateral character, all three roads act also as transnational corridors connecting Croatia with Austria or Italy, respectively.

To make the connection between Croatia and Italy safer and more efficient by diverting traffic from local roads, from 2005 onwards, the possibility to build a new motorway between actual motorways A1 (E61; in Slovenia) and A7 (E63; in Croatia) has been studied. Since the planned motorway is going to cross extensive karst areas or superficial drainage areas that are drained through karst caves (Reka River-Škocjanske jame Caves and Pivka River-Postojna Cave System) where important regional aquifers, high (underground) biodiversity and high density of natural monuments (all caves in Slovenia are protected as natural monuments of national importance) can be found, the Karst Research Institute ZRC SAZU was involved in several phases of planning (study of variants, environmental impact assessment) to evaluate and reduce impacts especially on caves to acceptable level. As many caves drain water from the project area, the protection of caves is thus also resulting in the better protection of karst water.

As all proposed motorway variants at least partly extend over the Reka River catchment area, the study is interesting due to the evaluation of the impact on the Škocjanske jame Caves as caves of international importance. As such the Škocjanske jame Caves were evaluated separately from the viewpoint of criteria relevant for the recognition of the Škocjanske jame Caves as a world heritage site recognized from 1986 by Unesco. Beside this reputation, the Škocjanske jame Caves were declared as Ramsar wetland and also included into MAB (Man and Biosphere) programme in 1999 and 2004, respectively.

25.1 Geomorphological and Hydrogeological Situation Along the Proposed Motorway

To make planning easier and more independent, the proposed motorway variants were grouped in the southern and northern group; within the southern group two variants (vS1, vS2) in one corridor are proposed, while the northern group consists of four variants (vN1, vN2, vN3, vN4) in three corridors (Fig. 25.1). The southern group is situated mostly on impermeable silicates from which water is drained superficially into the Škocjanske jame Caves. Only the most southern part is karstic with several hundred-metre-long active and relict stream caves of contact karst draining waters underground into the Kvarner Bay (Croatia). In the most southern part, the northern group possesses similar characteristics as the southern group while the northern part is karstic. Intensive karstification is evident especially on higher plateaus with high cave density and caves as long as

M. Knez and T. Slabe (eds.), *Cave Exploration in Slovenia*,
Cave and Karst Systems of the World, DOI 10.1007/978-3-319-21203-6_25

7.7 km (Vodna jama v Lozi). Regarding to tracing tests, the northern part was recognized as complex bifurcation area draining waters toward Malenščica spring (Planina Polje/Planinsko polje), Vipava spring (Vipava Valley) and Timavo spring (the Adriatic Sea). Southern and northern motorway variants drain water into the cross-boundary karst aquifers shared with Croatia (aquifer of Čičarija and Brgud lowland) and Italy (aquifer of the Karst plateau).

25.1.1 Possible Impacts Due to Motorway Construction and Methodology

Several changes are introduced during motorway construction that can result in an increased impact on karst caves:

- relief modification (the levelling of surface implementing roadcuts and embankments),
- land use changes with modification of runoff (natural vegetation with slow runoff and high retention → sealed surfaces with fast runoff and no retention),
- permanent (regular traffic) and occasional (winter road salting, accidents with spillage of dangerous liquids) changes in the chemical composition of water.

Impacts can be permanent (e.g., roadway) or temporary (during construction). While some changes are mainly reversible (e.g., building and removal of temporary construction platform), direct impact on caves can be generally considered as irreversible.

To define methodology, possible impacts were classified into two groups and two subgroups:

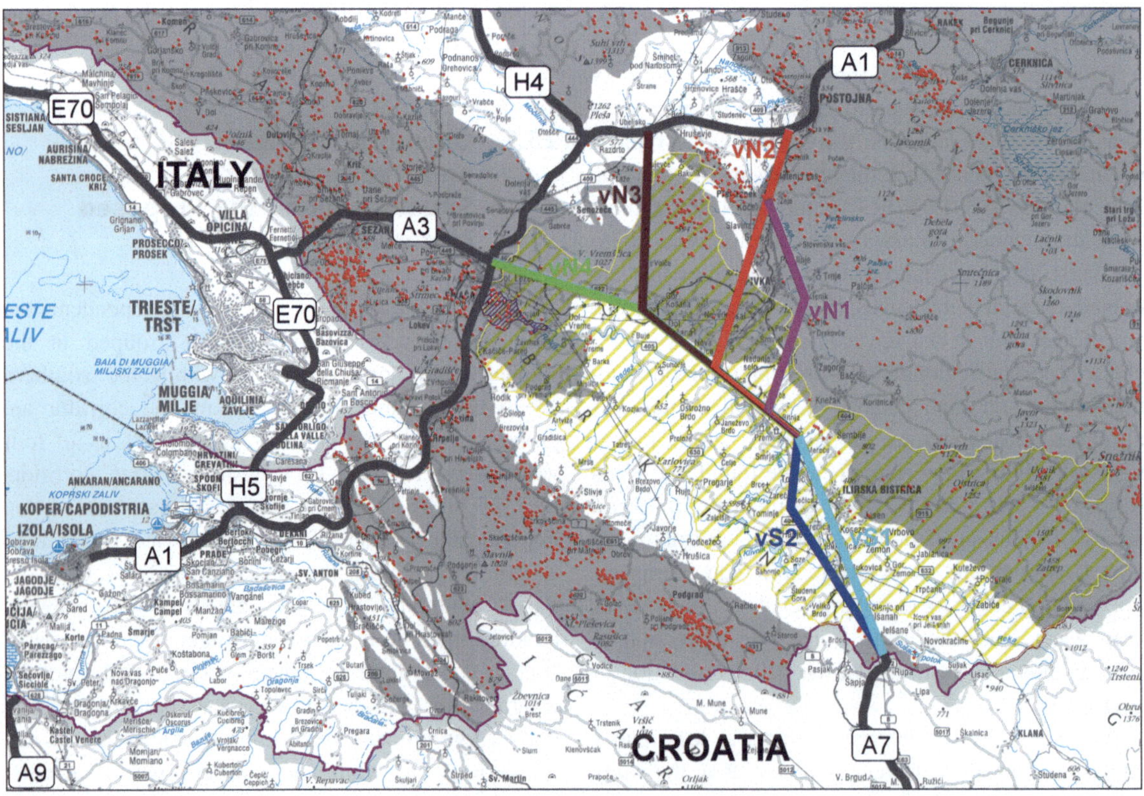

Fig. 25.1 Built motorways (*dark grey lines*) and planned versions of motorways (*coloured lines*) in relation to the Karst (*grey polygons*; only for Slovenia), caves (*red dots*), Reka River catchment area (*yellow pattern fill*; only for Slovenia) and Škocjan Caves Regional Park (*violet pattern fill*). Data sources: The Surveying and Mapping Authority of the Republic of Slovenia (topographic map), Geological Survey of Slovenia (extension of carbonate rocks on the surface)

- direct impact,
- indirect impact:
  - impact by vibration,
  - impact by waterflow.

Spatial analyses were done using ArcMap 10 software in scale 1:5000 on which plans and profiles of motorway variants were prepared. To define direct impacts and indirect impacts as a result of vibrations, distance between planned motorway and cave passages should be used. However, due to the lack of digital plans for all caves, coordinates of the entrance together with cave length was used from the Slovene Cave Registry holding data for 11,393 caves (March 2015); using both data and nearest distance between the cave entrance and motorway axis, factor A (A = d–D) was calculated representing the difference between the distance of cave entrance from the nearest motorway axis (d) and the length of cave passages (D; Fig. 25.2). Positive A (d > D) indicates the absence of direct impact since, even theoretically, cave passages are too short to reach the motorway. Negative A (d < D) indicates possible direct impacts if cave passages are directed toward the motorway; to define real conflict, plan (and to some extent also extended profile) of caves with negative A was georeferenced and threatened passages were defined. As coordinates of cave entrances can be inaccurate and the construction area several tens of metres away from the motorway axis, caves with A between 0 and 100 m were also taken into consideration; this and caves with negative A were found in the field and entrance coordinates were defined with a hand-held GPS navigator. Although we know that some coordinates of cave entrances can be inaccurate for several hundreds of metres, the highly increased number of caves made such an extensive examination economically ineligible. Regarding to past experiences with motorway construction, indirect impacts by vibration were expected in caves with A below 100 m during motorway construction; real indirect impact by vibration relies on the distance of cave passage from the construction area, the type of earth removal (use of dozers and excavators, hammers or explosives-rock blasting) and the specific vulnerability of cave formations.

To define the indirect impact on caves by water hydrological function of caves (draining authigenic or allogenic water) together with superficial and underground flow direction was taken into account. Although tracer tests provide information for many locations on a regional scale (Habe 1968; Gospodarič et al. 1970; Habič 1989a, b, c, d; Kogovšek 1999; Petrič 2009 after Schulte 1994; Petrič and Šebela 2004; Ravbar 2007), they were useful at least to estimate the general underground water flow direction (Fig. 25.3). Impact assessment covered both, changes in chemical composition as well as hydrodynamic changes. While downstream impact seems logical, less known but in some cases also important upstream changes (e.g., due to drainage of the tunnel) were took into consideration. Impact by water can also occur due to the concentration of meteoric wastewater, sinking while dry caves can be fed by treated wastewater. And vice versa, once dry caves can be (occasionally) fed by a concentrated flow of meteoric water from the

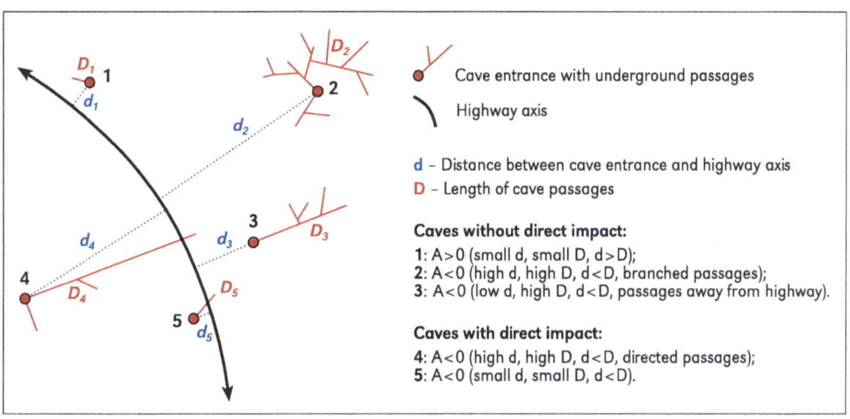

Fig. 25.2 Schematic presentation of methodology used to define possible direct impact using existing (cave length) and calculated data (distance of the cave entrance from the highway axis)

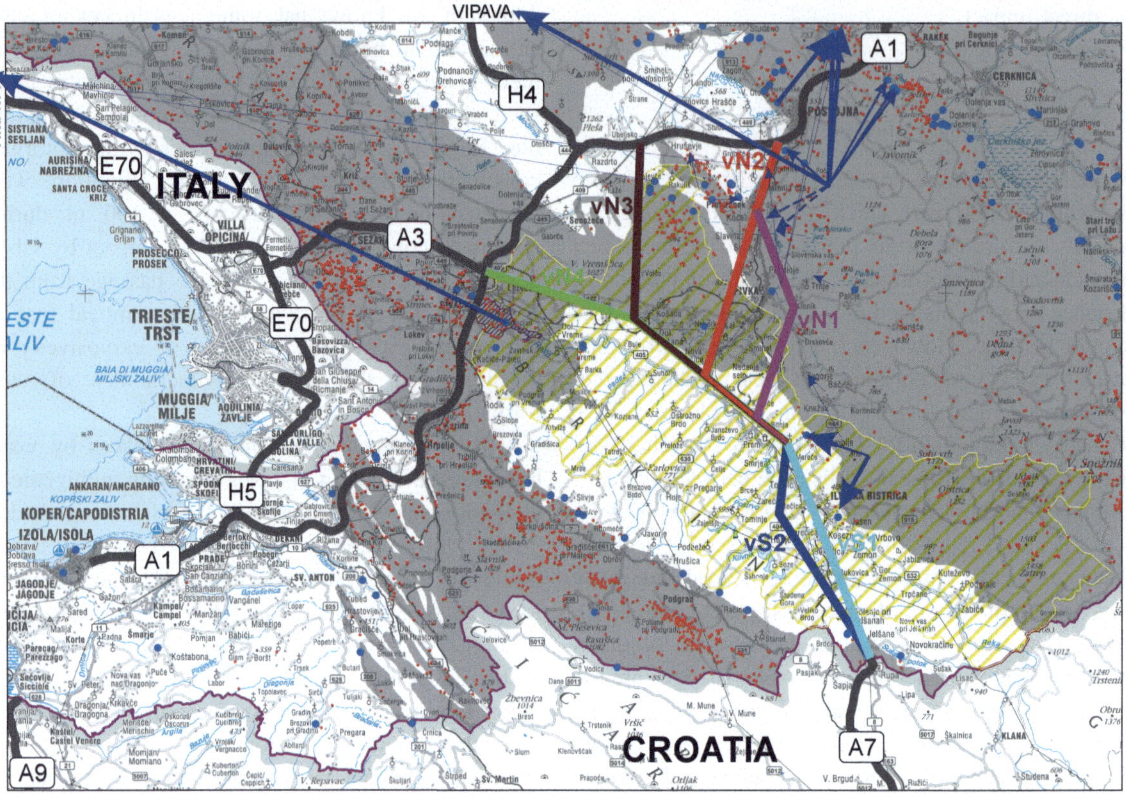

Fig. 25.3 Traced underground links indicating main directions of the underground water flow and bifurcation area in the Upper Pivka valley. *Blue dots* represent stream caves. Data sources: the same as in Fig. 25.1

roadway; to evaluate possible contamination in the vadose zone with water draining from the motorway, flow direction of percolation water was defined with a 45° angle downwards from the construction area.

The evaluation of the impact on the Škocjanske jame Caves took into consideration criteria that were recognized in the area of the Škocjanske jame Caves by Unesco. Therefore, special care was dedicated to impacts on significant on-going geological processes in the development of landforms (criteria viii) and impacts on the exceptional natural beauty's aesthetic value (criteria vii; Škocjan Caves 2015)—not only caves, but also broader contact karst phenomena (collapse dolines, ponor) and the visible cultural landscape. In practice, following critical evaluation of possible impacts was taken into consideration: direct and indirect changes of morphology, contamination and hydrodynamic changes of sinking and underground water and visual degradation of the area, as seen at least from major viewpoints.

25.2 Results

25.2.1 Direct Impact

Regarding to negative A factor, 9 caves can be subject to direct impact. However, the majority of those caves are relatively far away from the motorway, but have extensive cave passages developed on a relatively small area (type 2; e.g., Kačna jama) or directed predominantly away from the proposed motorway (type 3; e.g., Postojna Cave System). In the case of 88 m long and 36 m deep Brezno na Lovcah shaft (Reg. No. 7575), negative A factor indicated possible impact, yet according to the original plan of the cave, cave passages extended away from the motorway. Verification of entrance location showed 324 m of error; regarding to new coordinates of the entrance, the latter was situated directly at the axis of motorway variant vN3 (Fig. 25.4). A visit to the cave indicated

Fig. 25.4 Location of the entrance, the plan of Brezno na Lovcah and the course of the motorway before (*light red*) and after research and optimization (*red*)

error in the direction of cave passages and a new mapping of the cave was done.

Among the 11 caves with slightly positive A factor (between 0 and 100 m), slightly more than 50 % (6) caves had coordinates identical to the newly defined (less than 30 m of difference), which did not significantly change A factor. In addition to these unaffected caves, one cave was destroyed during A1 motorway construction. Three caves had too imprecise coordinates of the entrance or are blocked at the entrance, making it impossible to locate them in the perimeter of 50 m around the official coordinates. According to access description, one cave (Jama v Repljah) was found 215 m from the official coordinates that increased A factor. Since the 18 m long cave passages are parallel to the motorway, impact is not increased. However, due to subsequent safety spatial optimization, the distance of the cave entrance from the motorway axis was reduced to 50 m. A relatively short distance together with the location of the cave entrance at the proximity of the tunnel portal where the constructional platform is planned can result in direct impact.

25.2.2 Indirect Impact by Vibration

Indirect impact by vibration is, due to the 16 m distance between cave passages and the planned motorway, expected only in the case of Brezno na Lovcah shaft. Falling of the weakly adhered stalactites and thin stalagmites is expected in the case of deep rock blasting. Although the distance between the motorway and Jama v Repljah Cave is relatively short (50 m), the cave cannot be indirectly impacted due to lack of fragile speleothems. Gabranca Cave is located too far and too deeply to suffer from vibrations.

25.2.3 Downstream Indirect Impact by Waters

Downstream indirect impact by waters is possible by sinking streams (the Reka, Pivka and Nanoščica rivers, Drščevnik Stream at Dolenje Polje) or by percolation water that is drained from the planned motorway through stream caves. The Adriatic Sea and Black Sea catchment areas contain 32 and 14 such caves (without caves downstream of Planina Polje), respectively. Additional 14 possibly impacted caves can be identified in the bifurcational area between the catchment areas. For all possibly impacted caves, catchment areas were estimated; results for the Adriatic Sea, Black Sea and the bifurcational catchment area indicate 18, 14 and 5 caves, respectively. Within the Adriatic Sea catchment area, following groups of caves were identified: spring-ponor cave Gabranca, 2 ponor caves in Reka riverbed, 6 ponor and through-flow caves in the Slovene part of the Karst aquifer, 7 spring caves at the Vipava spring and 2 ponor caves in the area of Dolenje Polje. Within the Black Sea catchments area, two main groups of caves at risk exist: caves within Postojna-Planina Cave System and caves in the area of Rakov Škocjan. Despite several tracing tests, the most complex situation is in the bifurcational area between catchment areas, where sporadic water-table caves that can act as temporary springs were identified as possibly endangered.

Due to different motorway proximity, the impact on specific stream caves varies a lot—from destructive to barely noticeable. In addition, the mitigation of spillage and protection of ponor caves is much easier since the interception of pollutants can be much more easily solved at the surface stream while through-flow and spring caves are at higher risk as the interception of pollution at the karst surface is estimated as negligible (Kogovšek 2011b; Kogovšek and Petrič 2011). However, during high discharge (above 50 m^3/s) oil spillage at The Reka River cannot be intercepted due to the highly turbulent flow and due to safety reasons. Among the most endangered caves is the 214 m deep ponor-spring cave (estavelle) Gabranca that is fed by:

- the temporary surface stream Sušica (possible contamination during intensive precipitation following low-middle water levels, when the cave acts as a sink),
- the regional phreatic aquifer accessible through the deepest sump (possible contamination and hydrodynamic impact through regional aquifer), and
- local percolation water that drains through the cave permanently (possible contamination, hydrodynamical and hydrochemical impact on the stream that deposits flowstone).

While only general mitigation measures are effective for caves fed by regional flows, spatial optimization of motorway variant vN2 near Gabranca was proposed and accepted to reduce impacts on the cave by local percolation water (next subchapter).

25.2.4 Upstream Indirect Impact by Water

Due to the fact that the motorway lies substantially higher than the stream caves, upstream indirect impact by water is not expected.

25.2.5 Hydrodynamical Impact

Besides the Gabranca Cave, the hydrodynamical impact on caves can be neglected. This is true even for the Škocjanske jame Caves, where, according to different motorway variant, from 0.09 to 0.17 % of land will be sealed with asphalt (roadway, petrol stations and resting places). Taking into account the actual surface of the urban and transportation area, the motorway will increase those areas for only 2.5 to 5.0 %. Despite this fact and despite avoiding flood plains of the Reka River, the cumulative effect of all fast runoff surfaces should be taken into account, since built areas increase flood peaks and reduce retention capacities of soils and vadose part of the karst aquifer. However, due to the high permeability of the vadose zone and increased retention in the epiphreatic zone as a result of water level rise, motorway construction on the karst surface is expected to have a lower impact on hydrodynamics in comparison with non-karst areas. Just the opposite, impact on water quality on karst is expected to be worse due to very limited self-cleansing capability in the karst surface.

25.3 The Karstological/ Speleological Mitigation and Optimization of the Planned Motorway

Mitigation measures and optimization are considered as activities to eliminate or reduce conflicts between planned activity and environmental issues—in our case, caves. Optimization can be spatial (the horizontal and vertical adaptation of the motorway) as well as non-spatial (the adaptation of the roadcut slopes, the transformation of free roadway with a tunnel or viaduct, the intensive treatment of roadway wastewater, preventive measures for fast and effective action in case of oil spillage). In both cases, the impact on caves can be eliminated or reduced. However, spatial optimization is limited with the movement of the motorway for few tens of metres otherwise usually other problems appear and modification can be treated as a new motorway variant. In our case, optimization was used at four locations among which two are described in following paragraphs.

Without modification, planned motorway variant vN3 would directly eliminate 90 % of the Brezno na Lovcah Shaft (Fig. 25.4) while the rest 10 % would be highly devastated indirectly by vibrations. Since spatial optimization could solve the problem of direct and indirect impacts, movement of the motorway for 40 m eastward has been proposed and accepted. However, due to limited preciseness of GPS and use of 1:5000 scale for the positioning of the motorway and definition of the roadcut limit, detailed activities are planned

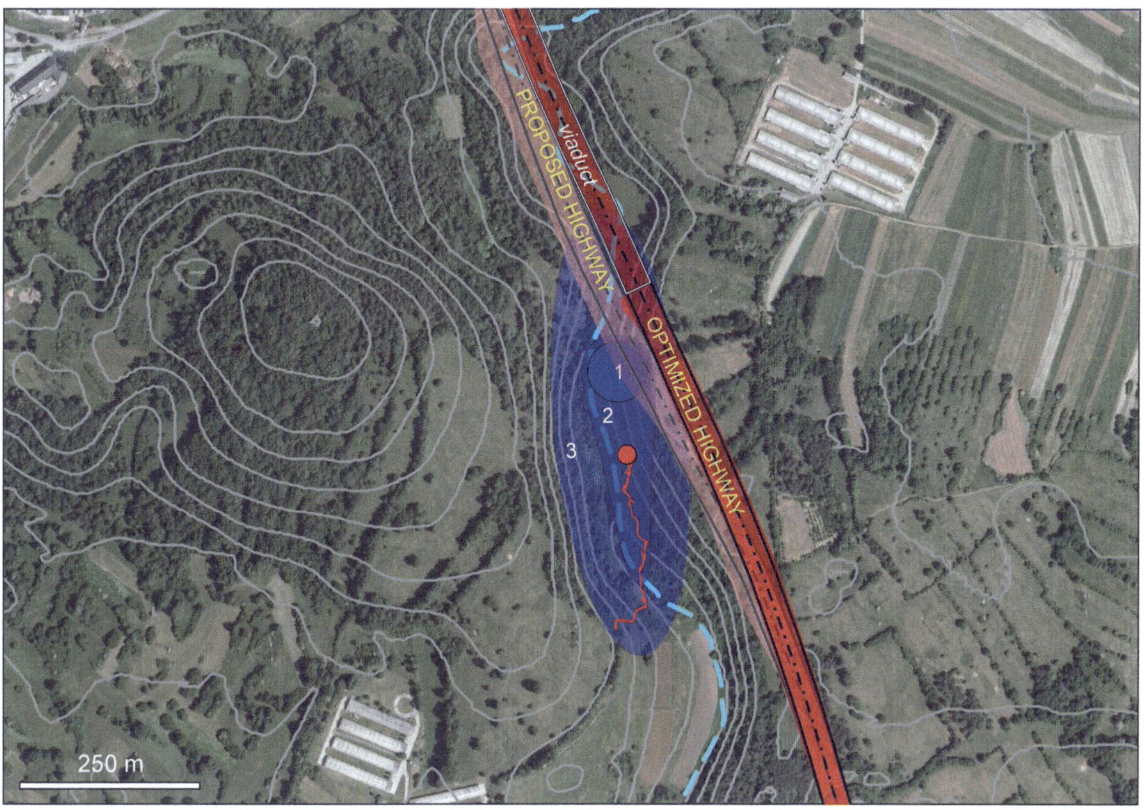

Fig. 25.5 The entrance to the Gabranca Cave (*red dot*) with underground passages (*red polygon*), occasional superficial flow of Sušica (*blue dashed line*) and local authigenic recharge zones (*blue*; *1* the core recharge area, *2* the underground feeders recharge area, *3* the wider bifurcational area from where the flow into the cave is probable but less important) in relation to the proposed (*light red*) and optimized course of the vN2 motorway (*red*)

in the future to ensure absence of impacts. In the case of shorter distance, low-intensity rock blasting is proposed.

In case of the Gabranca cave, the local authigenic recharge area was defined on the basis of discharge and speleological-hydrogeological data (25° strata dip followed by water channel). We found out that spatial optimization can mitigate the direct impact on the core local authigenic catchment area with the lowering of the roadway for 2 m to reduce the embankments and movement of the motorway for 35 m toward NE (Fig. 25.5); greater movement would open new demographic, agricultural and traffic-safety problems. To further minimize the impact area located on the possible local authigenic catchment area (zone 3 in Fig. 25.5; especially concerning $pCO_2 \approx 1.4\ \%$ in percolation water that was concerned also in the study of Milandre Cave (Switzerland) by Jeannin and coworkers (2013)), 20 m high and 500 m long viaduct was proposed and accepted by the contracting authority. To minimize the impact on the local superficial catchment area, interception and cleaning of roadway water was adopted. Since the impact on regional authigenic infiltration cannot be solved by spatial optimization, interception, cleaning and sinking of roadway water through dense array of artificial sinking areas along the motorway was proposed in a general way. The definition of the Gabranca's regional authigenic recharge area and density of the artificial sinking area array will be, due to the lack of appropriate data, studied in a specific study in the future.

25.4 Conclusion

The easiest protection of known caves (and indirectly karst water) during motorway construction can be carried out during the planning phase. Due to a fast planning procedure that does not allow long-lasting studies, the crucial factor that determines effectiveness is the extent and preciseness of data set—especially on the entrance location and the location of cave passages. To determine the impact on stream caves, knowledge on catchment areas is inevitable. However, scarce water tracing on a local scale, the lack of sampling in caves and hydrogeological complexity rarely provide a sufficient amount of data especially within the authigenically recharged bifurcational area (like the upper karst basin of the Pivka River).

Methodology used in this study to define the direct impact and indirect impacts of vibrations on caves has been proved useful on the level of general impact assessment. It is of crucial importance to upgrade by digitizing and georeferencing existing cave maps and field work, during which the correct location of cave entrances (and underground passages) can be defined. Regional water tracing takes at least several months and is as such, especially in complex hydrogeological situations or in several hundreds (tens) of km^2 big aquifers, too time-consuming and expensive to be performed to a relevant extent.

Spatial optimization and mitigation measures can importantly reduce local impacts on caves, especially direct and indirect impacts from vibrations. When dealing with the regional scale, spatial optimization proved ineffective in reducing the impact that water has on caves (e.g. in the case of Gabranca Cave). Effective mitigation measures can be interception and treatment of water in artificial oil and sediment interceptor that can be in desired density located along the motorway. However, past bad experiences along the A2 motorway between Unec and Postojna have shown that maintenance is crucial in reducing contaminants (Kogovšek 2011a); otherwise immediate random infiltration along the roadway can be even better from the perspective of water quality. In the case of actual water treatment, one can expect an occasionally high concentration of some very soluble contaminants in roadway water (like NaCl used during winter) and high chemical oxygen demand-COD (Kogovšek 2011a) whose interception is not economically eligible. Since spillage of dangerous chemicals cannot be avoided, it has to be anticipated and limited by the implementation of preventive and fast curative measures—especially in karst areas.

Part VI

Construction and Use of Motorways with Regard to Karst Waters

26 Biological Assessment of Habitats and Fauna in the Škocjanske Jame Caves and Reka River in the Motorway Construction Area of Influence

Škocjanske jame Caves are in the karst area of the Škocjan Caves Regional Park. The name, i.e. Škocjanske "caves" and not "cave", explains that it is the system of a large number of caves. As already mentioned, the Škocjanske jame Caves are on the list of World Heritage by Unesco and also on the Ramsar list of wetlands.

The Reka River runs above ground for nearly 52 km from its spring, located in impermeable flysch terrain near Klana, to Škocjanske jame Caves, where it sinks into the karst underground (320 m a.s.l.). The main tributaries are the karst springs of the Bistrica River near Ilirska Bistrica. From Bistrica and almost to the swallow hole of Škocjanske jame Caves, the Reka River again runs above ground, where it gains surface waters from the flysch terrain (Pipan 2000a).

The importance of Škocjanske jame Caves is therefore in their position, in their function as a sink and as a part of the water system of the karst aquifer. The importance is shown also in a unique heritage, which includes a diverse cave fauna, and fauna in the surface and underground river. In this paper the diverse aquatic fauna of Škocjanske jame Caves is presented.

26.1 Description of Methods and Techniques for Biological Sampling of Invertebrates in Aquatic Ecosystems

The biological assessment of water quality in running waters can be defined as the systematic use of biotic responses to evaluate changes in the aquatic environment. Assistance in this process is provided by bio-indicators, i.e. aquatic organisms at various trophic levels, such as bacteria, cyanobacteria, algae, macrophytes, macroinvertebrates and fish, whose presence or absence indicates the ecological state of the biotope. The biotic basis of monitoring is formed by populations and communities, and within them indicator species for specific abiotic conditions and chemistry (Toman 1999).

Sampling of biological material is an important part of faunistic as well as biodiversity studies, in various aquatic and terrestrial environments. Many sampling methods and techniques are well known and standardised to obtain optimal quantitative and qualitative data. In environments or habitats which are in remote places or difficult to access, standard sampling methods are often not usable. Researchers in such situations modify the known methods or introduce new methods that are often more suitable, convenient, and cheaper.

For studying the communities of aquatic invertebrates a simple method of filtering water through a 60 to 80 μm net can be used. The well-known planktonic or hand nets (Fig. 26.1) can be used for sampling the macroinvertebrates, whose body size exceeds 0.5 mm. For sampling smaller organisms, the so-called micro- and meiofauna with a body size between 0.06 and 0.5 mm, and for sampling in small aquatic habitats (pools, ponds), the standard sampling methods are not appropriate.

Pipettes are the basic laboratory equipment and the simplest device for collecting water from very small water bodies, crevices, and small pools at the top of flowstones and from other cave formations (Fig. 26.2). Water from the pipette can be collected directly into

M. Knez and T. Slabe (eds.), *Cave Exploration in Slovenia*,
Cave and Karst Systems of the World, DOI 10.1007/978-3-319-21203-6_26

small plastic bottles, into which the proper amount of the fixative is added before transport to the laboratory.

A cut plastic bottle (the cut is slightly inclined on the longitudinal axis) is very useful for sampling in small pools with clay or other sediment on the bottom and can be used on the surface or underground (Fig. 26.3). With the cut plastic bottle the bottom of the aquatic body is stirred in order to move the animals attached to the bottom into the water column. The collected water and aquatic fauna is then filtered through the filtering bottle (Fig. 26.4). The filtering bottle has holes covered with a net (60 or 80 μm) on two sides. The water from the bottle exits the net, while the animals are retained in the bottle. Afterwards the sample from the filtering bottle is emptied into a container for transport and for sorting in the laboratory.

Because epikarst (see Fig. 26.9) is almost impossible to sample directly, epikarst fauna must be explored indirectly by taking samples of percolation water (Pipan 2005). Unlike most subterranean habitats, continuous, long-term sampling of epikarst communities is possible. Epikarst communities are best sampled by collection and filtration of drip water over extended periods of time. The epikarst fauna in Škocjanske jame Caves was sampled indirectly by taking samples of percolation water that drips directly from the ceiling. The water from trickles was directed into a funnel and then into a collecting container with plankton netting (Fig. 26.5). The plastic container had

Fig. 26.1 Collecting of aquatic fauna using a fine mesh net

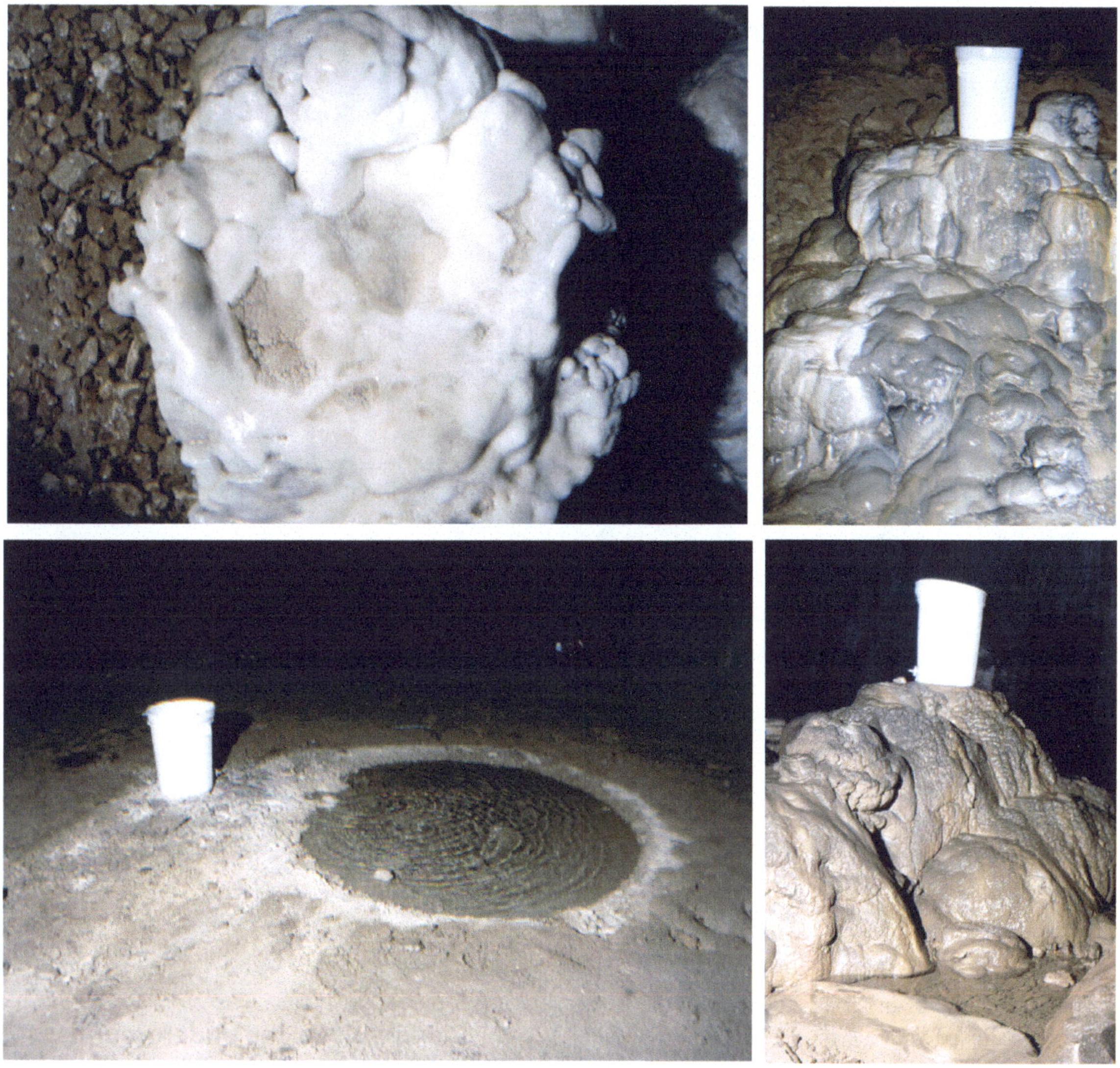

Fig. 26.2 Different types of pools filled by percolating water

holes on two sides, covered with a net of 60 μm mesh size. The collected animals and a small amount of water remained in the filtering bottle, while most of the water exited into a container. This water was used for measurements of physical and chemical parameters.

The easiest way to collect epikarst fauna is from pools filled with water which seeps down the walls or drips directly from the cave ceiling. By sampling such pools, the epikarst fauna is sampled and the fauna from hypogean streams is excluded. Pools are sampled by aspiration of the water (Fig. 26.6) filtered through the collecting container described above.

Aquatic fauna can be sampled using aquatic baited traps. Traps are usually modified from a plastic water bottle. The top of the bottle is cut off, inverted, and sealed onto the bottom of the bottle with duct tape. A small piece of bait (yoghurt) is placed into the bottle before the trap is deployed. At the appropriate site, each trap is submerged in the cave stream or pool and weighed down with a rock. In situations where the aquatic habitat is too shallow, a miniature version of the trap is used. The content of each bottle is washed through a fine mesh aquarium net several times. All organisms caught in the net are preserved immediately in 70 % alcohol.

Fig. 26.3 Collecting of water from a pool and filtering through a filtering bottle

The community of macroinvertebrates in the Reka River was sampled with a standard handnet, using the cross-section method, known in literature as "kick sampling", with transection sampling (Pipan 2000a). The sampling of each site was done at three transects of the riverbed, lying about 10 m apart. At each transect (subsample) we took four kick samples, at 25, 50, 75 and 100 % of the width of the riverbed, which we then joined into one subsample (Fig. 26.7). After repeating this at each transect we obtained three subsamples, each consisting of four kick samples. Where the width of the riverbed was less than 1 m (the width of four nets), the transects were set diagonally against the current. Sampling was always started at the lower transect and then proceeded upriver.

26.2 Biodiversity in the Underground Reka River

The underground river consists of a conduit of channels with larger intermediate chambers and local structures. Under the active stream there are large phreatic (permanently flooded) channels (Peric et al. 2007). The underground Reka River is only partially accessible to man. We can follow the river in Škocjanske jame Caves in the channel named Hankejev kanal across the length of 1 km. In other caves in Slovenia and Italy of this transboundary aquifer which are connected by the Reka River and which were included in the study (Kačna jama, Jama 1 v

Fig. 26.4 Filtering bottle with mesh screens on the sides that allow water to pass through and retain all organisms entering via the drips

Kanjaducah, Brezno v Stršinkni dolini, Labodnica, Jama Lazzaro Jerko, Pozzo dei Colombi), the river can be reached only by using a caving technique.

In the surface river 21 groups of macroinvertebrates were found which appear subsequently in the subterranean river, accidentally washed away as a drift, and later vanish from the cave environment (Pipan 2000b). They are not adapted to the unfavourable underground conditions and become a source of food for obligate cave-dwelling animals (stygobionts). Fauna in the epigean part of the river consists predominantly of species characteristic for Southern Europe. The most abundant are larvae and adult specimens of insects, followed by snails (*Zospeum spelaeum spelaeum*), oligochaets (*Haber monfalconensis*), nematodes, and different groups of crustaceas such as copepods, isopods (*Asellus aquaticus cavernicolus*, *Trichoniscus stammeri*), amphipods (*Niphargus timavi*), and decapods (*Troglocaris* sp.) (Culver and Sket 2002). Fish are represented by salmonids and ciprinids. Among the rare stygobionts which can be found in the Reka River are *Marifugia cavatica* (Polychaeta) and *Dendrocoelum spelaeum* (Turbellaria). *Proteus anguinus* (Amphibia) (Fig. 26.8) ranges from Italy to Monte Negro and in the Reka River reaches the most northwestern distribution.

26.3 Fauna of Percolating Water in Škocjanske Jame Caves

In addition to the fauna in the subterranean Reka River, the fauna in the percolating water which drips directly from the cave ceiling or seeps down the walls

Fig. 26.5 Epikarst drip filtration device

is likewise very rich. From the so-called epikarst zone (Fig. 26.9), which represents the upper layer of the karst underground, water drips from the surface. The fauna of the percolating water in Škocjanske jame Caves is represented by a wide variety of invertebrates from 11 groups, including Turbellaria, Nematoda, Gastropoda, Oligochaeta, Acarina, and among Crustacea are Ostracoda, Copepoda, Isopoda and Amphipoda, Collembola and the larvae of Diptera. The most abundant group in the epikarst are copepod crustaceans (Pipan 2005). These are from 0.2 to 0.8 mm long crustaceans, widespread in a variety of habitat types, e.g. marine, freshwater, groundwater and polar. Although parasitic copepods are also known, only free-living copepods have been found in subterranean habitats. They are representatives of four orders, but in percolating water individuals of only two orders (Cyclopoida and Harpacticoida) are known (Figs. 26.10 and 26.11).

In Škocjanske jame Caves we found 32 species of copepods, which belong to 16 genera and 3 families. Thirteen species and subspecies belonging to 7 genera are from the group Cyclopoida, whereas 9 genera and 19 species belong to Harpacticoida. Many of them (24) were brought into the cave environment by the percolating water, while the rest were found in the underground river (Pipan 2005). Twelve species from the epikarst are stygobionts—are obligate, permanent residents of aquatic subterranean habitats. Five species from 4 genera (*Bryocamptus*, *Moraria*, *Parastenocais* and cf. *Stygepactophanes*) are new to science. One species *Elaphoidella karstica* is endemic to Škocjanske jame Caves. This species was described after a specimen found in a trickle of percolating water, which is currently the only known habitat of its kind in the world.

26.4 Terrestrial Fauna of Škocjanske Jame Caves

Škocjanske jame Caves provide a habitat to many animals, especially to invertebrates that inhabit terrestrial habitats (Fig. 26.12), among which many are so well adapted to life without light and to situations where food is quantitatively and qualitatively poorer, that they become real underground animals. They typically spend their entire lives, from birth to death, and even during reproduction, underground. These organisms are adapted to aphotic habitats, high humidity, and a stable temperature and they cannot survive on the surface. They are troglobionts. Among these obligate, permanent residents of terrestrial subterranean habitats in Škocjanske jame Caves are isopods—*Alpioniscus* (*Illyrionethes*) *strasseri*, *Androniscus stygus tschameri*, *Titanethes* (*T.*) *dahli*, snails *Zospeum spelaeum spelaeum*, millipedes *Typhloiulus* (*Stygiiulus*) *illyricus*, collembolas—*Onychiurus canzianus*, *O. variotuberculatus*, *Oncopodura cavernarum* and beetles—*Anophtalmus schmidti trebicanus*,

Fig. 26.6 Device for collecting fauna from small amounts of water from pools and cracks. After aspiration, the water and sediment is passed through a filtering bottle (see Fig. 26.4)

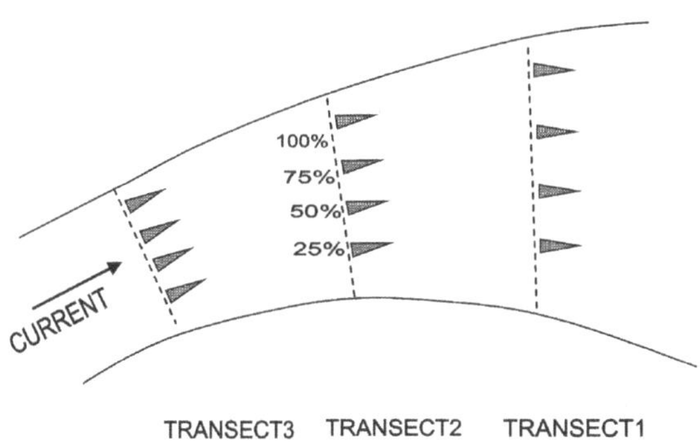

Fig. 26.7 Kick sampling strategy at the sampling site (Dall et al. 1995)

Bathysciotes khevenhuelleri tergestinus (Culver and Sket 2002).

26.5 Water Quality of the Reka River

Among the indices with which we assessed water quality in the Reka River, the assessments using a modified saprobic index and the average Chandler biotic score corresponded to each other the most (Pipan 2000a). Both methods are based on the same principle (including the diversity and number of macroinvertebrates, and their organic pollution tolerance), which

Fig. 26.8 Young *Proteus anguinus* with still visible eyes

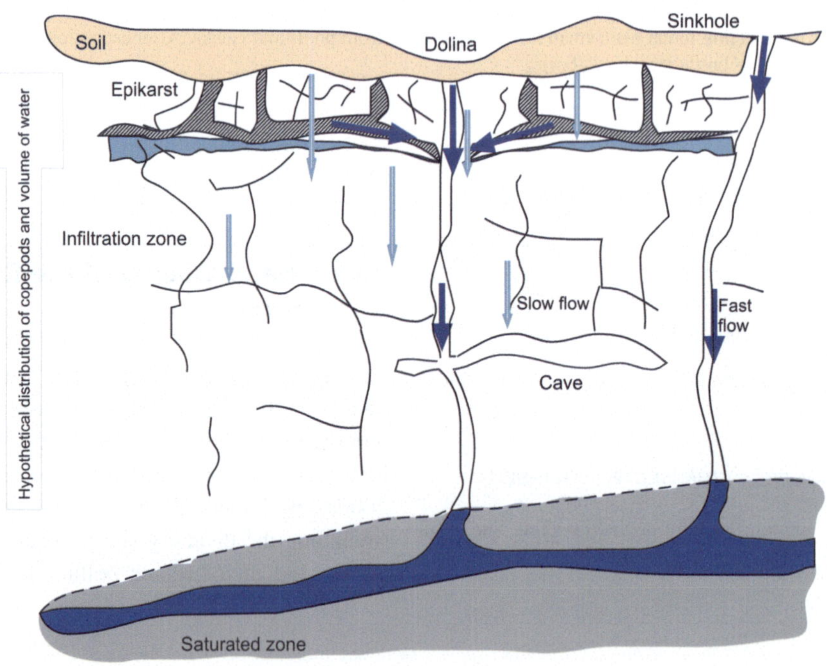

Fig. 26.9 Conceptual model of epikarst

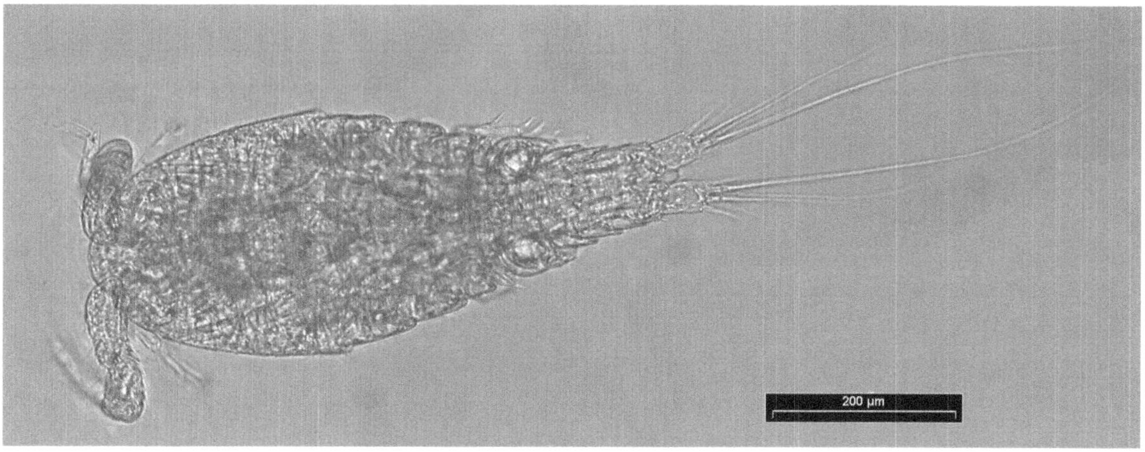

Fig. 26.10 Copepod crustacean (*Speocyclops infernus* from a group of Cyclopoida) from percolation water

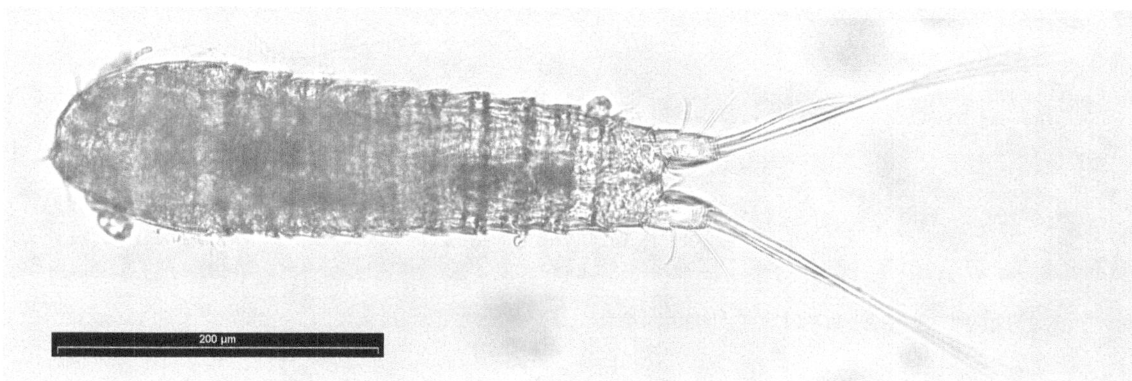

Fig. 26.11 Copepod crustacean from percolation water (*Morariopsis scotenophila* from a group of Harpacticoida)

further confirms our conclusions. However, the index values and the score at the spring might have been slightly higher, or better, even though the lower values are probably the result of consistent ecological conditions in which the more sensitive organisms are better competitors, and not the result of organic pollution. The assessment of extreme conditions (clean—heavily polluted) is usually not problematic; incorrect assessments are more often made within transitional zones (quality classes II or II–III) and most of Slovene waters are of this type (Toman 1999).

Biological investigations showed that in the Reka River optimal food exchange with slightly increased trophic activity takes place, but it does not have a negative effect on the community structure of macroinvertebrates (Pipan 2000a). The reason for this condition is the relatively rapid flow of the river, which does not allow the settling of organic matter, changes to the water level, etc. The Reka River, at its present biotic optimum, does not have a significantly negative impact on the underground stream of Škocjanske jame Caves, into which it flows, as was the case in the late 1980s.

Fig. 26.12 Pitfall trap for sampling terrestrial fauna

Yet there is still some anxiety because of an accumulation of a surplus of food in the underground river. Food consumption processes underground are different to those above ground because of the absence of primary production, and because of the difference between the circulation of substances above and below ground.

27 Impact of Motorways on Karst Waters

Impervious surfaces of road networks accumulate contaminants and pollutants, which are washed off during rain and snowstorm events into nearby waters and lands. Especially in areas with a developed traffic network, runoff from motorways is a significant source of pollution. Various protection and remediation measures were developed and implemented for preventing or at least diminishing its negative influences.

In karst areas, motorway runoff has an amplified impact on ground water, compared to other types of landscapes. The soil layer is commonly thin or non-existent and thus soil infiltration treatment is practically inexistent. Stormwater runoff may immediately flow into the aquifer through subsurface conduit networks, fractures, sinkholes and sinking streams, which makes karst groundwater especially vulnerable to pollution.

27.1 Motorways as a Source of Pollution

27.1.1 Pollutants from Road Runoff

There are three major sources of pollution associated with the road network: vehicles (emissions of motor vehicles, spilled and released oil, tires' particles, de-icing agents), road characteristics and paint markers, and atmospheric depositions influenced by the adjacent land use (Opher and Fidler 2010). Pollution from vehicles is either constant due to road runoff or periodic (catastrophic) due to the spills of pollutants during traffic accidents. The latter represent a great hazard as they instantly release a large amount of the pollutant, whereas in the long run the constantly present pollution under normal conditions may also discharge a large amount of hazardous anthropogenic substances into the environment. The main pollutants and their sources are (Pintar et al. 1998):

- inorganic pollutants:
  - lead (Pb): added to leaded petrol (exhaust fumes); it is no longer in use today, but the environment is still contaminated by it;
  - nickel (Ni), vanadium (V): added to diesel fuels (exhaust fumes);
  - cadmium (Cd), zinc (Zn), copper (Cu), iron (Fe), chromium (Cr): added to car tyres (wear and tear of car tyres);
  - manganese (Mn), chromium (Cr), nickel (Ni), iron (Fe): components of brake linings (wear and tear of brake linings);
  - molybdenum (Mo), bromine (Br), antimony (Sb): added to motor oils;
- organic pollutants:
  - low-volatile lipophilic substances—mineral oils and grease (TLS): lubricating and protective oils in cars;
  - polycyclic aromatic hydrocarbons (PAHs): incomplete combustion of various fuels, atmospheric deposits and wear and tear of the top layer of asphalt.

Heavy metals, polycyclic aromatic hydrocarbons (PAHs) and perfluorooctane sulphonate (PFOS) are

M. Knez and T. Slabe (eds.), *Cave Exploration in Slovenia*,
Cave and Karst Systems of the World, DOI 10.1007/978-3-319-21203-6_27

among the most toxic and carcinogenic compounds in the road runoff, which have a negative effect on humans and the aquatic ecosystem (Enserink et al. 1991, Maltby et al. 1995a, Nakayama et al. 2005). Motorway runoff results in an increase of total hydrocarbons, aromatic hydrocarbons (dominant phenanthrene, pyrene, and fluoranthene) in the sediment, and in an increase of heavy metals (dominant Zn, Cd, Cr, and Pb) and some anions concentrations in the water (Maltby et al. 1995b).

Besides pollutants directly associated with traffic, the impermeable surfaces can collect and drain a considerable quantity of organic waste, nitrogen quantity of organic waste, nitrogen and phosphorous, herbicides, pesticides and faecal pathogens (Scholz and Grabowiecki 2007). Nearby land use and storm events are key factors to introduce faecal pollution. The highest concentration of faecal indicator bacteria commonly occurs during the early phases of stormwater runoff, and the peak concentrations appear usually before the peak flow (Tiefenthaler et al. 2011). All these pollutants can be washed off from roads and can adversely impact waters, the aquatic environment and subsurface.

27.1.2 Motorway Runoff Impact on Biota

Stormwater runoff from urban motorways at the beginning of a storm event is generally more toxic than that collected later in a storm episode. A study where different indicators for toxicity were used (the water flea *Ceriodaphnia dubia*, the fathead minnow *Pimephales promelas*, green algae *Pseudokirchneriella subcapitatum*, the purple sea urchin *Strongylocentrotus purpuratus*, and the luminescent bacteria *Photobacterium phosphoreum* using Microtox™) showed that approximately 90 % of the toxicity is expressed during the first 30 % of storm duration. Toxicity was attributed mostly to Cu and Zn in 90 % (Kayhanian et al. 2008).

Impact of motorway runoff on macroinvertebrate assemblages in aquatic environments was frequently detected, but not on epilithic algae (Maltby et al. 1995a). Similarly it was reported by Boisson et al. (2005) that a small load of pollution from motorway runoff associated with low traffic did not notably affect physical and chemical parameters of the surface water, and periphyton expressed in biomass (chlorophyll *a*) and net primary production and respiration. An additional study showed that biomass and photosynthetic activity could increase when runoff was diluted (Boisson and Perrodin 2006). In runoff water from motorways, the majority of pathogenic bacterial indicators were associated with smaller particles, <50 μm (*Escherichia coli*, enterococci, faecal streptococci, total coliform, faecal coliform) (Zhang and Lulla 2006).

Organic material from the runoff can serve as a substrate for microbes (Datry et al. 2004). Comprehensive analyses of sediment cores affected by the motorway runoff at Plymouth, Massachusetts, USA, showed high total bacterial diversity and a dynamic nature of the communities (Rotaru et al. 2012). The majority of bacteria belonged to Proteobacteria, followed by Actinobacteria, Firmicutes and Chloroflexi. When contamination—acetate-based de-icing agents—was removed, a shift in the bacterial community was clearly evident and the community of predominant iron reducers (Geobacteriaceae) switched to a more diverse community (Holmes et al. 2005). Bacterial communities in the sediment were largely affected by the levels of iron (Fe^{2+}) and dissolved oxygen. There were also other factors influencing community structure which are generally relevant also for soil bacteria, nitrogen, total phosphorus, pH, clay loam and soil heterogeneity (Rotaru et al. 2012). In addition, the analyses showed a tendency of bacterial grouping based on distance from the infiltration basin and depth in the sediment core (Rotaru et al. 2012).

27.1.3 Protection and Remediation Measures

Soil infiltration treatment successfully removes most of pollutants, such as organic matter and dissolved organic carbon (DOC), phosphorus, PAHs, some heavy metals and oestrogenic activities. Some compounds (NO_3^-, Mn, Ni, alkaline earth metals, PFOS

Fig. 27.1 Oil separator is a treatment facility for mechanical treatment of road drainage: sedimentation of suspended material and oil separation. It can be used for trapping spillages caused by traffic accidents

and perfluorooctane sulphonamide-FOSA) were less successfully removed and they can potentially reach groundwater (Murakami et al. 2008).

High discharges of runoff waters represent a big problem. The simplest method of draining water from roads is dispersed draining by spilling water over their borders or draining runoff stormwater through individual outlets or through gutters for collecting and draining stormwater from these surfaces. In the case of motorways, which represent a greater burden to the environment, stormwater is drained at specific spots through appropriate outlets (Fig. 27.1). The solution to remove initial pollutants should include front-end treatment systems such as oil interceptors, spillage containment facilities and wetland or lagoons (Shutes et al. 1997). Along some motorway sections, stormwater from motorway runoff is directed and captured in collection basins. They usually consist of two parts: the first serves as a settling tank where sludge is formed and where oil and grease can be mechanically removed, and a second compartment where wastewater stays for few days for additional settling and degradation. Later on, liquid is released into the surface watercourse. Gotvajn and Zagorc-Končan (2009) reported that motorway stormwater runoff in Slovenia was commonly not very polluted, except during heavy rain after a longer dry period. In such cases, an addition of a biological activator can enhance degradation, up to 60 %. A biological activator is a mixture of enzymes, bacteria and nutrients and its primary use is to activate and improve the biodegradability of hydrocarbons in crude oil. In the case of motorway runoff, the application of a biological activator should be carefully dosed as improper application can increase the final total pollution level (Gotvajn and Zagorc-Končan 2009).

Constructed wetlands (CWs) after a settlement pond offer good removal of pollutants at the outlet (Bulc and Slak 2003), especially during storm events. CWs are accepted as sustainable sanitation systems because they prevent disease, protect the environment, and are cost-effective and simple (Langergraber 2013). CWs are aquatic ecosystems with static or flowing water with emergent, floating and submerged aquatic vegetation. In a study from the Netherlands, a system containing CWs demonstrated very good retention efficiencies for PAHs (90–95 %). In the case of heavy metals, most of their concentrations were lower than the standards at the outflow, except for Cu and Zn. High concentrations at the effluent from the CWs was attributed to de-icing salts and due to this fact it was suggested that roadwater should bypass the CWs during the application of de-icing salts on roads (Tromp et al. 2012). The proper selection and design of constructed wetlands is of great importance to achieve the proper degradation efficiency of runoff treatment (Shutes et al. 2004).

27.2 Particularity of Motorway Runoff in Karst Areas

Due to the permeability of fissured and soluble carbonate rocks in karst areas any contamination on the surface denotes a contamination of karst underground waters. Karst aquifers are supplied by precipitation percolating into the karst interior directly from the surface (Fig. 27.2), or potentially through karst sinking streams. Underground water flows back out onto the surface through karst springs, which are an important source of drinking water. Although karst rocks stand for only 7–12 % of the Earth's surface, almost a

Fig. 27.2 Revealed karst rock in the roadcut of the motorway

quarter of the world's population is supplied with drinking water from karst aquifers (Ford and Williams 2007). This share is even higher in Slovenia as karst water sources provide drinking water for over half of the population.

Traffic is one of the pollutants on the karst surface that jeopardize the quality of karst water. Water flowing off the roadway quickly percolates through the well-karstified karst surface into the underground. Accidents that cause spills of great quantities of hazardous and harmful substances are a particular hazard, as these substances enter karst aquifers directly through fissures across the vadose zone or flow into the streams and reach the karst aquifers via these. Equally dangerous are accidents that can occur at petrol stations or in warehouses for oil derivatives.

For proper management of motorway runoff in karst areas, all the environmental impact from other landscapes should be carefully considered. The presence of soil and vegetation is important for natural treatment processes. To enlarge the reaction surface and retention time for degradation processes, different constructed infiltration systems that include straw bales, sand bags, gravel etc., can be adapted to karst. CWs is another plausible solution for the treatment of motorway runoff in karst (Zhou and Beck 2005). There are varieties of CWs with potential use in karst, such as floating treatment wetlands (Gill et al. 2014).

Specifically in karst, it is crucial to prepare a stormwater runoff management plan for each particular section before roadway construction takes place. This should include an inspection for sinkhole collapse risk and groundwater drainage patterns along the proposed route, for example by tracing experiments (Zhou and Beck 2005).

In recent decades we have often monitored the aftermath of spills from traffic accidents in the Slovene karst, mainly of oil derivatives (Knez et al. 1994; Kogovšek 1996; Kogovšek and Petrič 2002a). Only in areas in which the courses and characteristics of the underground flow have already been determined based on previous research (including tracing experiments) could we predict with sufficient reliability which karst sources were threatened and prepare a plan for monitoring their quality and for taking appropriate action.

Below is a presentation of two accidents that caused spills of oil derivatives, which triggered parallel research and the monitoring of the presence of such pollutants at karst springs. Tracing was conducted in both areas, either in advance or afterwards; thus the use of results of all the research conducted and their comparison with the flow of water-soluble and water-insoluble substances has provided a better understanding of the characteristics of the transfer of oil derivatives in karst.

27.2.1 Pollution of a Karst Water Source Due to the Spillage of Gas Oil During a Traffic Accident

During a traffic accident on a road near Obrov in southwest Slovenia (Fig. 27.3) on 12 October 1994 almost 16 m^3 of D2 gas oil spilled out of a tank. The spilled cargo flowed off the surface very quickly (as estimated in 15 to 20 min), thus leading to the conclusion that in the case of such accidents in karst areas the spilled substance cannot be pumped in time and prevented from flowing into the karst. As soon as a hazardous substance enters a karst aquifer, remediation is no longer possible. A spillage and fast percolation into the karst interior took place at the second protection zone of the Rižana spring, which has been captured for water supply (Fig. 27.4), approximately 1 km SW of the sinking stream in the Jezerina blind valley, for which tracing experiments discovered a definite connection with the springs of the Rižana and Osapska Reka rivers (Krivic et al. 1989).

In the tracing experiment in Jezerina the tracer rhodamine appeared 18 days after injection, following a precipitation event, at the karst spring of Rižana. Maximum concentration appeared after another 3 and a half days. In one month after the appearance of the tracer, 10.6 % of the rhodamine flowed through the Rižana, with a flow velocity of 35 m/h. Rhodamine appeared in Osapska Reka in a water wave following a precipitation event, 21 days after being poured in. The maximum concentration was four times smaller than that in Rižana, yet the peak was more pronounced. The flow velocity in the direction of the Osapska Reka River was 35 m/h, same as in the direction of Rižana. At the springs of Ara and Sv. Ivan in Croatia the rhodamine appeared after 11 days, prior to a water wave following a precipitation event and in very low concentrations. The calculated flow velocity to the Ara spring was around 53 m/h, and 72 m/h to the spring of Sv. Ivan. Due to the low tracer concentrations these

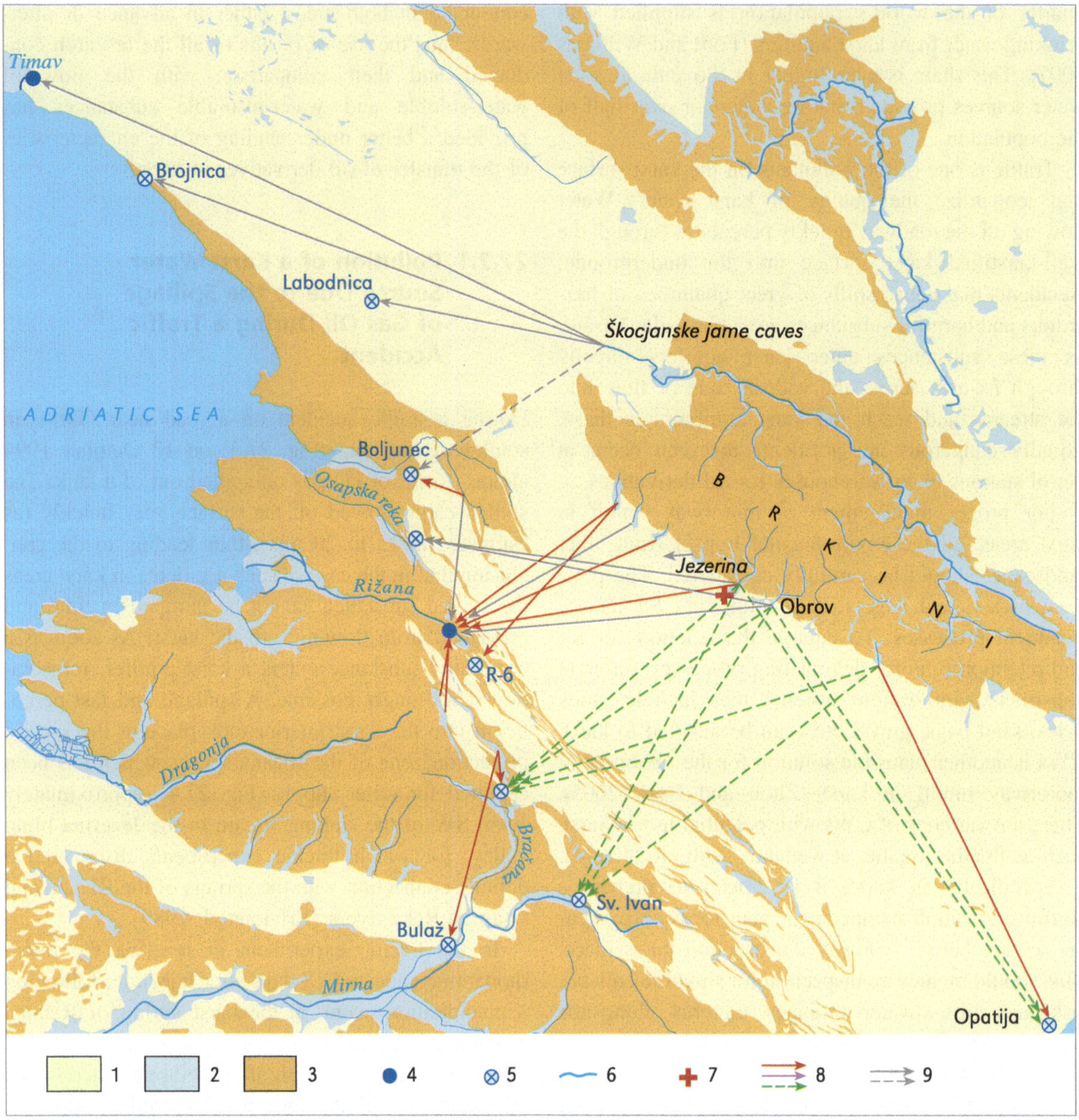

Fig. 27.3 Location of the spillage at Obrov and the directions of the groundwater flow, as determined with tracer tests (Krivic et al. 1987, 1989). Legend: *1* karst-fissure aquifer, *2* porous aquifer, *3* very poorly permeable rocks, *4* major spring, *5* spring, *6* surface flow, *7* location of the traffic accident spillage, *8* main, side, or unreliable direction of underground flow, as determined with tracer tests after 1976, *9* confirmed or unreliable direction of underground flow, as determined with tracer tests before 1976

two connections could not be confirmed with certainty, but they are possible in light of the determined velocities (Krivic et al. 1989).

Based on the results of preliminary tracing, it was estimated in the case of the accident near Obrov that the spreading of pollution would be mostly affected by precipitation or increased river flows and that the content of gas oil would have to be monitored more thoroughly in the Rižana catchment area, and occasionally also in the springs of Osapska Reka, Ara and Sv. Ivan. Analyses of samples were conducted by the Sanitary-Chemical Laboratory of the Koper Institute of Social Medicine and Hygiene (Sanitarno-kemični laboratorij Zavoda za socialno medicino in higieno

Fig. 27.4 The Spring of the Rižana River, captured in order to provide the coastal population with drinking water

Koper) using the gas chromatography method after extraction with hexane, as ordered by the water supply company Rižanski vodovod Koper.

After the spillage of gas oil during the traffic accident no precipitation occurred for 12 days. After a smaller precipitation event the flow of the Rižana River increased somewhat and on 26 October 1994, 14 days after the accident, the presence of gas oil was recorded, the highest during the one-month monitoring (Fig. 27.5). After two days the concentration of gas oil dropped and on 29 October, after abundant precipitation (70 mm) and the flow rate of the Rižana River increasing to 22.8 m^3/s, it again increased. Three days later it dropped below the method detection limit

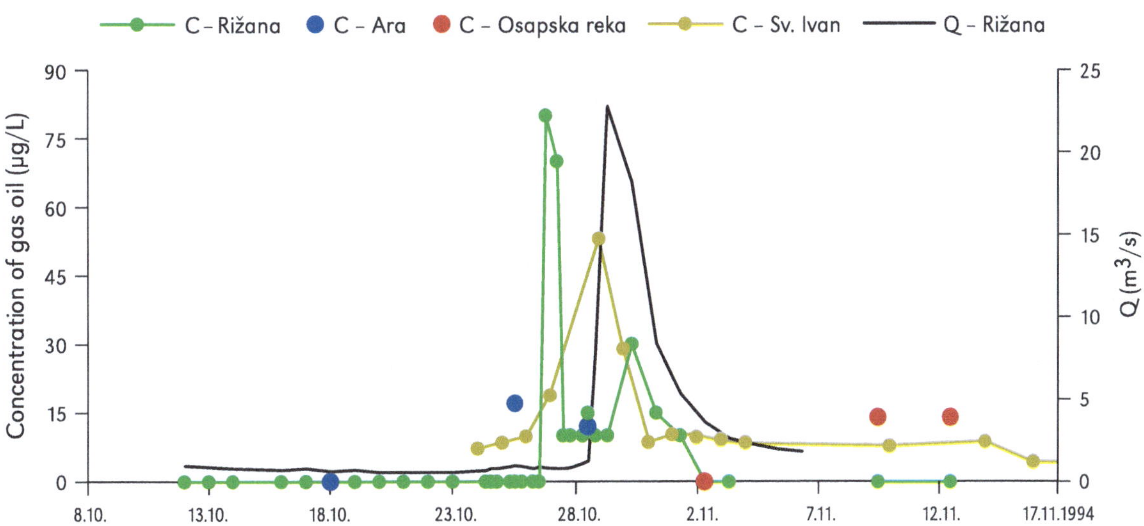

Fig. 27.5 Results of monitoring the emergence of gas oil in karst springs after the traffic accident at Obrov

(Kogovšek 1995b). Due to the detected pollution the Rižana spring was for a while excluded from the water supply network.

After the accident near Obrov the travelling velocity of the oil and water towards Rižana amounted to 45 m/h. The distinct transport of gas oil (80 μg/l) increased after abundant precipitation and a greatly increased flow, but afterwards the concentration quickly dropped below the detection limit.

After each subsequent precipitation event the oil was most likely pushed through the aquifer, however, sampling and measurements were no longer being carried out. It is possible that measurements would not have detected it, since greatly increased flows signify great dilution. Based on the available data, the mean daily discharges of the Rižana River and analyses of hydrocarbon concentration, we calculated the amount of gas oil that flowed out through the Rižana in the one month following the accident. It amounted to merely 88 kg during the measurements in the final week of October, which stands for only 0.5 % of the spilled quantity. This clearly indicates prolonged retention and slow outflow of such substances through karst aquifers, which can be estimated at a decade or more. Only a few samples were taken at the springs of Osapska Reka and Ara, which indicated a potential presence of gas oil in lower concentrations. From 24 October 1994 onwards water was also being analysed at the Sv. Ivan spring, which has been captured for drinking water supply (Vlahović 2000). The contents of mineral oils and total grease increased noticeably between 27 and 30 October. Since there is no record of the contents prior to the accident, we cannot deduce with certainty that they were increased by the spillage. This proves the importance of a regular monitoring of karst springs, especially those that have been captured for drinking water supply.

It must be pointed out that chlorination has a negative impact during the preparation of drinking water if the water contains organic substances, since these react to the chlorine, creating halogenated hydrocarbons which are carcinogenic. Their maximum permissible limit is 30 μg/L; after the accident near Obrov, a concentration of 10 μg/L of total trihalomethanes was measured at the end points of the Rižana water distribution system, which was at that time still using a chlorination process for water disinfection (Ožbolt 1994).

27.2.2 Pollution of a Karst Water Source Due to the Spillage of Gas Oil from a Warehouse for Oil Derivatives

During pumping at a warehouse for oil derivatives near Ortnek in SE Slovenia on 13 October 1998 an unknown quantity of gas oil was spilled down drainage pipes into a nearby stream, which flows into the Tržiščica Stream (Genorio 1999). The warehouse is situated in a non-karst area; however, after running on the surface for about 4 km the Tržiščica sinks at a contact with limestone through the Tentera cave into the karst underground (Fig. 27.6). After the accident this is exactly how the spilled gas oil entered the karst aquifer.

This karst aquifer supplies several karst springs; one of these, Globočec, has been captured for supplying the area of the Suha krajina region with drinking water (Fig. 27.7). Gas oil first appeared in this spring in a small concentration of 0.013 mg/L a good eight days after the spillage and roughly three days after heavier rain. As early as 8 h later the concentration dropped below 0.005 mg/L (the highest permissible limit for drinking water is 0.01 mg/L). After abundant precipitation on 5 November samples were taken 2 days later and merely contained the characteristic scent. Subsequent regular analyses showed pure water all the way until the end of November. In December and later on from January to March 1999 low temperatures prevailed, with relatively modest precipitation in the form of snow, which was insufficient for substances to be transferred in the karst. It was only after heavier precipitation in April 1999 that slightly increased values were again recorded for 5 h. No increase was recorded later on. When we visited the ponor of the Tržiščica in late August 1999, a year after the gas oil had been spilled, there was still a strong scent of gas oil present, as it had been adsorbed into the sediment in the riverbed.

Fig. 27.6 The Tržiščica Stream sinks from the surface into a karst aquifer through the Tentera sink cave

The planned expansion of the warehouse for oil derivatives also signified a greater risk of karst waters being polluted. In order to be able to prepare a plan of suitable protective measures we conducted a tracing experiment, which is an effective method for determining underground water connections in karst or for determining the potential directions in which pollution spreads from a specific spot. The uranine tracer was injected into the Tržiščica Stream at the Tentera ponor, into which oil derivatives would have flowed from the warehouse in the event of a spillage (Fig. 27.8).

The tracing experiment showed that under the hydrological conditions of decreasing flows from medium to low water levels the water of the Tržiščica Stream flows mainly into the Tominčev studenec and Javornikov izvir springs along the Krka River near Dvor (Figs. 27.8 and 27.9). It was estimated that by the end of May 2000 some 2/3 of the injected uranine flowed out through these two springs. To a significantly smaller extent the tracer also flowed into the Debeljakov izvir spring and in the direction of the Podpeška jama cave (Kogovšek and Petrič 2002b). Under the described hydrological conditions we did not establish a connection to the Globočec spring, even though it had been sampled most thoroughly as a captured source of water.

Water flowed into the permanent Tominčev studenec spring with the apparent velocity of 144 m/h, and with 137 m/h into the intermittent Javornikov izvir spring, as calculated based on the appearance of maximum concentration. The rather high velocities of water flows indicate a fast transfer of pollution to the springs near Dvor should the Tržiščica Stream be polluted. We recorded as much as 3 times higher concentrations of uranine in Javornikov izvir than in Tominčev studenec, which indicates a more concentrated outflow

Fig. 27.7 The Globočec Spring, captured in order to provide the population of the Suha krajina region with drinking water

or smaller dilution along the underground flow in the direction of this spring. In Podpeška jama uranine appeared only after heavier precipitation, which occurred after two months of a constant decrease in discharge, when the springs Javornikov izvir and Debeljakov izvir even dried up.

It can be concluded that the underground flow of water from the ponor of the Tržiščica Stream is directed mainly towards Javornikov izvir and Tominčev studenec near Dvor and that it depends on the hydrological conditions. It has been deduced that during high water levels a connection exists between the Tržiščica and the captured Globočec spring. This connection is poor and only a smaller part of the Tržiščica flows in this direction during high waters; however, in the event of pollution from hazardous substances (such as oil derivatives) that would be enough to contaminate the captured spring. In the case of the above-mentioned accident from 1998 this spring had to be excluded from the network for a longer period of time.

Only periodic, short-term presence of gas oil after each precipitation event indicates that substances which do not dissolve in water or mix with it may stay in the karst underground for a long period of time. Comprehensive data on the velocities and above all on

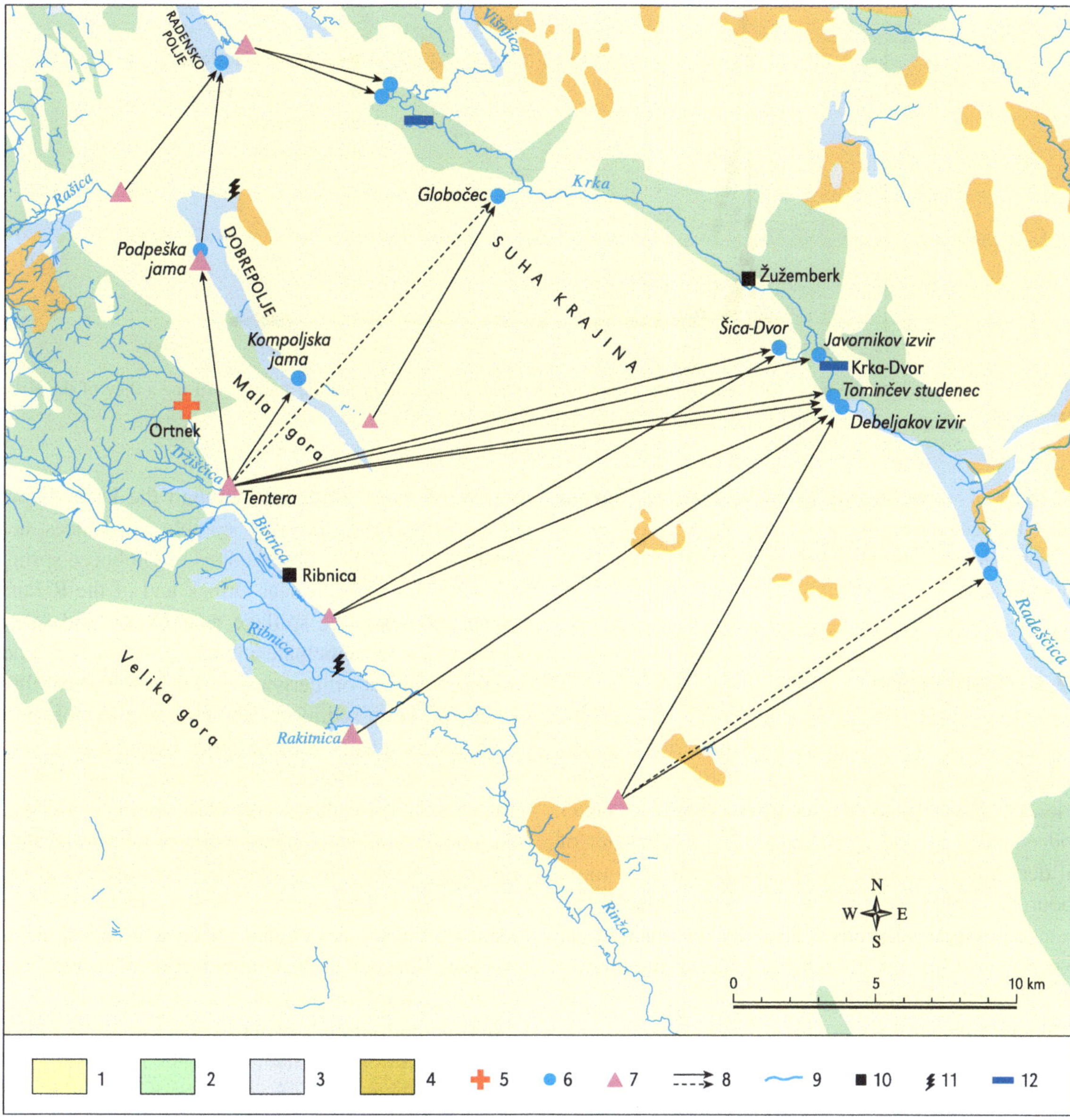

Fig. 27.8 Directions of the underground water flow, as determined with a tracer test by injecting the tracer into the Tržiščica Stream in the Tentera sink cave and with other tracer tests in the wider area of the Suha krajina region. Legend: *1* karst aquifer, *2* fissure aquifer, *3* porous aquifer, *4* very poorly permeable rocks, *5* location of the gas oil spillage, *6* spring, *7* ponor, *8* main and side direction of underground flow, as determined with tracer tests, *9* surface flow, *10* settlement, *11* precipitation station, *12* hydrologic station

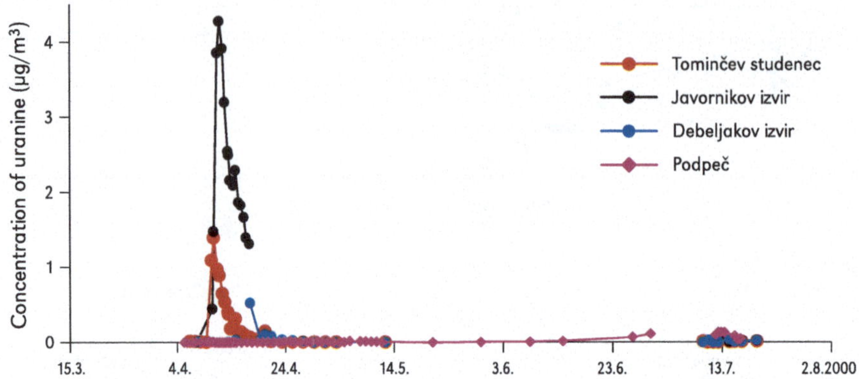

Fig. 27.9 Appearance of the tracer injected into the ponor of the Tržiščica Stream in the karst springs of the valley of the Krka River

the share of water flowing from the Tržiščica Stream into Globočec during high and very high waters could be obtained only by carrying out tracing under such conditions.

27.3 Conclusion

Analyses of accidents with spills of oil derivatives and their comparison with the results of hydrogeological research have gradually contributed to a better understanding of the characteristics of the transfer of oil derivatives in karst. We could find out even more about the differences in the way water-soluble and water-insoluble substances flow by conducting an experiment using a water-soluble tracer at the time of the accident, simultaneously as water-insoluble oil derivatives are spilled.

Accidents that cause greater quantities of hazardous substances to flow into the karst, for whatever reason, endanger our environment, karst waters, and even in smaller quantities also the quality of the karst springs that have been captured for drinking water supply. Especially dangerous are spills of oil derivatives, since we know too little about the flow of such substances that do not dissolve in water and are lighter than it. Based on observations of accidents with spills of oil derivatives in karst areas conducted thus far, it is known that these derivatives flow across the same paths as stormwater does when it flows from the surface into the karst interior. The first appearance in springs is mostly influenced by precipitation. The conducted comparisons indicate that the subsequent transfer of these substances through the karst differs substantially from the way soluble substances are transferred. Based on observations of Globočec spring following the accident near Ortnek and of the Rižana Spring following the spillage near Obrov and other similar cases, we anticipate a longer retention time and washing away of oil derivatives due to the possibility of adsorption on sediments and retention in siphons.

Should accidents become more frequent in a specific area the substantial collection/accumulation of substances in the recharge area of the spring could lead to a more permanent contamination of the spring, thus preventing its use. As a reminder, the example of the spring of the Krupa River in SE Slovenia is still vivid today; as a water source this spring will be lost to us for a long time due to the contamination of its recharge area with hazardous polychlorinated biphenyls.

It is certainly much easier to solve concrete cases, when a hazardous substance is spilled and we must foresee the directions and velocities of the outflow or assess which karst springs will become contaminated, if research into the directions and velocities of the flow of underground waters have already been conducted in the area in question. Hydrogeological and hydrochemical research, but above all tracing experiments, enable us to determine the underground water paths. In doing so we must be aware that in the event of larger quantities of liquids spilled in an instant the flow differs from that during precipitation, when liquid is being entered over a longer period of time and dispersedly. Research into the flow of water through soil and the vadose zone (Kogovšek and Šebela 2004; Kogovšek 2010) has shown that this is exactly where water and other potential liquids may be retained the longest on their way towards the more permeable parts of the aquifer, where the flow is substantially faster. Hence pollution may appear in a karst spring with greater delay. In the event of accidents it thus makes sense to conduct observation for a longer period of time, especially after subsequent intense and abundant precipitation events.

Even though concentrations of the contaminant in a contaminated karst spring are low and appear only periodically, it has been ascertained that in the event of contamination with hazardous substances (such as oil derivatives) that is enough to have to exclude the water source from use. It must be especially pointed out that in the event of greater increases in the flow of the contaminated spring, when the concentration of oil derivatives has dropped below the detection limit, they are still being transferred significantly. In such cases the relatively high method detection limit has proved to be a limitation factor for determining mineral oils. In the case of the captured spring, which has been polluted and where chlorine is being used to disinfect the water, halogenated derivatives are being created, which are carcinogenic.

We should know the recharge area of each spring that has been captured for supplying the population with drinking water in order to be able to protect the quality of water and take appropriate actions should unexpected contamination occur. However, that is not enough to ensure clean water. It is exceptionally important that we build roads appropriately or equip them with retention and cleaning facilities that prevent or at least reduce the direct outflow of hazardous substances into the karst.

Transfer of Contamination from Motorways Towards Karst Water Sources: The Example of the Malenščica Karst Spring

28

A large part of motorways in Slovenia runs across the karst aquifers which are an important source of drinking water supply. Contamination washed off road surfaces by precipitation water can be a great threat to the quality of water sources. When the water, and with it harmful substances, enters the karst underground, it flows away rapidly through permeable karst channels and fissures towards the springs. Its cleansing ability, either natural or artificial, is very low, and the degree of endangerment of water sources is thus greater.

Modern motorway construction has addressed this issue by building various protective facilities. The most frequent such facilities in the Slovenian karst terrain are oil separators with outflows from motorways leading into them. In practice, unfortunately, their operation has shown various deficiencies which has rendered the prevention or at least limitation of contamination rather unsuitable. Presented below is a study of the potential impact of contamination at the chosen motorway section on the regionally significant water source. The conclusion provides a few general findings about the impact of motorways on the karst water sources and the necessary measures for their protection.

28.1 Measures for Limiting the Negative Impact of Motorways on Water Sources

The majority of the Slovenian motorway connections was built in the past twenty years. Road drainage, water treatment, and final disposal of precipitation waste water from the roads were entirely unregulated. It was not until 2005 that the Decree on the emission of substances in the discharge of meteoric water from public roads (Decree 2005) was adopted. One of the key reasons for the delay was the fact that the European Union had not had this area uniformly regulated. The above Decree lays down evaluation, measuring, threshold values, and measures for decreasing the emission of substances with the discharge of meteoric waste water in relation to the traffic load, natural conditions on the route of the roads, and water vulnerability. It also lays down the criteria which determine when the precipitation waste water needs to be captured in a retarding basin before draining. In the case of the karst surface, these criteria are the strictest (Table 28.1).

When the threshold values for precipitation waste water parameters at the outflow from the retarding basin are exceeded (Table 28.2), water treatment is required. Treatment is required for the water of the so-called critical rainfall, i.e. to the initial, purging rainfall when the water is the most contaminated. According to the scope of precipitation waste water treatment, there are several different special-regime areas (e.g. water protection areas, nature conservation areas etc.) where the requirements regarding treatment due to the increased level of the protection regime increase.

Lacking proper national legislation, the Motorway Company of the Republic of Slovenia (DARS) adopted an internal rule in 1995 which entailed instructions for building designers regarding the drawing up of technical documentation for draining of meteoric water from the surface of motorway roads; the rule was amended in 1999 (Rismal et al. 1995, 1999). The rule stipulates the assessment of contamination by the

M. Knez and T. Slabe (eds.), *Cave Exploration in Slovenia*,
Cave and Karst Systems of the World, DOI 10.1007/978-3-319-21203-6_28

Table 28.1 Threshold values which determine that the precipitation waste water needs to be captured in a retarding basin before draining

Aquifer type	Daily mean traffic flow
Karst aquifer	over 6000
Porous or fissure aquifer	over 12,000
Precipitation waste water is drained directly into a river or sea	over 12,000
Rocks with water permeability under 10^{-6} m/s	over 40,000

Table 28.2 Threshold parameter values for precipitation waste water

Parameter	Threshold values (mg/L)	
	Draining directly into the waters	Draining directly into the sewage system
Suspended material	80/160[a]	[b]
Sedimentable material	0.5/10[a]	10
Cadmium	0.1	0.1
Copper	0.5	0.5
Zinc	2.0	2.0
Total chromium	0.5	0.5
Nickel	0.5	0.5
Total hydrocarbons (mineral oils)	10/50[a]	20
Semi-volatile aromatic hydrocarbons (BTX)	0.1	0.1
Adsorbable organic halogens (AOX)	0.5	0.5
Phenols	0.1	10
Polycyclic aromatic hydrocarbons (PAHs)	0.00006	–

[a]Lower value applies to water protection areas according to regulations regulating the water protection areas
[b]Threshold concentration of suspended material in precipitation water is determined with the value stipulated in the environmental permit where this value does not yet affect the sewage system or the sewage plant

outflow of precipitation water from the motorways, defines criteria, design, and draining of the water off the road surfaces and from outflow treatment facilities, settling lagoons, and containment basins, as well as providing alternative protection proposals. Guidelines for the establishment of the method for protecting groundwater in motorway areas provide for an additional regulation of groundwater protection (Ajdič et al. 1999). They stipulate the method of treatment (mechanical and biological treatment, reed lagoons).

According to the Decree on the emission of substances in the discharge of waste water from petrol stations, facilities for the maintenance and repairs of motor vehicles and car-washes (Decree 1999), the waste water from service stations has to be treated at a sewage plant which ensures removal of mineral oils and petrol (oil separators) and operates in accordance with the technical requirements. For total hydrocarbons (mineral oils), the threshold parameter values for waste water from stations for filling motor vehicles and tanks with liquid fuel is 5 mg/L for the outflow into water and 10 mg/L for the outflow into the sewage system.

The Rules on minimum technical and other requirements for parking lots and places of vehicles maintenance (Rules 1996) apply to parking lots for cargo vehicles and buses. The Rules stipulate that such surfaces need to be paved so that they are impermeable to water and petroleum products; they also stipulate that meteoric water needs to be drained through oil (and fuel) separators of suitable design.

Technical requirements for the construction of oil (and fuel) separators are laid down in the European standard SIST EN 858-1 (2002). It stipulates the method of construction, operation, and testing, as well as marking and operation control. The rule distinguishes between two classes of oil separators—Class I with maximum permissible content of residual oil of 5 mg/L (coalescing separators), and Class II with maximum permissible content of residual oil of 100 mg/L (gravitational separators). The rule also provides a detailed description of the quality requirements for all built-in materials, concrete, shut-off safety system, testing and so on.

28.2 The Motorway in the Recharge Area of the Karst Water Source of Malenščica

The possibility of contamination of the karst water due to the impact of the motorway in its hinterlands was demonstrated in detail on an example of the karst water source of Malenščica, which is captured to supply drinking water to a population of 21,000, mainly in the area of Postojna and Pivka (Fig. 28.1). On the southern edge of Planinsko polje (southwest region of Slovenia), there are two permanent springs, i.e. Malenščica and Unica which both flow into the Unica River (Fig. 28.2). The flow rates of the Malenščica spring range from 1.1 to 11.9 m^3/s, with the mean flow rate of 6.7 m^3/s. The flow rates of the Unica spring, which were measured in the 2007–2009 period, were ranging between 0.04 and 70 m^3/s, with the mean flow rate of 13.3 m^3/s. The recharge area of both springs are intertwined and expand across an area of over 740 km^2. Prevalent here are the Mesozoic carbonate rock, mostly well-karstified limestone, and to a lesser extent somewhat less permeable dolomites. The surface waters accumulate in the Pivka Basin on the Eocene flysch; at the contact area with the karst terrain, the waters sink and flow away underground towards the springs in Planinsko polje.

The recharge area of both springs are cut by the motorway section between Vrhnika and Razdrto, which was built all the way back in 1974. Since the motorway section for the most part (41 km) runs across the karst landscape, the meteoric water was already then routed from the road surface and into the oil separators. The latter were constructed in order to retain potential traffic accident spillages of petroleum products and fluids that are lighter than water. Up until then, the meteoric water from the road surface had been flowing directly into the karst terrain, and the only issue that was considered was the terrain's ability to swallow certain amounts of water. However, during motorway construction there was already a hint of great vulnerability of the karst terrain and the possibility that a spillage of harmful substances could compromise the quality of the karst water sources. A more detailed description of the oil separators is given in the chapter "Impact of motorways on karst waters".

In the research on the motorway's impact on the karst waters in the example of the Malenščica water source, the composition of water flowing off the road and into the oil separator was first analysed, followed by the analysis of the composition of water flowing from the oil separator into the karst aquifer. The conclusion that this water can be highly contaminated brought up a question, how this contamination can spread through the karst aquifer and which springs are consequently in danger. The answer to this question was sought by analysing the results of various hydrogeological researches and by implementing a tracer test.

28.3 The Composition of Water Flowing off the Motorway

28.3.1 Initial Periodic Measurements

First, the current samples of inflow into the oil separator A at Stara vas near Postojna (Fig. 28.3) were measured, into which the meteoric water from approx. 2200 m long motorway section is flowing. Towards Postojna, the road begins to climb which most likely means higher fuel consumption and wear, and consequently also higher risk of contamination. The composition of water draining off the road surface depends also on precipitation quantity which is washing away and diluting the contamination. Varying intensity and quantity of precipitation means lesser or greater dilution of contamination off the road surface.

From March 1992 to January 1993, nine samples were taken from the inflow to the oil separator (Fig. 28.4) in various conditions: after a long period of drought, at the time of salting the roads, and during and after heavy rainfalls (Table 28.3).

The acquired samples were determined according to their electrical conductivity (EC), turbidity, chemical (COD—dichromatic method) and biological oxygen demand (BOD_5), oil and chloride content,

Fig. 28.1 The Malenščica Spring in Planinsko polje, captured in order to provide the population of 21,000 with drinking water

cadmium and lead content, and sulphate, nitrate, and o-phosphate content.

Initial analyses and measurements (Fig. 28.5) have shown the type and the extent of contamination in the outflow of water from the motorway in various conditions (Kogovšek 1993). The highest values for EC, COD, BOD_5, and turbidity were measured during winter. The weather in the beginning of December 1992 (sample 8) was dry, but the oil separator was still receiving melted snow. Since this was the time of salting the road surfaces, higher EC (12.7 mS/cm) and higher content of chlorides (4.2 g/L) were expected.

On 12 January, after a month of drought, another sample was taken (sample 9) at the beginning of the period of light rain. The maximum values of the measurements were recorded for all parameters. The turbidity was as high as 290 NTU, EC was measured at 33 mS/cm, and chloride content at 13.9 g/L, with high calcium content which indicated the use of $CaCl_2$ for de-icing of road surfaces. The COD was as high as 2500, and BOD_5 was 84 mg O_2/L. The COD/BOD_5 ratio was 30, the highest among all measurements, which indicates a substantial amount of persistent contaminants compared to the amount of biodegradable contaminants. Other results showed 16 mg of NO_3^-/L, 440 mg of SO_4^{2-}/L, and 0.016 mg/L of cadmium and 1.1 mg/L of lead.

The above measurements in given sampling conditions showed that the high EC corresponds particularly with the high concentration of chlorides due to the salting of road surfaces in winter, which can start as early as October and last until April. High chloride content had no effect on grass growth at the side of the road; it did, however, cause the pine tree needles to dry. Turbidity is higher after longer periods of drought when the solids that have accumulated on the road surface begin to wash off. Higher turbidity corresponds with the higher COD and BOD_5 which indicates that the source of this contamination are the solids on the road surface. The COD and BOD_5 values often exceeded the values that had been determined for the outflows from treatments plants into rivers (COD = 160 mg O_2/L and BOD_5 = 30 mg O_2/L); however, dilution still takes place. On the other hand, the water from oil separators in karst flows directly into the karst aquifer.

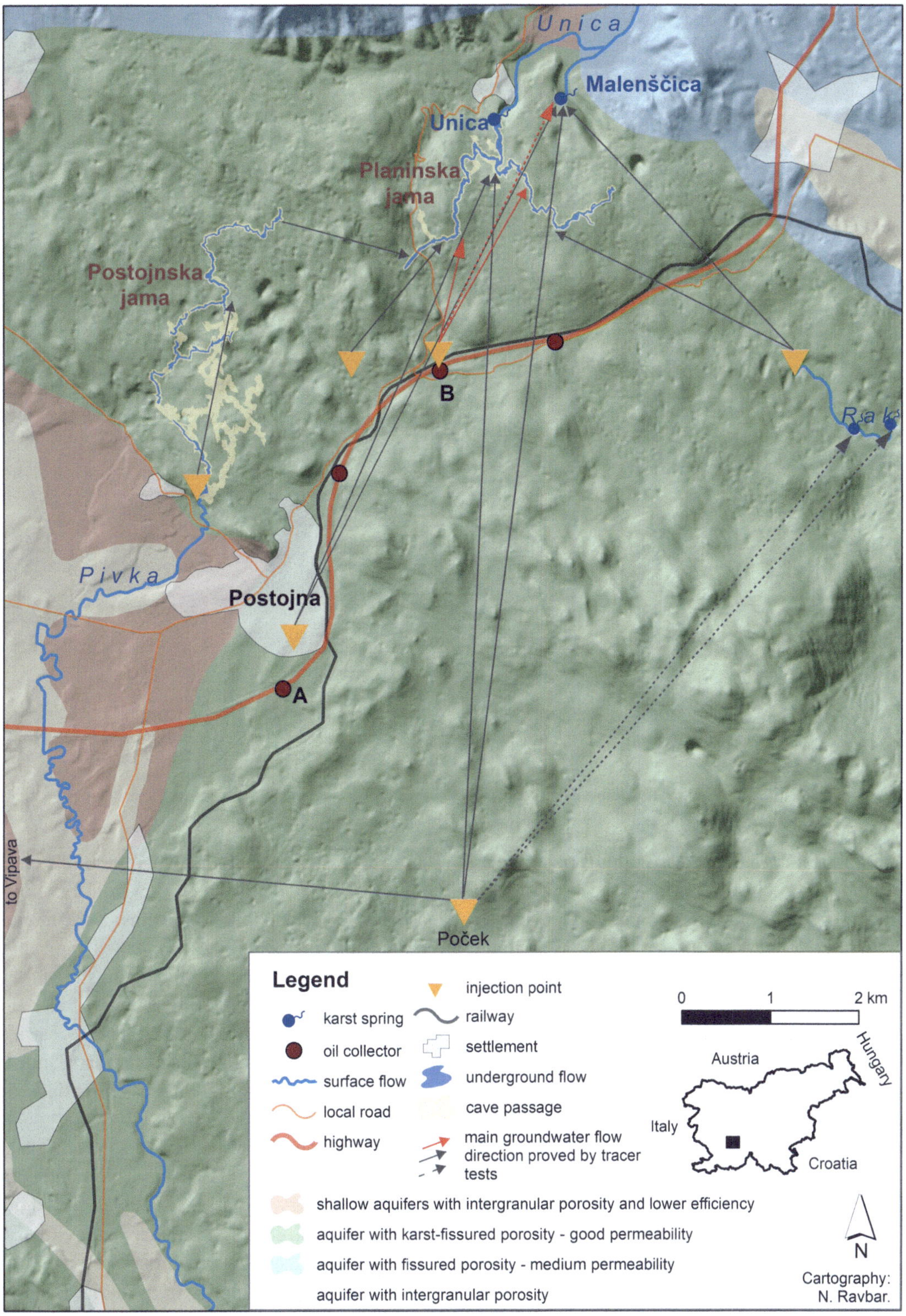

Fig. 28.2 Hydrogeological map of the area

Fig. 28.3 Oil separator at the village of Stara vas

The composition of the water coming off the road surface changes in close relation to the precipitation conditions. The highest values of contaminants were measured after longer periods of drought at initial washing-off of the road surface during light precipitation when the dilution tends to be lower. Heavy precipitation has a significant impact on dilution.

The results described above have shown that it would be prudent to determine how water composition changes depending on the precipitation quantity, or to determine the quantity of precipitation required to wash the majority of contaminants off the road surface.

28.3.2 Sampling During Precipitation Events

In order to better understand the changes in water quality at different hydrological conditions, the next phase addressed in detail the dynamics of the changing composition of the water that flows off the motorway during different precipitation events. Such monitoring required that the sampling was started right after the onset of precipitation and continued at various short-time intervals. Normally, up to four samples were taken. All the measurements were carried out using the same methods and at the same motorway section as with previous periodical samplings described above.

In total, seven water waves were analysed (Kogovšek 1995c). The characteristic values of individual parameters for all 18 samples are shown in Tables 28.4 and 28.5. The higher values for EC are quite noticeable, as well as higher chloride content, higher turbidity, and higher COD during winter-time compared to the summer measurements when higher metal values (lead and cadmium) were observed. Two typical waves are presented in detail below.

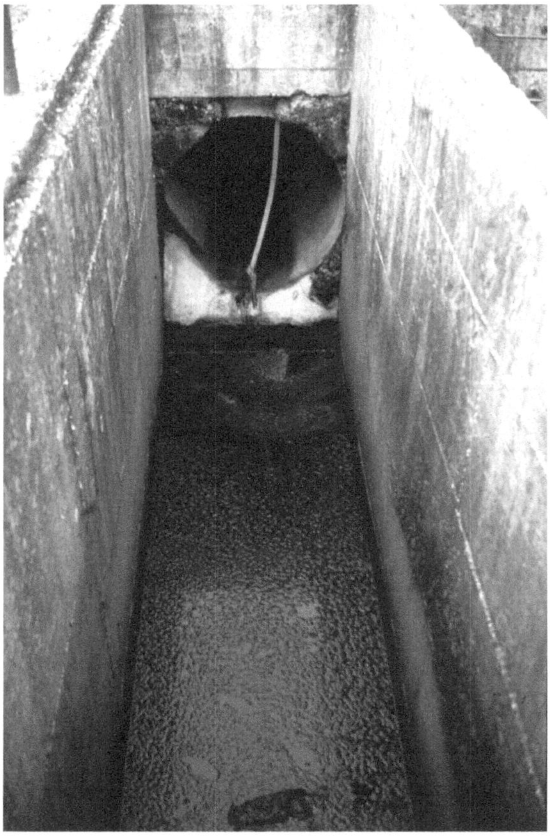

Fig. 28.4 Sampling location at the inflow leading to the oil separator

Table 28.3 Conditions during sampling

Seq. No.	Date	Conditions at sampling
1	17.3.1992	At the time of salting the road surfaces
2	24.3.1992	A few days after precipitation when some of the water was already drained
3	22.5.1992	Light precipitation during water level drop
4	1.6.1992	A drizzle after 20 days of dry weather
5	12.6.1992	Most abundant inflow during observation after several days of precipitation
6	28.9.1992	After a long summer drought with little periodic precipitation
7	20.10.1992	After heavy precipitation (170 mm) and with a well washed-out road
8	10.12.1992	Dry, snow was melting
9	12.1.1993	Onset of light rain

28.3.2.1 Water Wave on 24 March 1993

The water wave that took place on 24 March 1993 was monitored after a long period of winter drought, when the bigger part of December 1992 had gone by without any precipitation, whereas in January, February, and in the beginning of March the precipitation was a mere 18.5 mm. After three months of drought, it started raining on 24 March 1993 at 7 AM; during the day, the rain turned into drizzle.

The first sample was taken when the inflow into the oil separator was at its highest. The measured turbidity was very high (464 NTU), whereas COD was high (340 mgO_2/L) and BOD_5 somewhat low, which indicates that the contaminants that were accumulating on

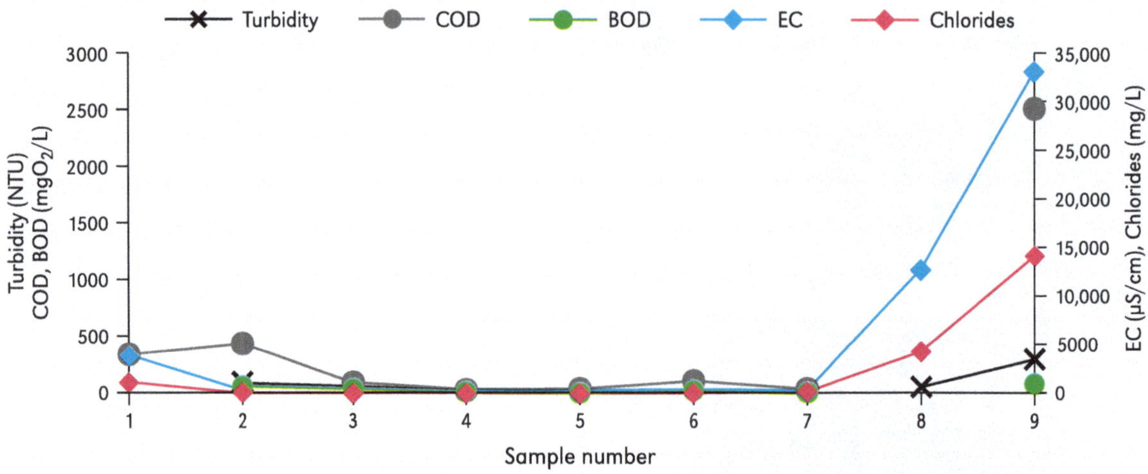

Fig. 28.5 Measurements of turbidity, EC, chlorides, COD, and BOD_5 in the period from March 1992 to January 1993

Table 28.4 Measurement results during winter

	EC	Turbidity	COD	BOD_5	Chlorides	Sulphates	Pb	Cd	Nitrates
	μS/cm	NTU	mg O_2/L		mg/L		μg/L		mg/L
Max	7,810	780	480	45	1,980	158	1,790	76	19
Min	173	57	33	5	30	3	180	11	3
Mean	1,920	285	150	17	520	47	680	50	8

Table 28.5 Measurement results during summer

	EC	Turbidity	COD	BOD_5	Chlorides	Sulphates	Pb	Cd	Nitrates
	μS/cm	NTU	mg O_2/L		mg/L		μg/L		mg/L
Max	216	93	274	70	35	39	11,100	250	10
Min	50	23	19	4	7	11	2,280	65	2
Mean	126	50	113	21	18	23	3,750	167	5

the road surface for a long period of time without being washed off (Fig. 28.6) were now washed off. The high value for EC (3500 μS/cm) was predominantly the result of high chloride content (1200 mg/L) due to the salting of road surfaces and the result of high sulphate content (108 mg/L). The lead content was also relatively high. This initial condition changed significantly during subsequent sampling. As the inflow of water into the oil separator was slowly dropping, a steep initial and later on a slower drop of all parameters was observed. The second sample which was taken three hours after the first sample, i.e. after another 10 mm of rain, showed that the water was still quite contaminated.

28.3.2.2 Water Wave on 3 June 1993

The April and May of 1993 were relatively dry (only 97 mm of rain). A good two weeks before the monitoring of the second wave started on 3 June there was no precipitation. On the day of observation, there was 30.5 mm of rainfall. The heaviest rainfall took place between 7 and 8 AM with 15.5 mm of rain, between 8 and 9 AM the rainfall dropped to 3 mm, and at 2 PM it was reduced to 1 mm of rain in intervals.

The first sampling carried out at 8.30 AM during the highest observed inflow of water into the oil separator showed relatively low values of EC and chlorides since the road surface had been already thoroughly washed off after the salting of roads in

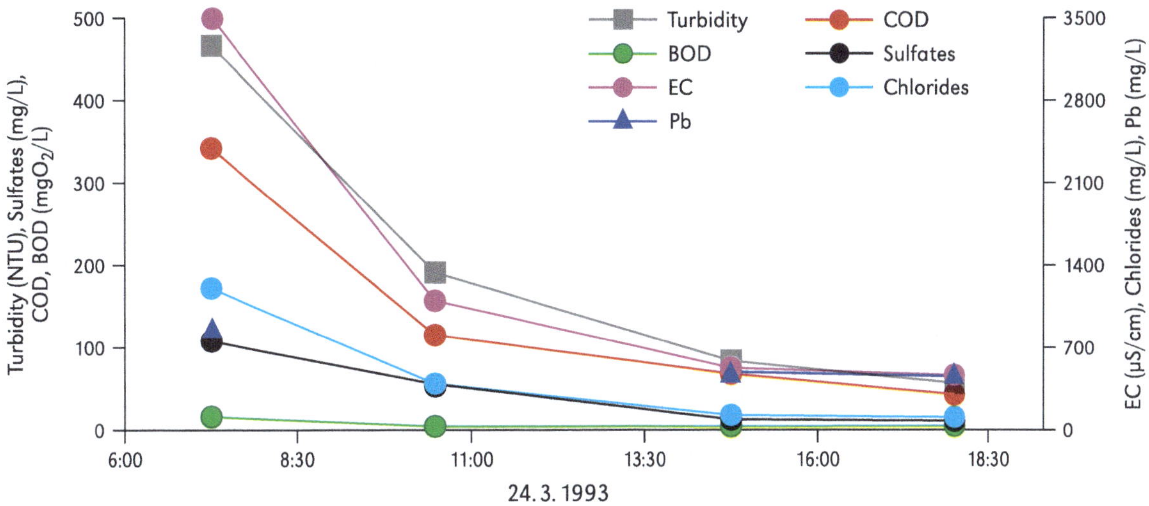

Fig. 28.6 The composition of the water draining from the motorway after the precipitation event of 24.3.1993. After a relatively good and long-lasting washing of the road surface (with a total rainfall of 32 mm) it has been determined that the water flowing into the oil separator still had elevated turbidity (57 NTU), KPK (44 mg O_2/L) and lead values (460 µg/L)

winter (Fig. 28.7). Turbidity was also relatively low (93 NTU), whereas BOD_5 (70 mg O_2/L) was strikingly high, COD relatively high (220 mg O_2/L), and the lead and cadmium content high as well. The second sampling which was carried out at an inflow that was two times lower, the measured values of all parameters (excluding lead) were significantly lower. Just over an hour later, the third sampling was carried out when the inflow was at its minimum, i.e. about four times lower than at the second sampling. It is clear that nearly all parameters were slightly higher, which can be explained with the washing-off of the remaining contaminants on the road surface and with the lower dilution compared to the second sample. Since the heaviest rainfall took place between 7 and 8 AM, but the first sample was taken at 8.30 AM, it is likely that the most contaminated water had flown away before the first sampling was carried out. This finding confirms how significant it is that sampling is planned ahead since the conditions change rapidly and the time of taking a sample has a major impact on the determined quality.

Similar measurements were carried out in other motorway sections as well. For the inflow of water into the concrete sedimentation basin at the section of the motorway Čebulovica–Sežana near the Divača cemetery in the years 1997 and 1998, it was determined that the intensity and quantity of precipitation affect the concentrations of contaminants in the water (Pintar et al. 1998). Again, the major contaminants turned out to be lead, total organic carbon (TOC), and suspended matter to which most of the metals are tied to. The measured values seldom exceeded the permissible limit for the outflow of waste water into open rivers.

Afterwards, during the precipitation events on 4 and 5 October 2001, 20-min samples were taken at the same sedimentation basin (Kompare et al. 2002a). It was determined that neither inflow nor outflow from the facility exceeded the maximum permissible concentrations. However, it did turn out that in order to assess the effectiveness of environment protection, the concentrations themselves are not enough, but the quantity of substances that flow into the karst terrain need to be assessed as well since they also depend on the quantity of the water that flows out.

28.4 The Composition of Water Flowing from the Oil Separator

In parallel to the monitoring of the water at the inflow into the oil separator A at the village of Stara vas, the water in the drainage canal, flowing out from the oil separator (after sedimentation) directly into the karst

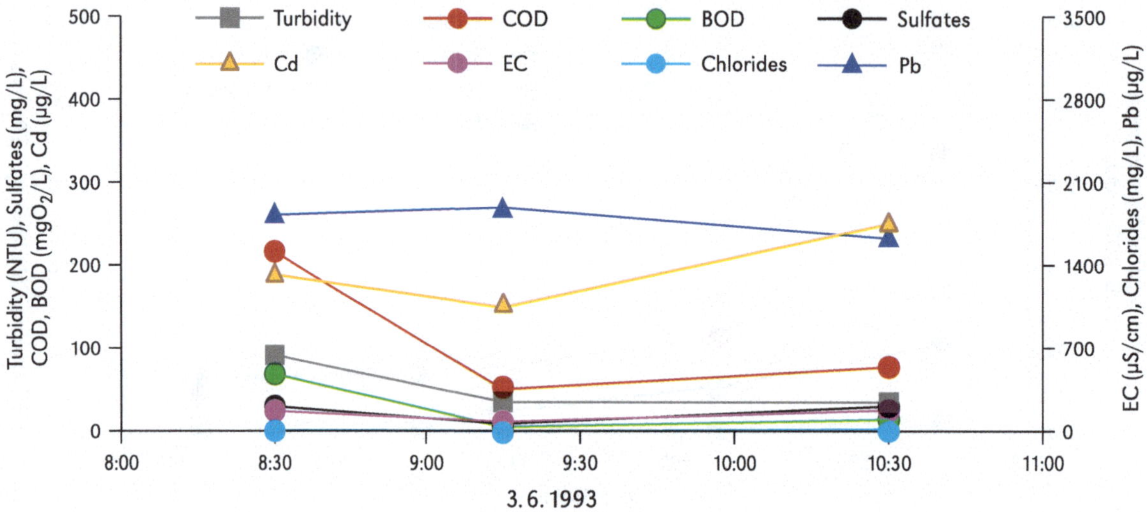

Fig. 28.7 The composition of the water draining from the motorway after the precipitation event of 3.6.1993

terrain, was analysed as well. The results showed that, during periods of drought, the soluble matter in the oil separator undergoes enrichment due to water evaporation.

Lower values for turbidity and COD in the outflow are due to the solids in the oil separator during lower inflow of water when the mixing of the sediment does not occur. The oil separator thus also acted as a sedimentation basin. The mixing of the sediment is more intensive during increased inflow of water, which means that the sediment flows into the karst terrain in form of a suspension which puts more pressure on the environment. Therefore, it is important to carry out regular maintenance works in the oil separator and to regularly remove the sediment.

28.5 Transfer of Contaminants Through the Karst Aquifer

The area in question is made of fissured and karstified carbonate rock. The precipitation water and water from the karst surface flow rapidly and directly into the underground. Earlier researches of this area have shown that intensive and abundant precipitation water can flow through 150 m thick limestone along more permeable conduits in as little as six hours (Kogovšek and Habič 1981). These major conduits are usually accompanied by a network of poorly permeable fissures which convey smaller amounts of water significantly slower; if contaminated, this means that the water accumulates in the underground and takes longer to resurface through the karst springs. Thus the velocity of the water flow and of the water-soluble matter through the vadose zone (from the surface to the continuous water flows in the karst terrain) ranges from 25 m/h to less than 1 cm/h (Kogovšek 2010). Less abundant precipitation, typical of dry summer periods or dry winter conditions, together with the contaminants, remain in the rock near the karst surface, sometimes up to several months, and it takes abundant and intensive precipitation after a period of drought to squeeze the accumulated water and the contaminants deeper into the karst terrain. In such cases, the contamination is detected only after several months, which can lead to a false conclusion that the contamination never occurred, mainly because of the lack of knowledge about the situation and because the monitoring was too short.

Slower flow through the vadose zone can lead to a certain degree of self-cleaning, but only of light or biodegradable organic matter. In the karst, dilution plays an important role since, due to the increased inflow of water after intensive precipitation events, it often screens out the contamination.

As the contamination reaches the continuous underground water flows, further transfer becomes faster. The researches on the flow of the karst water in the Slovenian Karst conducted thus far, mainly by means of tracer tests, have shown that the flow velocities range from 40 to 200 m/h (on average 100 m/h), meaning that the spreading of contamination is relatively quick and takes place across big distances from the point of contamination all the way to the karst springs (Kogovšek 2000; Kogovšek and Petrič 2002a).

28.5.1 Tracer Test in the Area of the Oil Separator B at Postojna

Tracer tests are one of the most effective research methods for studying the characteristics of water flow and transfer of substances in the karst. By using artificial tracers, dissolved in water and injected into the karst aquifer, it is possible to simulate the spreading of water-soluble contamination through the karst underground from the source to the springs. In the past, several tracer tests were carried out in the recharge area of the springs Malenščica and Unica. The tests were used to determine the characteristics and the direction of the underground water flow (Fig. 28.2). However, based on the collected results, it was not possible to exactly determine towards which spring the main outflow from the observed motorway section at Postojna is flowing and the degree with which it compromises the Malenščica spring, which is the most important drinking water supply in this area.

Therefore, a tracer test was carried out on 18 November 2008 in the area of the oil separator B at Postojna, approx. 3.5 km from the source of the Malenščica spring. At 10.00 AM, at the outflow from the oil separator leading into the sinking stream, an aqueous solution containing 1 kg of the uranine tracer was injected and washed out with approx. 11 m^3 of water from the tank (Figs. 28.8 and 28.9). Following the period of high water levels in the beginning of

Fig. 28.8 Injection of uranine into the sinking stream of the oil separator B at Postojna

Fig. 28.9 Outflow of uranine from the sinking stream of the oil separator B at Postojna

November, the water level at the time of injection was dropping. The flow rate of the Malenščica spring was approx. 7 m^3/s, and of the Unica spring approx. 9 m^3/s. The samples at the source of the Malenščica spring were taken using the ISCO 6700 sampler at various intervals, most frequently every 2 h. At the same time, the fluorescence was measured in situ using the »Fibrooptic Fluorimeter LLF-M« at 30-min intervals. The sampling at the source of the Unica spring was done manually. The sampling interval was set to 2 h at the beginning of the test, afterwards it was less frequent. The fluorescence in the acquired samples was measured in the laboratory using the »Elmer Luminiscence Spectrometer LS 30« (E_{ex} = 491 nm, E_{em} = 512 nm, detection limit 0.005 mg/m^3).

The first traces of uranine in the Unica spring appeared on 24 November 2008 (Fig. 28.10). On 25 March, after a period of less intensive precipitation, the flow rates slightly increased and the concentration of uranine gradually rose to 0.16 mg/m^3. On the other hand, heavy precipitation at the end of November and in the beginning of December resulted in very high flow rates which did not subside until the end of the month. Along with the increased flow rates, the concentration of uranine increased significantly as well, reaching its highest value of 0.6 mg/m^3 on 30 November at 12.00 AM, with the flow rate of the Unica spring at 24 m^3/s. After 8 h, the concentration dropped to its initial value, and after 5 days it dropped below the detection limit. Afterwards, several small peaks were recorded. At the same time, the flow rate of the Malenščica spring was rising as well, but the uranine in this spring was not detected until the morning of 2 December. After 24 h, with the flow rate of 9.2 m^3/s, the concentration reached its peak value of 0.11 mg/m^3 (Gabrovšek et al. 2010).

During the initial period with no rain, the water flow through the vadose zone was very slow and the

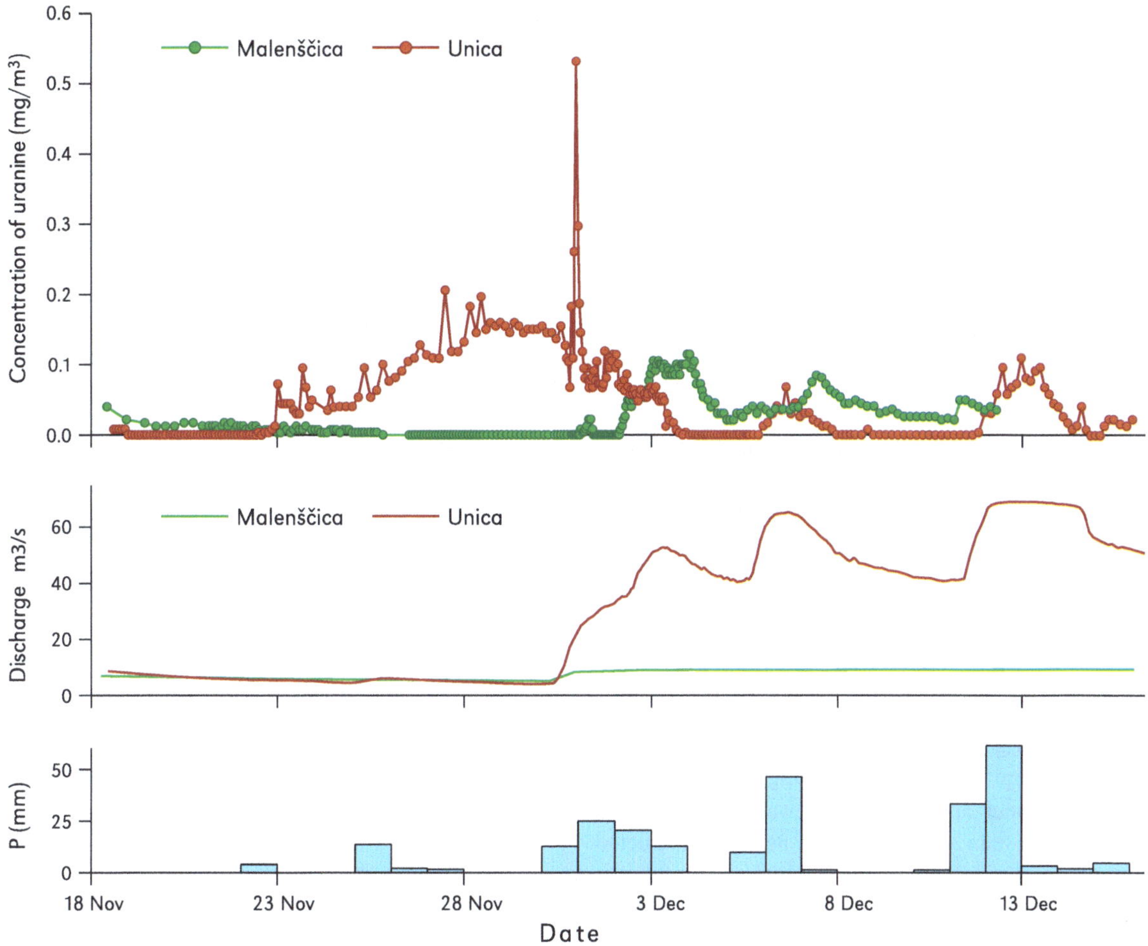

Fig. 28.10 Uranine breakthrough curves, discharges of springs and precipitation measured in Postojna in November 2008

uranine appeared in the springs in very low concentrations. It was not until intensive precipitation at the end of November 2008 that the uranine was pushed through the karst system; however, due to the dilution effect during rapid increase of flow rates, these concentrations were rather short-lived. Under these conditions, the main flow direction was determined as leading from the oil separator B towards the source of the Unica spring, whereas the link with the source of the Malenščica spring was rather weak.

Based on the results of the tracer test, it was possible to determine that the risk for deteriorating the quality of the karst water source of the Malenščica spring due to the potential contamination at the observed motorway section was relatively small, but not to be discounted. Therefore, it is necessary to monitor the quality of the water flowing out from the oil separator, since its significant deterioration can have a negative impact on the quality of the water of the Malenščica spring. One also needs to consider the fact that there are several other oil separators in the recharge area of this spring (Fig. 28.2); for these, no detailed research was carried out, plus there is a chance that they are even more directly connected with the observed water source.

28.6 Conclusion

The monitoring of the water flowing off the motorway was carried out more than two decades ago when traffic density was much lower due to the political circumstances in the Balkans. Even back then it was clear that the concentrations of some of the parameters had exceeded the permissible values. The results have shown particularly higher concentrations of chlorides and sulphates during winter-time which, at the same time, means higher EC and higher turbidity and COD compared to the results during summer-time when higher content of lead and cadmium was observed. Provided that the precipitation quantity of a single rainfall is high enough, the road surface is gradually washed away which was reflected in the increasing quality of successive samples. This means that the highest quantity of contamination flows off at initial washing-off, with precipitation intensity as an important factor to this. However, the parameter values do not drop below the set limit, even after heavy washing-off of the road surface following abundant and intensive precipitation; on the other hand, this is not to be expected since steady traffic results in constant contamination.

The findings about the contamination of water flowing off the motorway surfaces have had an impact on the construction of retarding basins and treatment plants for later motorway construction in Slovenia. Thus, several generations of such facilities have been built which are constantly being upgraded and updated. Subsequent analyses of the water off the road surface at certain section of the newly constructed motorways have only seldom shown values that exceeded the maximum permissible concentrations (Pintar et al. 1998), and the measurements during the precipitation event in October 2001 recorded no exceeded values (Kompare et al. 2002a). The comparison of all the measurements that were carried out has shown that the load on individual sections of the motorway can be different and varies. Furthermore, the quality is normally also affected by the conditions during sampling. Therefore, in order to determine the actual state and show a real picture of the degree of contamination, frequent and systematic measurements are required.

During smaller inflows into the oil separator, the outflow from the oil separator into the karst terrain was usually of better quality than the inflow, and of significantly poorer quality during abundant inflows, since the sediment that had accumulated in the oil separator was also flushed out, and the majority of the organic contaminants and metals are tied to the sediment. It is thus necessary to ensure regular maintenance of oil separators and to properly clear out the accumulating sediment.

The realization that the areas with higher annual precipitation quantity experience greater dilution of the contaminated water off the road surfaces indicates that the concentrations are not an adequate criterion for contamination that flows off the motorway surfaces and into the environment. The calculations of the input of individual contaminants, which are the product of concentrations and quantities of water flowing off the road surfaces, make more sense. This is particularly true of the karst landscape since the water flows out directly into the environment and has no chance of cleansing itself through the underground on its way to the karst springs, where the latter are often captured to provide drinking water supply.

For individual oil separators on motorway sections on the karst, tracer tests are a sure method to determine and evaluate the link with drinking water catchments. Where a direct link with high flow rates was determined, it would have to be provided for effective treatment of waste water before it is released into the karst terrain. On the other hand, if no such detailed research was carried out in the recharge area of the karst water sources, it would still be necessary to intercept and properly treat at least initial quantities of the most contaminated outflow water from the road surfaces.

Regarding the Planning and the Construction of Slovenian Motorways in the Karst Region

One of the biggest projects currently being finalized in Slovenia is linking the country with modern motorways. The Karst makes up for almost half of Slovenia's landscape and more than half of the water supplied for drinking comes from karst aquifers. Slovenia is the home of the classical Karst region, which provided its name to numerous world languages for the type of landscape that develops on carbonate stone and where the science of karstology began to develop. Comprising an important part of our natural and cultural heritage, the sensitive karst landscape demands that we possess good knowledge and serious effort for its preservation.

Since 1994, Slovene karstologists have cooperated in the planning and construction of motorways in the karst regions. When choosing the route of the motorways and railway lines, one must first consider the integrity of the karst landscape and the recommendations for bypassing the more important surface karst phenomena (dolines, polje, collapse dolines, karst walls) and the already known caves. Special attention was given to the impact of the construction and use of motorways on karst waters. Motorways should be impermeable. Water from the road surface is first collected in oil separators, cleansed, and then released into the karst terrain. Included in the study were also the contaminants in the water that flows daily off the motorways.

Construction work has provided a series of important discoveries about the formation of the karst and its development on various bedrock, in different conditions, and through various processes. We have studied the karst along motorways between Razdrto, Karst Edge, and Fernetiči (southwest Slovenia), the central part of the Dolenjska karst region (south Slovenia), and the young karst in the Vipava Valley (southwest Slovenia). Their cross-section thus provided the most important areas of the Slovenian Karst. This way, a great deal of knowledge about surface karst phenomena and the epikarst where the earthwork cut deeper into the surface, as well as the knowledge about the tunnels and within them about the vadose zones and the palaeokarst was acquired. Common to all of these are the important traces of the karst development, especially in the form of numerous old caves. In fact, more than 350 new caves have opened.

In the karst region of western Slovenia we have studied surface karst phenomena such as dolines and karren. The numerous newly uncovered caves have revealed the cavernosity of individual parts of the aquifer. These include old caves that are now dry since they remained high above the water table, and caverns with water that flows from the permeable karst surface and into the aquifer. A good portion of the caves is filled with alluvia. Along other speleological features, the shape of cave networks and their parts and subsoil rock forms, these sediments have helped uncover important periods in the formation, development, and age of the caves. Their age was also determined with the aid of palaeomagnetic dating. The oldest sediments probably filled the caves after the Messinian crisis and are therefore more than 5 million years old. Therefore, they can be ranked among our oldest caves, and their age exceeds our previous estimates. The opening of

M. Knez and T. Slabe (eds.), *Cave Exploration in Slovenia*,
Cave and Karst Systems of the World, DOI 10.1007/978-3-319-21203-6

unroofed caves also provided important information, as we discovered that these traces of the development of the aquifer helped form the karst surface much more than we previously suspected. Unroofed caves, which are the consequence of the lowering of the karst surface, comprise a new form that we have added as a unique form to the international list of karst forms. On the karst surface that is predominantly overgrown with forest, traces of human activity have been discovered, the history of their former intensive exploitation, primarily for agriculture and water supply.

Construction works on the low and largely alluvium-covered Dolenjska karst area revealed predominantly the subsoil shaping of the karst. Here, subsoil stone forests and underground shafts were found; this was the very first record of these forms in the Slovenian Karst in all their glory and eloquence. Large areas of stone forests with their characteristic subsoil rock forms have revealed the wealth of the subsoil shaping of karst surfaces. The water that was penetrating through the soil and alluvium uniquely shaped various carbonate rocks. Consolidated underground streams of percolating water have been carving underground shafts, whereas hollows that are similar to empty shafts are more or less completely filled with alluvia.

Earthworks during motorway construction through the Vipava Valley provided a number of unique discoveries. Here, the karst formed in young breccia that have developed from the consolidated slope rubble below Mount Nanos. With the percolation of water from the surface, forms began to develop similar in shape to that of the dolines. Many different types of caves were discovered. Smaller caves have formed in the most consolidated parts of breccia and are thus typical karst caves, whereas larger caves have formed at the contact with the flysch that lies below the breccia. Smaller streams of water run through them. Fissure caves are the result of the tension in the slope breccia that lies on inclined flysch bedrock. The majority of water flows along this contact area.

In short, the results of regular research of karst features that were revealed during motorway construction have broadened our knowledge of the natural and cultural heritage and provided more insight into the field of karstology. In any case, they are worth a second glance and need to be presented collectively and preserved for further study. At the same time, they provide a starting point for planning life in the Karst and for protecting the karst landscape (Knez et al. 2015).

References

Ajdič M, Brenčič M,Ločniškar A, Petkovšek A, Prestor J, Žmavc J, Rismal M, Veselič M, Hauck J (1999) Smernice za določitev načina zaščite podzemne vode na območju avtocest. Postopki določanja načina zaščite podzemne vode v odvisnosti od stopnje občutljivosti vodonosnika s tehničnimi ukrepi. DARS, Ljubljana, p 20

Archives of the Karst Research Intitute ZRC SAZU

Bavdek A (1998) Poročilo o arheoloških izkopavanjih vrtač na trasi AC Divača-Kozina. Notranjski muzej Postojna, Postojna, p 84

Bavdek A (2003a) AC Divača-Kozina, končno poročilo o obdelavi arheološkega gradiva z arheoloških izkopavanj vrtač. Notranjski muzej Postojna, Postojna

Bavdek A (2003b) Vrtače. In: Zemlja pod vašimi nogami, Vodnik po najdiščih, Zbirka Dnevi evropske kulturne dediščine, Ljubljana, pp 285–287

Bavdek A, Križman P (2000) Obdelava arheološkega gradiva, pridobljenega pri arheoloških izkopavanjih vrtač na trasi AC Divača–Kozina: delno poročilo. Notranjski muzej Postojna, Postojna, p 24

Beguš T, Kočevar M, Prestor J, Sotlar K (2003) Tunnels in karst and flysch in Slovenia. RMZ—Mater Geoenviron 50(1):17–20

Birkeland PW (1984) Soils and geomorphology. Oxford University Press, Oxford, p 372

Boisson JC, Perrodin Y (2006) Effects of road runoff on biomass and metabolic activity of periphyton in experimental streams. J Hazard Mater 132(2–3):148–164

Boisson JC, Boisson C, Perrodin Y (2005) The impacts of road runoff on periphyton in a small upland stream during summer. Aquat Ecosyst Health Manag 8(4):415–425

Bosák P, Ford DC, Głazek J (1989) Terminology. In: Bosák P, Ford DC, Glazek J, Horaček I (eds) Paleokarst. A systematic and regional review (developments in earth surface processes, Book 1). Academia, Praha, Elsevier Science, Amsterdam, pp 25–32

Bosák P, Pruner P, Zupan Hajna N (1998a) Palaeomagnetic research of cave sediments in SW Slovenia. MS, Institute of Geology, Academy of Sciences of the Czech Republic and Karst Research Institute Slovenian Academy of Sciences and Arts, Praha–Postojna, p 144

Bosák P, Pruner P, Zupan Hajna N (1998b) Palaeomagnetic research of cave sediments in SW Slovenia. 6th International Karstological School Classical Karst: Alpine Karst, Guide-booklet for the excursions and abstracts of papers. Speleological Association of Slovenia and Karst Research Institute ZRC SAZU, Postojna, pp 35–37

Bosák P, Pruner P, Zupan Hajna N (1998c) Paleomagnetic research of cave sediments in SW Slovenia. Acta Carsolog 27(2):151–179

Bosák P, Mihevc A, Pruner P (eds) (1999a) Cave fill in the Črnotiče Quarry, SW Slovenia: palaeomagnetic, mineralogical and geochemical study (preliminary report). MS, Geol ust AV ČR and Karst Res Inst SAZU, Praha–Postojna, p 108

Bosák P, Mihevc A, Pruner P, Melka K, Venhodová D, Langrová A (1999b) Cave fill in the Črnotiče Quarry, SW Slovenia: palaeomagnetic, mineralogical and geochemical study. Acta Carsolog 28(2):15–39

Bosák P, Knez M, Otrubová D, Pruner P, Slabe T, Venhodová D (2000a) Palaeomagnetic research of a fossil cave in the highway construction at Kozina, SW Slovenia. Acta Carsolog 29(2):15–33

Bosák P, Pruner P, Mihevc A, Zupan Hajna N (2000b) Magnetostratigraphy and unconformities in cave sediments: case study from the Classical Karst, SW Slovenia. Geologos 5:13–30

Bosák P, Mihevc A, Pruner P (2004) Geomorphological evolution of the Podgorski Karst, SW Slovenia: contribution of the magnetostratigraphic research of the Črnotiče II site with *Marifugia* sp. Acta Carsolog 33(1):175–204

Bressan G, Snidarcig A, Venturini C (1998) Present state of tectonic stress of the Friuli area (eastern Southern Alps). Tectonophysics 292(3–4):211–227

Brodar S (1958) Črni Kal, nova paleolitska postaja v Slovenskem Primorju (Zusammenfassung: Črni Kal, eine neue paläolitische Station im Küstengebiet Sloweniens). Razprave 4, Razred za prirodoslovne vede SAZU, Ljubljana, pp 269–364

Bulc T, Šajn Slak A (2003) Performance of constructed wetland for highway runoff treatment. Water Sci Technol 48(2):315–322

Bulić J, Jurišić-Polšak Z (2009) Macropalaeontology and stratigraphy of lacustrine Miocene deposits at Crnika beach on the Island of Pag (Croatia). Geol Croat 62(3):135–155

Buser S (1968) Osnovna geološka karta SFRJ 1:100.000, list Gorica. Savezni geološki zavod, Beograd

Buser S (1969) Osnovna geološka karta SFRJ 1:100.000, list Ribnica. Savezni geološki zavod, Beograd

M. Knez and T. Slabe (eds.), *Cave Exploration in Slovenia*,
Cave and Karst Systems of the World, DOI 10.1007/978-3-319-21203-6

Buser S (1973) Osnovna geološka karta SFRJ 1:100.000, tolmač lista Gorica. Savezni geološki zavod, Beograd

Buser S (1976) Tektonska zgradba južnozahodne Slovenije. In: Drobne K et al (eds) 8. jugoslovanski geološki kongres, Bled 1.–5. oktober 1974, 3, Ljubljana, pp 45–58

Buser S, Grad K, Pleničar M (1967) Osnovna geološka karta SFRJ 1:100.000, list Postojna. Savezni geološki zavod, Beograd

Cander H (1995) Interplay of water-rock interaction efficiency, unconformities, and fluid flow in a carbonate aquifer: Floridan aquifer system. In: Budd DA, Saller AH, Harris MP (eds) Unconformities and Porosity in Carbonate Strata. AAPG Memoir 63, pp 116–137

Caves Registry. Karst Research Institute ZRC SAZU, Postojna, Speleological Association of Slovenia, Ljubljana

Choquette PW, Pray LC (1970) Geologic nomenclature and classification of porosity in sedimentary carbonates. AAPG Bull 54:207–250

Clauzon G (1973) The eustatic hypothesis and pre-Pliocene cutting of the Rhône valley. In: Ryan WBF, Hsü KJ (eds) Initial reports of the deep sea drilling project, vol 13, pp 1251–1256

Clauzon G (1979) Le canyon messinien de la Durance (Provence, France): une preuve paléogéographique du bassin profond de dessiccation. Palaeogeogr Palaeoclimatol Palaeoecol 29(1/2):15–40

Clauzon G, Puig JM, Guendon JL (1997) Géomorphologie quantitative et paléo-géomorphologie dans les karsts du domaine Mediterranéen. In: Manifestations karstiques induites par le creusement messinien: exemples rhodano-duranciens, Livret-guide d'excursion. Université de Provence, Aix-Marseille, p 33

Cucchi F, Forti F, Ulcigrai F (1994) Valori di abbassamento per dissoluzione di superfici carsiche. Acta Carsolog 23:55–61

Culver DC, Sket B (2002) Biological monitoring in caves. Acta Carsolog 31(1):55–64

Čar J (1982) Geološka zgradba požiralnega obrobja Planinskega polja (summary: geologic setting of the Planina polje ponor area). Acta Carsolog 10:75–105

Čar J, Juren A (1980) Šmihelska tektonska krpa (abstract: The Klippe of Šmihel). Geologija 23(2):279–283

Dall PC, Friberg N, Lindegaard C, Toman MJ (1995) A practical guide of biological assessment of stream water quality. In: Toman MJ, Steinman F (eds) Proceedings of the Tempus workshop on biological assessment of stream water quality (theory, application and comparison of methods). University of Ljubljana, Ljubljana, pp 97–125

D'Ambrosi C, Legnani F (1965) Sul problema delle sabbie silicee del Carso di Trieste. Boll Soc Adriat Sci Nat 53 (2):211–220

DARS (2015) Družba za avtoceste v Republiki Sloveniji/ Motorway Company in the Republic of Slovenia: http:// www.dars.si/ (20 Feb 2015)

Datry T, Malard F, Gibert J (2004) Dynamics of solutes and dissolved oxygen in shallow urban groundwater below a stormwater infiltration basin. Sci Total Environ 329:215–229

Debeljak I, Košir A, Otoničar B (1999) A preliminary note on dinosaurs and non-dinosaurian reptiles from the Upper Cretaceous carbonate platform succession at Kozina (SW Slovenia). Razprave 40, Razred za naravoslovne vede SAZU, Ljubljana, pp 3–25

Debeljak I, Košir A, Buffetaut E, Otoničar B (2002) The late cretaceous dinosaurs and crocodiles of Kozina (SW Slovenia): a preliminary study. Mem Soc Geol Ital 57:193–201

Decree (1999) Decree on the emission of substances in the discharge of waste water from petrol stations, facilities for the maintenance and repairs of motor vehicles and car-washes. Official Gazette of the RS, No. 10/99, Ljubljana

Decree (2005) Decree on the emission of substances in the discharge of meteoric water from public roads. Official Gazette of the RS, No. 47/05, Ljubljana

Drobne K (1977) Alvéolines paléogènes de la Slovénie et de l'Istrie. Mem Suisses de Paléont 99:124

Dular A (1993a) Geološko-geomehansko poročilo za izvedbeni projekt trase AC na odseku Čebulovica-Divača in za razcep Divača. Elaborat ZRMK, Ljubljana

Dular A (1993b) Geološko-geomehansko poročilo za izvedbeni projekt trase AC na odseku Divača-Dane. Elaborat ZRMK, Ljubljana

Durn G, Aljinović D (1995) Teška mineralna frakcija u terra rossama istarskog poluotoka, Hrvatska. Abstracts of first Croatian geological congress, 31, Zagreb

Durn G, Ottner F, Tišljar J, Mindszenty A, Barudžija U (2003) Regional subaerial unconformities in shallow-marine carbonate sequences of Istria: sedimentology, mineralogy, geochemistry and micromorphology complexes. In: Vlahović I, Tišljar J (eds) 22nd IAS meeting of seedimentology. Field trip guidebook, Opatija, pp 209–254

Enserink EL, Maas-Diepeveen JL, Van Leeuwen CJ (1991) Combined effects of metals: an ecotoxicological evaluation. Water Res 25:679–687

Fisher R (1953) Dispersion on a sphere. Proc R Soc Lond A 217:295–305

Foos AM (1991) Aluminous lateritic soils. Eleuthera, Bahamas: a modern analog to carbonate paleosols. J Sed Petrology 61 (3):340–348

Ford DC, Williams PW (2007) Karst hydrogeology and geomorphology. Wiley, Chichester, p 562

Freytet P, Plaziat JC (1982) Continental carbonate sedimentation and pedogenesis: late cretaceous and early tertiary of southern France. Contrib Sediment 12:213

Gabrovšek F, Kogovšek J, Kovačič G, Petrič M, Ravbar N, Turk J (2010) Recent results of tracer tests in the catchment of the Unica River (SW Slovenia). Acta Carsolog 39(1):27–37

Gams I (1959) Problematika regionalizacije Dolenjske in Bele krajine (résumé: Le problème de la délimitation régionale de la Basse-Carniole et de la Carniole-Blanche). Geografski Vestn 31:79–95

Gams I (1962a) Morfografski pregled novomeških pokrajin. In: Gams I, Savnik R (eds) Dolenjska zemlja in ljudje. Dolenjska založba, Novo mesto, pp 31–38

Gams I (1962b) Meritve korozijske intenzitete v Sloveniji in njihov pomen za geomorfologijo (summary: measurements of corrosion intensity in Slovenia and their geomorphological significance). Geografski Vestn 34:3–20

Gams I (1962c) Nekatere značilnosti Krke in njenih pritokov. In: Gams I, Savnik R (eds) Dolenjska zemlja in ljudje. Dolenjska založba, Novo mesto, pp 92–110

Gams I (1965) H kvartarni geomorfogenezi ozemlja med Postojnskim, Planinskim in Cerkniškim poljem (summary: on the quaternary geomorphogenesis of the area among the

karst poljes of Postojna, Planina and Cerknica). Geografski Vestn 37:61–101
Gams I (1971) Podtalne kraške oblike (summary: subsoil karst forms). Geografski Vestn 43:27–45
Gams I (1974) Kras. Slovenska matica, Ljubljana, p 360
Gams I (1987) Razvoj reliefa na zahodnem Dolenjskem, s posebnim ozirom na poplave (Relief evolution in the western part of Lower Carniola, Dolenjsko, with special reference to floods). Geografski Zbornik 26:63–96
Gams I (1990) Depth of Rillenkarren as a measure of deforestation age. Stud Carsologica 1(2):29–36
Gams I (1991) Sistemi prilagoditve primorskega dinarskega krasa na kmetijsko rabo tal (summary: systems of adapting the Littoral Dinaric Karst to agrarian land use). Geografski Zbornik 31:5–106
Gams I (1998) Kras. In: Gams I, Vrišer I (eds) Geografija Slovenije. Slovenska matica, Ljubljana, pp 55–90
Gams I (2004) Kras v Sloveniji v prostoru in času (summary: karst in Slovenia in space and time). Založba ZRC, Ljubljana, p 515
Genorio M (1999) Varovanje kraških vodnih virov: primer havarije plinskega olja v zajetju Globočec. Diplomska naloga, Visoka šola za zdravstvo, Ljubljana, p 65
Geršl M, Stepišnik U, Šušteršič S (1999) The "unroofed cave" near the bunker (Laški Ravnik). Acta Carsologica 28(2):77–90
Gill LW, Ring P, Higgins NMP, Johnston PM (2014) Accumulation of heavy metals in a constructed wetland treating road runoff. Ecol Eng 70:133–139
Głazek J (1993) Kryzys messyński (wyschniecie Morza Środziemnego w gornym miocenie) i jego wplyw w paleogeografie Europy. Streszczenia referatow wygloszonych na posiedzeniach Oddzialu Poznaiiskiego (1991–1992), Pol. Tow. Geol, Inst. Geol. UAM. Poznan, pp 13–15
Gospodarič R (1985) O speleogenezi Divaške jame in Trhlovce (summary: on the speleogenesis of Divaška jama and Trhlovca cave). Acta Carsolog 13(1984):5–36
Gospodarič R, Habe F, Habič P (1970) Orehovški kras in izvir Korentana (summary: the Karst of Orehek and the source of the Korentan). Acta Carsolog 5:95–108
Gotvajn AŽ, Zagorc-Končan J (2009) Bioremediation of highway stormwater runoff. Desalination 248(2):794–802
Gregorič V (1967) Minerali glin v nekaterih talnih enotah Slovenskega primorja (summary: clay minerals in some soils in Littoral Slovenia). Geologija 10:247–269
Gregorič V (1969) Nastanek tal na triadnih dolomitih (summary: soil formation on the triassic dolomites). Geologija 12:201–230
Habe F (1968) Poročilo o barvanju Orehovških ponikev. In: Gospodarič R, Habe F, Habič P (eds) Vodni viri za oskrbo Postojne. Elaborat, Karst Research Institute ZRC SAZU, Postojna, pp 46–48
Habič P (1968) Kraški svet med Idrijco in Vipavo. Prispevek k poznavanju razvoja kraškega reliefa (the Karstic region between the Idrijca and Vipava rivers. A contribution to the study of development of the Karst relief). Dela 21, Razred za prirodoslovne in medicinske vede SAZU, Ljubljana, p 243
Habič P (1974) Poročilo o kraških pojavih na AC Senožeče–Divača–Sežana. Elaborat, Karst Research Institute ZRC SAZU, Postojna, p 20
Habič P (1982) Pregledna speleološka karta Slovenije (summary: the synoptical speleological map of Slovenia). Acta Carsolog 10:5–22
Habič P (1987) The Renčelica doline near Sežana. Man's impact in Dinaric Karst, Guide book, Ljubljana, pp 115–117
Habič P (1989a) Sledenje kraških voda v Sloveniji: zgodovinski pregled in raziskave po letu 1965 (summary: about karst waters tracing in Slovenia). Geografski Vestn 61:3–20
Habič P (1989b) Sledenje v zaledju Timava. Karst Research Institute ZRC SAZU, Postojna
Habič P (1989c) Kraška bifurkacija Pivke na jadransko črnomorskem razvodju (summary: pivka karst bifurcation on Adriatic-Black Sea watershed). Acta Carsolog 18:233–264
Habič P (1989d) Študij sledilnih metod v Sloveniji (2. faza). Karst Research Institute ZRC SAZU, Postojna, p 36
Habič P (1992) Les phénomènes paléokarstiques du karst Alpin et Dinarique en Slovénie. In: Salomon JN, Maire R (eds) Karst et évolutions climatiques. Presses Universitaires de Bordeaux, Bordeaux, pp 411–428
Hill CA, Forti P (1997) Cave Minerals of the World, 2nd edn. National Speleological Society, Huntsville, Alabama, p 463
Holmes DE, Nevin KP, O'Neil RA, Ward JE, Adams LA, Woodard TL, Vrionis HA, Lovley DR (2005) Potential for quantifying expression of the Geobacteraceae citrate synthase gene to assess the activity of Geobacteraceae in the subsurface and on current-harvesting electrodes. Appl Environ Microbiol 71(11):6870–6877
Horáček I, Mihevc A, Zupan Hajna N, Pruner P, Bosák P (2007) Fossil vertebrates and paleomagnetism update of one of the earlier stages of cave evolution in the Classical karst, Slovenia: Pliocene of Črnotiče II site and Račiška pečina Cave (Fosilni vretenčarji in paleomagnetna posodobitev ene od zgodnjih faz razvoja jam na Klasičnem krasu, Slovenija: pliocen Črnotič II in Račiške pečine). Acta Carsolog 36 (3):453–468
Horvat A (2004) Srednjemiocenske kremenične alge Slovenije: paleontologija, stratigrafija, paleoekologija, paleobiogeografija (middle miocene siliceous algae of Slovenia: paleontology, stratigraphy, paleoecology, paleobiogeography). Založba ZRC, Ljubljana, p 255
Hrovat A (1953) Kraška ilovica: njene značilnosti in vpliv na zgradbe. Državna založba Slovenije, Ljubljana, p 91
Hsü KJ (1973) Quando il Mediterraneo si disseccò. Le Scienze 56:19–29
Hsü KJ, Cita MB, Ryan WBF (1973) Origin of Mediterranean evaporites. In: Ryan WBF, Hsü KJ (eds) Initial reports of the deep sea drilling project, vol 13, pp 1203–1231
Hsü KJ, Montadert L, Bernoulli D, Cita MB, Erickson A, Garrison RE, Kidd RB, Mélières F, Müller C, Wright R (1977) History of the Mediterranean salinity crisis. Nature 267:399–403
Hudoklin A (1995) Poročilo o barvanju Temenice. Bilten '94 Dolenjskega krasa, Novo mesto, pp 16–17
Hydrometeorological Institute of the Republic of Slovenia 1995–1997, Ljubljana, pp 411–428
Ilešič S (1958) Problemi geografske rajonizacije ob primeru Slovenije (résumé: Sur les problèmes de délimitation et classification des régions géographiques d'après l'exemple de la Slovénie). Geografski Vestn 29–30:83–140

Jeannin PY, Häuselmann P, Lütscher M, Blant D, Meury PX (2013) Presentation of a water injection system to control the growth of speleothems at the Milandre Test-Site (JU, Switzerland). In: Filippi M, Bosák P (eds) Proceedings of the 16th International congress of speleology (21–28 July 2013, Brno, Czech Republic), vol 2, Czech Speleological Society and the Speleo2013, Prague, pp 408–412

Jelínek V (1966) A high sensitivity spinner magnetometer. Stud Geophys Geod 10(1):58–78

Jelínek V (1973) Precision A.C. bridge set for measuring magnetic susceptibility and its anisotropy. Stud Geophys Geod 17(1):36–48

Jones B, Khale CF (1985) Lichen and algae: agents of biodiagenesis in karst breccia from Grand Cayman Island. Bull Can Pet Geol 33:446–461

Jurkovšek B, Toman M, Ogorelec B, Šribar L, Drobne K, Poljak M, Šribar L (1996) Formacijska geološka karta južnega dela Tržaško-Komenske planote 1:50.000: kredne in paleogenske karbonatne kamnine (Geological map of the southern part of the Trieste-Komen Plateau 1:50,000: cretaceous and paleogene carbonate rocks). Geološki zavod Slovenije, Ljubljana, p 143

Kayhanian M, Stransky C, Bay S, Lau S-L, Stenstrom MK (2008) Toxicity of urban highway runoff with respect to storm duration. Sci Total Environ 389(2–3):386–406

Khumakov IS (1967) Pliocenovye i pleistocennovye otlozhenia doliny Nila v Nubii i verchnem Egipte. Tr Geol Inst AN SSSR 170:1–117

Khumakov IS (1971) Nekotorye voprosy geologicheskoy istorii Sredizemnomorskogo basseyna v konce miocena-nachala pliocena po novym dannym. Sov Geol 14(10):3–14

Kirschvink JL (1980) The least-squares line and plane and the analysis of palaeomagnetic data. Geophys J Int 62(3):699–718

Kladnik D (1996) Naravnogeografske členitve Slovenije (summary: natural-geographical divisions of Slovenia). Geografski Vestn 68:123–159

Klimchouk A (1995) Karst morphogenesis in the epikarstic zone. Cave Karst Sci 21(2):45–50

Klimchouk AB (2000) The formation of epikarst and its role in vadose speleogenesis. In: Klimchouk AB, Ford DC, Palmer AN, Dreybrodt W (eds) Spleogenesis. Evolution of karst aquifers. Huntsville, Alabama, pp 91–99

Knez M (1995) Pomen in vloga lezik pri makroskopskih raziskavah karbonatnih kamnin, v katerih so oblikovani freatični kanali. Annales 5(7):127–130

Knez M (1996) Vpliv lezik na razvoj kraških jam. Primer Velike doline, Škocjanske jame (summary: the bedding-plane impact on development of karst caves. An example of Velika dolina, Škocjanske jame caves). Zbirka ZRC 14, Ljubljana, p 186

Knez M, Slabe T (1999a) Unroofed caves met during the motorway construction near Kozina and their recognition on karst surface. Abstracts of the 7th International karstological school Classical Karst–Roofless Caves, Postojna, pp 30–31

Knez M, Slabe T (1999b) Unroofed caves and recognising them in karst relief (discoveries during motorway construction at Kozina, South Slovenia). Acta Carsolog 28(2):103–112

Knez M, Slabe T (2000) Jame brez stropa so pomembna oblika na kraškem površju: s krasoslovnega nadzora gradnje avtocest na krasu (Denuded caves as important feature on karst surface: recognition from supervision of motorway construction on karst). In: Gostinčar A (ed) Proceedings of the 5th Slovenian road and transport congress. Družba za raziskave v cestni in prometni stroki Slovenije, Ljubljana, p 29

Knez M, Slabe T (2001a) Oblika in skalni relief stebrov v Naigu kamnitem gozdu, JZ Kitajska (summary: shape and rock relief of the pillars in Naigu stone forest, SW China). Acta Carsolog 30(1):13–24

Knez M, Slabe T (2001b) The lithology, shape and rock relief of the pillars in the Pu Chao Chun stone forest (Lunan stone forests, SW China). Acta Carsolog 30(2):129–139

Knez M, Slabe T (2001c) Karstology and expressway construction. In: Proceedings of the 14th IRF road world congress, Paris (on CD-Rom). International Road Federation, Genève

Knez M, Slabe T (2002a) Unroofed caves are an important feature of karst surfaces: examples from the classical karst. Z Geomorphol 46(2):181–191

Knez M, Slabe T (2002b) Lithologic and morphological properties and rock relief of the Lunan stone forests. In: Gabrovšek F (ed) Evolution of karst: from prekarst to cessation (Carsologica 1). ZRC SAZU, Postojna–Ljubljana, pp 259–266

Knez M, Slabe T (2004a) Karstology and the opening of caves during motorway construction in the karst region of Slovenia. Int J Speleol 31(1/4):159–168

Knez M, Slabe T (2004b) Highways on karst. In: Gunn J (ed) Encyclopedia of caves and karst science. Fitzroy Dearborn, New York, pp 419–420

Knez M, Slabe T (2005) Caves and sinkholes in motorway construction, Slovenia. In: Waltham T, Bell F, Culshaw M (eds) Sinkholes and subsidence. Karst and cavernous rocks in engineering and construction. Springer, Praxis, Chichester, pp 283–288

Knez M, Slabe T (2006a) Krasoslovne raziskave pri gradnji avtocest preko slovenskega krasa (summary: karstological research during the construction of motorways crossing the Slovene Karst). Annales 16(2):259–266

Knez M, Slabe T (2006b) Dolenjska subsoil stone forests and other karst phenomena discovered during the construction of the Hrastje-Lešnica motorway section, Slovenia. Acta Carsolog 35(2):103–109

Knez M, Slabe T (eds) (2007) Kraški pojavi, razkriti med gradnjo slovenskih avtocest (summary: karst phenomena discovered during the construction of the Slovene motorways). Carsologica 7. ZRC Publishing, Postojna–Ljubljana, p 250

Knez M, Slabe T (2008) Kraški pojavi, odkriti med gradnjo slovenskih avtocest. Proceedings of the 10th Slovenian road and transport congress, Portorož, October 2008. DRC-Družba za raziskave v cestni in prometni stroki Slovenije, Ljubljana, pp 867–871

Knez M, Slabe T (2009a) Caves in breccia and flysch below Mount Nanos in the Vipava Valley (Slovenia). Annales 19 (1):85–94

Knez M, Slabe T (2009b) Karst features discovered during motorway construction. In: White WB (ed) Proceedings of the 15th international congress of speleology, July 2009, Kerrville, USA (2009 ICS Proceedings, vol 2, Symposia, Part 2). International Union of Speleology, pp 1139–1144

Knez M, Slabe T (2009c) Lithological characteristics, shape and rock relief of the Lunan stone forests. In: Ginés Á, Knez M, Slabe T, Dreybrodt W (eds) Karst rock features: karren sculpturing. Carsologica, vol 9. ZRC Publishing, Postojna–Ljubljana, pp 439–452

Knez M, Slabe T (2009d) Karstology and motorway construction. In: Bonacci O (ed) Abstract book of the international interdisciplinary scientific conference sustainability of the karst environment—Dinaric karst and other karst regions, Plitvice Lakes (Croatia), September 2009. Centre for Karst, Gospić, pp 84–85

Knez M, Slabe T (2010a) Expressway in breccia and flysch below Mount Nanos in the Vipava Valley (Slovenia). Geophysical research abstracts, vol 12, [1] p http://meeting-organizer.copernicus.org/EGU2010/EGU2010-14807.pdf

Knez M, Slabe T (2010b) Karstology and motorway construction. In: Bonacci O (ed) Sustainability of the karst environment—Dinaric karst and other karst regions (international hydrological programme 7, series on groundwater 2). Unesco, Paris, pp 107–114

Knez M, Slabe T (2010c) Uncovered caves and their conservation during the construction of motorways over Classical Karst and other types of Slovenian Karst. In: Andreo B (ed) Advances in research in Karst media (Environmental earth sciences, Book 1). Springer, Berlin, pp 401–406

Knez M, Slabe T (2010d) Planning and construction motorways and railroads crossing Classical Karst in Slovenia. In: Lakušić S (ed) Road and rail infrastructure. Proceedings of the first international conference on road and rail infrastructure CETRA, Opatija, Croatia, May 2010. Department of transportation, Faculty of Civil Engineering, University of Zagreb, Zagreb, pp 837–843

Knez M, Slabe T (2010e) Prevotljenost breče in fliša pod Nanosom v Vipavski dolini. In: Proceedings of the 10th Slovenian road and transport congress, Portorož, October 2010. DRC-Družba za raziskave v cestni in prometni stroki Slovenije, pp 1016–1022

Knez M, Slabe T (2010f) Over 350 new caves discovered during motorway construction in Slovenia. In: Abstracts [S. l.: International Geographical Union, 2010]

Knez M, Slabe T (2011a) Cave preservation during motorway construction in Slovenia. In: Prelovšek M, Zupan Hajna N (eds) Pressures and protection of the underground karst: cases from Slovenia and Croatia. Karst Research Institute ZRC SAZU, Postojna, pp 83–92

Knez M, Slabe T (2011b) Karstification in breccia and flysch (Mount Nanos, Slovenia). In: Polk JS, North LA (eds) Karst: hydrogeology and ecosystems: proceedings of the International conference on karst hydrogeology and ecosystems. Hoffman Environmental Research Institute, Department of Geography and Geology, Western Kentucky University, Bowling Green, 8–10 June 2011, pp 96–97

Knez M, Slabe T (2012a) Expressway construction on young karst in breccia (Vipava Valley, Slovenia). In: Lakušić S (ed) Road and rail infrastructure II. Proceedings of the 2nd International conference on road and rail infrastructures CETRA 2012, 7–9 May 2012, Dubrovnik, Croatia. Department of Transportation, Faculty of Civil Engineering, University of Zagreb, Zagreb, pp 773–779

Knez M, Slabe T (2012b) Planning, research and karstological monitoring of expressways crossing classical Karst (Slovenia). In: Proceedings of the world congress on engineering and technology (CET 2012), October 2012, Beijing, China (Creative Education 3:39–42)

Knez M, Slabe T (2012c) Kraški pojavi, razkriti med gradnjo avtoceste na nizkem krasu Dolenjske. Proceedings of the 11th Slovenian road and transport congress, Portorož, October 2012. DRC-Družba za raziskave v cestni in prometni stroki Slovenije, Ljubljana, pp 1–6

Knez M, Slabe T (2012d) Kraški pojavi, razkriti med gradnjo avtoceste na nizkem krasu Dolenjske. In: Abstracts of the reports from the 11th Slovenian road and transport congress, Portorož, October 2012. DRC-Družba za raziskave v cestni in prometni stroki Slovenije, Ljubljana, p 212

Knez M, Slabe T (2014) Subsoil stone forest discovered during the construction of the motorway (SE Slovenia). In: Lakušić S (ed) Road and rail infrastructure III. Proceedings of the 3rd international conference on road and rail infrastructures CETRA 2014, 28–30 April 2014, Split, Croatia. Department of Transportation, Faculty of Civil Engineering, University of Zagreb, Zagreb, pp 603–608

Knez M, Šebela S (1994) Novo odkriti kraški pojavi na trasi avtomobilske ceste pri Divači. Naše jame 36:102

Knez M, Kranjc A, Otoničar B, Slabe T, Svetličič S (1994) Posledice izlitja nafte pri Kozini (summary: the effect of the oil spill near Kozina). Ujma 8:74–80

Knez M, Otoničar B, Slabe T (2003) Subcutaneous stone forest (Trebnje, Central Slovenia). Acta Carsolog 32(1):29–38

Knez M, Slabe T, Šebela S (2004a) Karstification of the aquifer discovered during the construction of the expressway between Klanec and Črni Kal, Classical Karst. Acta Carsolog 33(1):205–217

Knez M, Slabe T, Šebela S (2004b) Karst uncovered during the Bič-Korenitka motorway construction (Dolenjska, Slovenija). Acta Carsolog 33(2):75–89

Knez M, Slabe T, Šebela S, Gabrovšek F (2008) The largest karst caves discovered in a tunnel during motorway construction in Slovenia's Classical Karst (Kras). Environ Geol 54(4):711–718

Knez M, Slabe T, Šebela S, Gabrovšek F (2009) Largest karst cave discovered in tunnel during construction of motorways in Slovenia's classical karst (Kras). Proceedings of the 9th international conference on tunnel construction and underground structures, Ljubljana, September 2009. Naravoslovno tehniška fakulteta, Oddelek za geotehnologijo in rudarstvo, Ljubljana, pp 121–126

Knez M, Petrič M, Slabe T (eds) (2012) Krasoslovje v razvojnih izzivih na krasu II. Gradnja, turizem, ekologija, varovanje (Karstology and development challenges on karst II.

Construction, tourism, ecology, protection). ZRC Publishing, Postojna–Ljubljana, p 199

Knez M, Petrič M, Slabe T, Šebela S (eds) (2015) The Beka-Ocizla cave system. Karstological railway planning in Slovenia. Cave and karst systems of the world, vol 1. Springer, Heidelberg, p 102

Kogovšek J (1993) Kakšna je sestava voda, ki odtekajo z naših cest (summary: water composition flowing off our roads). Ujma 7:67–69

Kogovšek J (1995a) The surface above Postojnska jama and its relation with the cave: the case of Kristalni rov. In: Cigna AA (ed) Proceedings of the international symposium Show caves and environmental monitoring, Frabosa Soprana, pp 29–39

Kogovšek J (1995b) Izlitja nevarnih snovi ogrožajo kraško vodo. Onesnaženje Rižane oktobra 1994 zaradi izlitja plinskega olja ob prometni nesreči v Obrovu. Annales 5 (7):141–148

Kogovšek J (1995c) Podrobno spremljanje kvalitete vode, odtekajoče z avtoceste in njen vpliv na kraško vodo (Detailed monitoring of the quality of the water that runs off the motorway and its impact on karst water). Annales 5 (7):149–154

Kogovšek J (1996) Onesnaževanje Krasa ob različnih nesrečah (summary: incidences of pollution in the Karst). Ujma 10:170–171

Kogovšek J (1999) Nova spoznanja o podzemnem pretakanju vode v severnem delu Javornikov, visoki kras (summary: new knowledge about the underground water drainage in the northern part of Javorniki Mountains, High Karst). Acta Carsolog 28(1):161–200

Kogovšek J (2000) Ugotavljanje načina pretakanja in prenosa snovi s sledilnim poskusom v naravnih razmerah (summary: how to determine the way of percolation and transport of substances by water tracing test in natural coonditions). Annales 10(1):133–142

Kogovšek J (2010) Characteristics of percolation through the karst vadose zone. Carsologica, vol 10. ZRC Publishing, Postojna–Ljubljana, p 168

Kogovšek J (2011a) Ogroženost kraških vodnih virov zaradi prometa v rednih razmerah (threats to the karst water sources from traffic in normal conditions). In: Knez M, Petrič M, Slabe T (eds) Krasoslovje v razvojnih izzivih na krasu I. Voda (Karstology and development challenges on karst I. Water). ZRC Publishing, Postojna–Ljubljana, pp 47–64

Kogovšek J (2011b) The impact of traffic on karst waters. In: Prelovšek M, Zupan Hajna N (eds) Pressures and protection of the underground karst. Cases from Slovenia and Croatia. Karst Research Institute, ZRC Publishing, Postojna, pp 119–127

Kogovšek J, Habič P (1981) Preučevanje vertikalnega prenikanja vode na primerih Planinske in Postojnske jame (summary: the study of vertical water pecolation in the case of Planina and Postojna caves). Acta Carsolog 9:129–148

Kogovšek J, Petrič M (2002a) Ogroženost kraškega sveta. In: Ušeničnik B (ed) Nesreče in varstvo pred njimi. Uprava RS za zaščito in reševanje Ministrstva za obrambo, Ljubljana, pp 170–183

Kogovšek J, Petrič M (2002b) Podzemno raztekanje vode iz ponora Tržiščice, JV Slovenija (summary: underground water flow from the Tržiščica sinking stream, SE Slovenia). Acta Carsolog 31(2):75–91

Kogovšek J, Petrič M (2011) Izlitja nevarnih snovi ogrožajo kraške vode (spillages of hazardous substances endanger karst waters). In: Knez M, Petrič M, Slabe T (Eds) Krasoslovje v razvojnih izzivih na krasu I. Voda (karstology and development challenges on karst I. Water). Karst Research Institute ZRC SAZU, ZRC Publishing, Postojna–Ljubljana, pp 65–82

Kogovšek J, Šebela S (2004) Water tracing through the vadose zone above Postojnska Jama, Slovenia. Environ Geol 45 (7):992–1001

Kogovšek J, Slabe T, Šebela S (1997) Motorways in Karst (Slovenia). In: Byerly DW (ed) Proceedings and a fieldtrip excursion guide of the 48th Highway geology symposium Knoxvill, Tennessee, May 1997. University of Tennessee, Department of Geological Studies, pp 49–55

Kompare B, Atanasova N, Babič R, Panjan J, Bulc T, Cerar U, Rodič P, Knez M, Kogovšek J, Petrič M, Pintar M (2002a) Odvodnja avtocest in zaščita voda: analiza delovanja čistilnega objekta na Krasu. In: Vilhar M (ed) Zbornik referatov 6. slovenskega kongresa o cestah in prometu Portorožu 23–25 oktobra 2002. Družba za raziskave v cestni in prometni stroki Slovenije, Ljubljana, pp 93–102

Kompare B, Atanasova N, Babič R, Panjan J, Bulc T, Cerar U, Rodič P, Knez M, Kogovšek J, Petrič M, Pintar M (2002b) Zaščita hidrosfere pred spiranjem in dreniranjem z avtocest. Zbornik seminarja Varstvo in kvaliteta pitne vode, Terme Olimia, pp 60–68

Košir A (2003) Litostratigrafska revizija zgornje krede in paleogena v jugozahodni Sloveniji. In: Horvat A (ed) Treatises and reports from the 16th meeting of Slovenian geologists (Geološki zbornik 17). Naravoslovnotehniška fakulteta, Oddelek za geologijo, Ljubljana, pp 92–98

Košir A, Otoničar B (1997) Diskonformnosti v zgornjekredno-paleogenskem zaporedju Jadransko-Dinarske karbonatne platforme v jugozahodni Sloveniji. In: Horvat A, Fister M, Zupančič N (eds) Treatises reports from the 13th meeting of Slovenian geologists, April 1997 (Geološki zbornik 11). Naravoslovnotehniška fakulteta, Oddelek za geologijo, Ljubljana, pp 15–16

Košir A, Otoničar B (2001) The evolution of Upper Cretaceous and Palaeogene synorogenic carbonate platforms in NW Dinaric foreland basin. In: Dragičević I, Velić I (eds) Abstracts of the 1st scientific meeting on carbonate platform or carbonate platforms of Dinarids, Zagreb, 1–2 Oct 2001. Rudarsko-geološko-naftni fakultet, Prirodoslovno-matematički fakultet, Institut za geološka istraživanja and Hrvatsko geološko društvo, Zagreb, pp 62–63

Košir A, Otoničar B (2002) Evolucija zgornjekrednih in paleogenskih sinorogenih karbonatnih platform v severozahodnih Dinaridih. In: Horvat A, Košir A, Vreča P, Brenčič M (eds) Book of abstracts of the 1st congress of Slovenian geologists, 9–11 oktober 2002. Črna na Koroškem, p 42

Košir A, Goričan Š, Debeljak I, Otoničar B, Turnšek D (1999) Vodnik po ekskurziji: ob 50-obletnici ustanovitve Paleontološkega inštituta Ivana Rakovca ZRC SAZU. Ljubljana, Založba ZRC SAZU, p 20

Kranjc A (1986) Transport rečnih sedimentov skozi kraško podzemlje na primeru Škocjanskih jam (summary: underground fluvial sediments transport as an example from

Škocjanske jame, Kras, Slovenia). Acta Carsolog 14–15:109–116

Kranjc A (1989) Recent fluvial cave sediments, their origin and role in speleogenesis, vol 27(1). Dela SAZU, Razred za naravoslovne vede, Ljubljana, p 167

Kranjc A (1990) Dolenjski kraški svet. Dolenjska založba, Novo mesto, p 240

Kranjc A (ed) (1997) Kras: Slovene Classical Karst. Založba ZRC, Ljubljana, p 254

Krivic P, Bricelj M, Trišič N, Zupan M (1987) Sledenje podzemnih vod v zaledju izvira Rižane (summary: water tracing in the Rižana spring ground water basin). Acta Carsolog 16:83–104

Krivic P, Bricelj M, Zupan M (1989) Podzemne vodne zveze na področju Čičarije in osrednjega dela Istre (summary: underground water connections in Čičarija region and in Middle Istria). Acta Carsolog 18:265–295

Ladišić B (1981) Hidrografsko speleološke odlike temeniškega podolja. Osmi jugoslovenski speleološki kongres, Beograd, pp 75–85

Langergraber G (2013) Are constructed treatment wetlands sustainable sanitation solutions? Water Sci Technol 67 (10):2133–2140

Maltby L, Forrow DM, Boxall ABA, Calow P, Betton CI (1995a) The effects of motorway runoff on freshwater ecosystems: 1. Field study. Environ Toxicol Chem 14 (6):1079–1092

Maltby L, Boxall BA, Forrow DM, Calow P, Betton CI (1995b) The effects of motorway runoff on fresh-water ecosystems. 2. Identifying major toxicants. Environ Toxicol Chem 14 (6):1093–1101

Marton E, Drobne K, Cimerman F, Ćosović V, Košir A (1995) Paleomagnetism of latest Maastrichtian through Oligocene rocks in Istria (Croatia), the Karst region, and S of the Sava fault (Slovenia). In: Vlahović I, Velić I, Šparica M (eds) Proceedings of the 1st Croatian geological congress, Opatija, October 1995. Institut za geološka istraživanja and Hrvatsko geološko društvo, Zagreb, pp 355–360

Melik A (1959) Posavska Slovenija. Slovenija. Geografski opis II, Opis slovenskih pokrajin, 3. zvezek. Slovenska Matica, Ljubljana

Melik A (1960) Slovensko primorje. Slovenska matica, Ljubljana, p 546

Melik A (1961) Fluvialni elementi v krasu (résumé: Les elements fluviatiles du karst). Geografski zbornik 6:333–362

Melik A (1962) Geografski pregled Dolenjske. Dolenjska zemlja in ljudje, Dolenjska založba, Novo mesto, pp 7–14

Mihevc A (1984) Nova spoznanja o Kačni jami (summary: new discoveries in the Kačna jama cave). Naše jame 26:11–20

Mihevc A (1991) Morfološke značilnosti ponornega kontaktnega krasa, izbrani primeri s slovenskega krasa. Magistrska naloga, Univerza v Ljubljani, Filozofska fakulteta, Ljubljana, p 202

Mihevc A (1994) Contact karst of Brkini hills (Brkinski kontaktni kras). Acta Carsolog 23:99–109

Mihevc A (1996) Brezstropa jama pri Povirju (abstract: the cave Brezstropa jama near Povir). Naše jame 38:65–75

Mihevc A (1998) Speleogeneza Matičnega krasa. PhD thesis, Univerza v Ljubljani, Filozofska fakulteta, Ljubljana, p 150

Mihevc A (1999a) The caves and the karst surface—case study from Kras, Slovenia. In: Karst 99: Colloque européen Des paysages du karst au géosystème karstique: dinamiques, structures, et enregistrement karstiques. Études de Géographie Physique, suppl 28, Aix-en-Provence, pp 141–144

Mihevc A (1999b) Roofles caves. Guide-booklet for the excursions for the 7th International Karstological School »Classical Karst—Roofless Caves«. ZRC SAZU Karst Research Institute, Postojna, pp 2–25

Mihevc A (1999c) Unroofed caves as geomorphic and speleologic features. Abstracts of the 7th International Karstological School »Classical Karst—Roofless Caves«. ZRC SAZU Karst Research Institute, Postojna, pp 33–34

Mihevc A (2001) Speleogeneza divaškega krasa (summary: the speleogenesis of the Divača Karst). ZRC Collection, vol 27, Založba ZRC, Ljubljana, p 180

Mihevc A (2007) The age of Karst relief in West Slovenia (Starost kraškega reliefa v zahodni Sloveniji). Acta Carsolog 36(1):35–44

Mihevc A, Zupan Hajna N (1996) Clastic sediments from dolines and caves found during the construction of the motorway near Divača, on the Classical Karst. Acta Carsolog 25:169–191

Mihevc A, Slabe T, Šebela S (1998) Denuded caves—an inherited element in the karst morphology; the case from Kras. Acta Carsolog 27(1):165–174

Mlakar I (1969) Krovna zgradba idrijsko žirovskega ozemlja (summary: nappe structure of the Idrija-Žiri region). Geologija 12:5–72

Murakami M, Sato N, Anegawa A, Nakada N, Harada A, Komatsu T et al (2008) Multiple evaluations of the removal of pollutants in road runoff by soil infiltration. Water Res 42 (10–11):2745–2755

Mylroie JE, Carew JL (1995) Karst development on carbonate islands. In: Budd DA, Saller AH, Harris PM (eds) Unconformities and Porosity in Carbonate Strata. AAPG Memoir, vol 63, pp 68–89

Nakayama S, Harada K, Inoue K, Sasaki K, Seery B, Saito N et al (2005) Distributions of perfluorooctanoic acid (PFOA) and perfluorooctane sulfonate (PFOS) in Japan and their toxicities. Environ Sci 12:293–313

Novak D (1970) Hidrogeološke značilnosti osrednje Dolenjske. Naše jame 11:17–24

Novak D (1994) Raziskave na Temenici leta 1994. Naše jame 36:109–110

Ogorelec B, Drobne K, Jurkovšek B, Dolenec T, Toman M (2001) Paleocene beds of the Liburnia formation in Čebulovica (Slovenia, NW Adriatic-Dinaric platform). Geologija 44(1):15–65

Opher T, Friedler E (2010) Factors affecting highway runoff quality. Urban Water J 7(3):155–172

Osborne RAL (2000) Paleokarst and its significance for speleogenesis. In: Klimchouk AB, Ford DC, Palmer AN, Dreybrodt W (eds) Speleogenesis. Evolution of karst aquifers. Huntsville, Alabama, pp 113–123

Otoničar B (2006) Razvoj paleokrasa v zgornjekrednih apnencih Matarskega podolja in Slavnika (Paleokarst development in the Upper Cretaceous limestones of Matarsko Podolje and Slavnik). PhD thesis, University of Ljubljana, p 309

Otoničar B (2007) Upper Cretaceous to Paleogene forbulge unconformity associated with foreland basin evolution (Kras, Matarsko podolje and Istria; SW Slovenia and NW Croatia). Acta Carsolog 36(1):101–120

Otoničar B (2008) Evolution of paleokarst related to a forebulge unconformity: an example from the late Cretaceous and Paleogene of Kras, Matarsko Podolje and Istria, southwest Slovenia and northwest Croatia. In: Sasowsky ID, Feazel CT, Mylroie JE, Palmer AN, Palmer MV (eds) Karst from recent to reservoirs (special publication), vol 14. Karst Waters Institute, Leesburg, Virginia, pp 146–156

Otoničar B, Košir B (1998) Palustrine carbonates of the Upper Cretaceous/Paleogene Liburnia Formation, Southwestern Slovenia. In: Canaveras JC, García del Cura MA, Soria J (eds) Abstracts. of the 15th international sedimentological meeting, April 1998, Alicante. IAS and Universidad de Alicante, Alicante, pp 593–594

Otoničar B, Košir B (2002) Vpliv geotektonskih dogodkov na sedimentna zaporedja karbonatnih depozicijskih sistemov (zgornja jura—eocen, severni del Jadransko-Dinarske karbonatne platforme). Book of abstracts of the 1st congress of Slovenian geologists, October 2002. Črna na Koroškem, pp 65–66

Ožbolt A (1994) Poročilo o spremljanju onesnaženja na izviru Rižane po izlitju plinskega olja D2. Poročilo Zavoda za socialno medicino in higieno, Koper

Peric B, Gabrovšek F, Pipan T (2007) Vzpostavitev monitoringa podzemnega toka Reke. Park Škocjanske jame, Škocjan

Perna G (1996) Il carsismo messiniano nell'area del Mediterraneo. Speleologia 34:5–12

Petrič M (2009) Pregled sledenja voda z umetnimi sledili na kraških območjih v Sloveniji (review of water tracing with artificial tracers on karst areas in Slovenia). Geologija 52 (1):127–136

Petrič M, Šebela S (2004) Vulnerability mapping in the recharge area of the Korentan spring, Slovenia. Acta Carsolog 33 (2):151–168

Pintar M, Ajdič M, Leskovšek H (1998) Kemizem padavinske vode z avtoceste pri Divači in v Vipavski dolini. Zbornik 4. kongresa o cestah in prometu, Portorož. Družba za raziskave v cestni in prometni stroki Slovenije, Ljubljana, pp 259–267

Pipan T (2000a) Biological assessment of stream water quality —the example of the Reka River (Slovenia). Acta Carsolog 29(1):201–222

Pipan T (2000b) Function feeding groups of macroinvertebrates in the Reka River (Slovenia). Acta Carsolog 29(2):293–301

Pipan T (2005) Epikarst—a promising habitat. Copepod fauna, its diversity and ecology: a case study from Slovenia (Europe). ZRC Publishing, Ljubljana, p 101

Placer L (1981) Geološka zgradba jugozahodne Slovenije (summary: geologic structure of southwestern Slovenia). Geologija 24(1):27–60

Placer L (1998a) Structural meaning of the Sava folds. Geologija 41:191–221

Placer L (1998b) Contribution to the macrotectonic subdivision of the border region between Southern Alps and External Dinarides. Geologija 41:223–255

Placer L (2002) Predhodna objava rezultatov strukturnega profiliranja Kraškega roba in Istre (AC Kozina-Srmin, Sečovlje)/Preliminary results of structural profiling of the Kras edge in Istria (Kozina–Srmin Motorway, Sečovlje). Geologija 45(1):277–280

Placer L (2005) Nekatere posebnosti hidrografske mreže v slovenski Istri. In: Horvat A (ed) 17th meeting of Slovenian Geologists, Ljubljana, April 2005. Geološki zbornik 18:91–92

Placer L (2008) Principles of the tectonic subdivision of Slovenia (Osnove tektonske razčlenitve Slovenije). Geologija 51(2):205–217

Placer L, Vrabec M, Celarc B (2010) The bases for understanding of the NW Dinarides and Istria Peninsula tectonics (Osnove razumevanja tektonske zgradbe NW Dinaridov in polotoka Istre). Geologija 53(1):55–86

Pleničar M (1954) Vrnik. Proteus 17:89–90

Pleničar M (1970) Osnovna geološka karta SFRJ 1:100.000, Tolmač za list Postojna. Savezni geološki zavod, Beograd, p 62

Pleničar M, Premru U (1977) Osnovna geološka karta SFRJ 1:100.000, Tolmač za list Novo mesto. Savezni geološki zavod, Beograd, p 61

Pleničar M, Polšak A, Šikić D (1969) Osnovna geološka karta SFRJ 1:100.000, List Trst. Savezni geološki zavod, Beograd

Pleničar M, Polšak A, Šikić D (1973) Osnovna geološka karta SFRJ 1:100.000, Tolmač za list Trst. Savezni geološki zavod, Beograd

Pleničar M, Premru U, Herak M (1976) Osnovna geološka karta SFRJ 1:100.000, List Novo mesto. Savezni geološki zavod, Beograd

Plut D (1984) Nekatere značilnosti vodnih virov novomeške občine, Dolenjske in Bele krajine. In: Zbornik 13. zborovanja slovenskih geografov, Ljubljana, pp 99–111

Premru U (1976) Neotektonika vzhodne Slovenije (summary: neotectonic evolution of eastern Slovenia). Geologija 19:211–249

Přihoda K, Krs M, Pešina B, Bláha J (1989) MAVACS—a new system of creating a non-magnetic environment for palaeomagnetic studies. Quad Geol Iberica 12:223–250

Pruner P, Bosák P (1999) Zprava k projektu KONTAKT ME 251 (1998) Česko-slovinska spoluprace, Research of karst sediments on the example of the Classical Karst, Slovenia. MS, Geol list AV ČR, Praha, pp 1–78

Radinja D (1962) Neka iskustva u proučavanju gruboklastičnog materijala sa morfometrijskom metodom. In: Zbornik VI. kongresa geografov FLRJ v LR Sloveniji od 27.9. do 5.10.1961. Savez geografskih društava FNRJ and Geografsko društvo Slovenije, Ljubljana, pp 235–243

Radinja D (1972) Zakrasevanje v Sloveniji v luči celotnega morfogenetskega razvoja (résumé: La karstification et l'évalution générale du relief en Slovénie). Geografski zbornik 13:197–243

Rakovec I (1958) Pleistocenski sesalci iz jame pri Črnem Kalu (summary: the pleistocene mammals from the cave Črni Kal in Northern Istria). Razprave SAZU 4, Razred za prirodoslovne vede, Ljubljana, pp 365–433

Ravbar N (2007) The protection of karst waters. A comprehensive Slovene approach to vulnerability and contamination risk mapping. ZRC Publishing, Ljubljana, p 254

Remškar B (2006) Jame v breči na južnem pobočju Trnovskega gozda. Naše jame 46:4–15

Rismal M, Saje-Rejc M, Šmid F, Hauck J, Ločniškar A (1995) Navodila projektantom za izdelavo tehnične dokumentacije:

odvodnjavanje meteornih voda iz avtocestnih površin. DARS, Ljubljana

Rismal M, Hauck J, Šmid F, Saje-Rejc M, Ločniškar A (1999) Navodila projektantom za izdelavo tehnične dokumentacije: odvodnjavanje meteornih voda iz avtocestnih površin. DARS, Ljubljana

Rossinsky V, Wanless HR (1992) Topographic and vegetative controls on calcrete formation, Turks and Caicos Islands, British West Indies. J Sed Petrol 62(1):84–98

Rotaru C, Woodard TL, Choi S, Nevin KP (2012) Spatial heterogeneity of bacterial communities in sediments from an infiltration basin receiving highway runoff. Microb Ecol 64:461–473

Rules (1996) Rules on minimum technical and other requirements for parking lots and places of vehicles maintenance. Official Gazette of the RS, No 33/96, Ljubljana

Savnik R (1962) Nekateri problemi kraške hidrografije na Dolenjskem. In: Gams I, Savnik R (eds) Dolenjska zemlja in ljudje. Dolenjska založba, Novo mesto, pp 15–30

Scholz M, Grabowiecki P (2007) Review of permeable pavement systems. Building Environ 42(11):3830–3836

Schulte U (1994) Geologische und hydrogeologische Untersuchungen im Karst von Orehek (Slowenien). Diplomarbeit, Universität Karlsruhe, Deutschland, p 105

Shutes RBE, Ellis JB, Revitt DM, Forshaw M, Winter B (2004) Urban and highway runoff treatment by constructed wetlands. In: Wong MH (ed) Developments in ecosystems, vol 1. Elsevier, pp 361–382

Shutes RBE, Revitt DM, Mungur AS, Scholes LNL (1997) The design of wetland systems for the treatment of urban run off. Water Sci Technol 35(5):19–25

SIST EN 858-1 (2002) Separator systems for light liquids (e.g. oil and petrol)—Part 1: Principles of product design, performance and testing, marking and quality control. http://www.en-standard.eu/ (20 Feb 2015)

Skoberne P, Peterlin S (eds) (1991) Inventar najpomembnejše naravne dediščine Slovenije, 2. del: osrednja Slovenija. Zavod Republike Slovenije za varstvo naravne in kulturne dediščine, Ljubljana

Slabe T (1995) Cave rocky relief and its speleogenetical significance. Zbirka ZRC 10, Založba ZRC, Ljubljana, p 128

Slabe T (1996) Karst features in the motorway section between Čebulovica and Dane. Acta Carsolog 25:221–240

Slabe T (1997a) Karst features discovered during motorway construction in Slovenia. Environ Geol 32(3):186–190

Slabe T (1997b) The caves in the motorway Dane-Fernetiči. Acta Carsolog 26(2):361–372

Slabe T (1998a) Karst features discovered during motorway construction between Divača and Kozina. Acta Carsolog 27 (2):105–113

Slabe T (1998b) Rock relief of pillars in the Lunan stone forests. In: Chen X et al (eds) South China Karst, vol 1. Založba ZRC, Ljubljana, pp 51–67

Slabe T (1999) Subcutaneous rock forms. Acta Carsolog 28 (2):255–269

Slabe T (2005) Two experimental modelings of karst rock relief in plaster: subcutaneous "rock teeth" and "rock peaks" exposed to rain. Z Geomorphol 49(1):107–119

Slabe T, Knez M (2004) Kraške podtalne skalne oblike (summary: karst subsoil rock forms). Annales 14 (2):259–266

Stampfli GM, Mosar J (1999) The making and becoming of Apulia. Mem Sci Geol 51(1):141–154

Stampfli GM, Mosar J, Marquer D, Marchant R, Baudin T, Borel G (1998) Subduction and obduction processes in the Swiss Alps. Tectonophysics 296(1–2):159–204

Steininger FF, Wessely G (2000) From the Tethyan ocean to the Paratethys sea: Oligocene to Neogene stratigraphy, paleogeography and paleobiogeography of the circum-Mediterranean region and the Oligocene to Neogene Basin evolution in Austria. Mitt Österr Geol Ges 92:61–77

Stepišnik U, Šušteršič S (1999) The "unroofed cave" near the bunker (Laški Ravnik). Acta Carsolog 28(2):77–90

Stoessell RK (1992) Effects of sulfate reduction on $CaCO_3$ dissolution and precipitation in mixing-zone fluids. J Sed Petrology 62(5):873–880

Šebela S (1996) Predhodne krasoslovne raziskave trase avtoceste Divača-Kozina. Annales 9:103–106

Šebela S (1999) Morphological and geological characteristics of two denuded caves in SW Slovenia. Acta Carsolog 28 (2):175–185

Šebela S, Mihevc A (1995) The problems of construction on karst—the examples from Slovenia. In: Beck BF, Pearson FM (eds) Karst geohazards, engineering and environmental problems in karst terrain. In: Proceedings of the 5th multidisciplinary conference on dolines and engineering and environmental impacts on karst. A.A. Balkema, Rotterdam, pp 475–479

Šebela S, Mihevc A, Slabe T (1999) The vulnerability map of karst along highways in Slovenia. In: Beck BF, Pettit AJ, Herring JG (eds) Hydrogeology and engineering geology of dolines and karst. In: Proceedings of the 7th multidisciplinary conference on dolines and the engineering and environmental impacts on karst. A.A. Balkema, Rotterdam, pp 419–422

Šifrer M (1970) Nekateri geomorfološki problemi dolenjskega krasa (résumé: Quelques problèmes géomorphologiques du Karst de Dolenjska, SE Slovénie). Naše jame 11:7–15

Škocjan Caves (2015) URL: http://whc.unesco.org/en/list/390 (10 Feb 2015)

Šušteršič F (1978) Nekaj misli o zasutih breznih in njihovem polnilu (summary: some considerations about the stuffed potholes and their filling). Naše jame 19:7–14

Šušteršič F (1985) Metoda morfometrije in računalniške obdelave vrtač (summary: a method of doline morphometry and computer processing). Acta Carsolog 13(1984):79–98

Šušteršič F (1998) Interaction between cave systems and the lowering karst surface, case study: Laški Ravnik. Acta Carsolog 27(2):115–138

Tiefenthaler L, Stein ED, Schiff KC (2011) Levels and patterns of fecal indicator bacteria in stormwater runoff from homogenous land use sites and urban watersheds. J Water Health 9(2):279–290

Toman MJ (1999) Kakovost tekočih voda—problematika biološkega ocenjevanja. Vestnik Univerze v Ljubljani 30 (3–4):17–18

Tromp K, Lima AT, Barendregt A, Verhoeven JTA (2012) Retention of heavy metals and poly-aromatic hydrocarbons from road water in a constructed wetland and the effect of de-icing. J Hazard Mater 203–204:290–298

Turk I, Bavdek A, Vidrih-Perko V, Culiberg M, Šercelj A, Dirjec J, Pavlin P (1992) Acijev spodmol pri Petrinjah,

Slovenija. Poročilo o raziskovanju paleolita, neolita in eneolita v Sloveniji 20:27–48

Vanstone SD (1998) Late Dinantian palaeokarst of England and Wales: implications for exposure surface development. Sedimentology 45(1):19–37

Vlahović T (2000) Kemizam vode kao indikator regionalnog kretanja podzemne vode u krškim vodonosnicima: izvor Sv. Ivan, Istra (chemistry of water as indicator of regional movement of groundwater in karstic aquifers: the Spring Sv. Ivan, Istria). In: Proceedings of the 2nd Croatian geological congress, Cavtat–Dubrovnik. Institut za geološka istraživanja, Zagreb, pp 827–832

Vrabec M, Fodor L (2006) Late Cenozoic tectonics of Slovenia: structural styles at the northeastern corner of the Adriatic microplate. In: Pinter N, Grenerczy G, Weber J, Stein S, Medak D (eds) The Adria microplate: GPS geodesy, tectonics and hazards. NATO science series IV, Earth and environmental sciences, vol 61. Springer, Dordrecht, pp 151–168

Zhang X, Lulla M (2006) Distribution of pathogenic indicator bacteria in structural best management practices. J Environ Sci Health Part A Toxic-Hazard Subst Environ Eng 41 (8):1421–1436

Zhou W, Beck BF (2005) Roadway construction in karst areas: management of stormwater runoff and sinkhole risk assessment. Environ Geol 47(8):1138–1149

Zupan N (1991) Flowstone datations in Slovenia. Acta Carsolog 20:187–204

Zupan Hajna N (1992) Mineralna sestava mehanskih sedimentov iz nekaterih delov slovenskega krasa (summary: mineral composition of mechanical sediments from some parts of Slovenian karst). Acta Carsolog 21:115–130

Zupan Hajna N (1997) Poročilo o I. fazi speleoloških raziskav na trasi AC Klanec–Križišče Ankaran. ZRC SAZU Karst Research Institute, Postojna

Zupan Hajna N (1998) Mineral composition of clastic sediments in some dolines along the new motorway Divača-Kozina. Acta Carsolog 27(1):277–296

Zupan Hajna N (1999) Comparison of some clastic sediments from the surface and caves of Divača Karst, SW Slovenia. In: Karst 99: Colloque européen Des paysages du karst au géosystème karstique: dinamiques, structures, et enregistrement karstiques. Études de géographie physique, suppl 28, Aix-en-Provence, pp 209–214

Zupan Hajna N (2004) The caves of the contact karst of Beka and Ocizla, in SW Slovenia (jame kontaktnega krasa Beke in Ocizle, JZ Slovenija). Acta Carsolog 33(2):91–105

Zupan Hajna N, Drole F (2003) Jama S 647 v levi cevi tunela Kastelec (abstract: the cave S 647 in the left tube of the Kastelec tunnel). Naše jame 45:69–79

Zupan Hajna N, Mihevc A (2008) (eds) Karst sediments: guide book and abstracts of the 16th International Karstological School »Classical Karst«. ZRC SAZU Karst Research Institute, Postojna, p 120

Zupan Hajna N, Mihevc A, Pruner P, Bosák P (2008) Paleomagnetism and magnetostratigraphy of Karst sediments in Slovenia. Carsologica, vol 8. Založba ZRC, ZRC SAZU, Ljubljana, p 268

The manufacturer's authorised representative in the EU is Springer Nature Customer Service Centre GmbH, Europaplatz 3, 69115 Heidelberg, Germany. If you have any concerns regarding our products, please contact ProductSafety@springernature.com

Printed and bound by CPI Group (UK) Ltd, Croydon, CR0 4YY
30/09/2026
02225562-0020